Magnesium Alloys Containing Rare Earth Metals

Advances in Metallic Alloys

A series edited by **J.N. Fridlyander**, *All-Russian Institute of Aviation Materials, Moscow, Russia* and **D.G. Eskin**, *Netherlands Institute for Metals Research, Del The Netherlands*

Volume 1
Liquid Metal Processing: Applications to Aluminium Alloy Production
I.G. Brodova, P.S. Popel and G.I. Eskin

Volume 2
Iron in Aluminum Alloys: Impurity and Alloying Element
N.A. Belov, A.A. Aksenov and D.G. Eskin

Volume 3
Magnesium Alloys Containing Rare Earth Metals: Structure and Properties
L.L. Rokhlin

Magnesium Alloys Containing Rare Earth Metals

Structure and Properties

L.L. Rokhlin
Baikov Institute of Metallurgy and Materials Science, Moscow, Russia

LONDON AND NEW YORK

First published 2003 by Taylor & Francis
11 New Fetter Lane, London EC4P 4EE

Simultaneously published in the USA and Canada
by Taylor & Francis Inc,
29 West 35th Street, New York, NY 10001

Taylor & Francis is an imprint of the Taylor & Francis Group

This book has been produced from camera-ready copy supplied by the authors
Printed and bound in Great Britain by TJ International Ltd, Padstow, Cornwall

British Library Cataloguing in Publication Data
A catalogue record for this book is available from the British Library

Library of Congress Cataloging in Publication Data
A catalog record for this book has been requested

ISBN 0-415-28414-7

Contents

Preface

Magnesium alloys are widely used in modern industry. High strength properties combined with quite low density make them very attractive as structural materials in applications where weight saving is of great importance. Such areas are, first of all, aircraft and space machinery, but ground transport, including automobiles, has applications for such alloys, too. Application of magnesium alloys in machines enables a decrease in fuel consumption improvements in their dynamic and other technical characteristics, making them more profitable. There are also other important applications of magnesium alloys as light structural materials. Besides, they are used as materials with particular physical or chemical properties.

The wide usage of magnesium alloys is promoted by their other outstanding features. Magnesium products may be made by different methods, common and die casting, rolling, forging and extrusion. They may be easily treated by machining and working. In general, they are quite corrosion resistant in many media. There are large natural sources of magnesium for extraction.

In recent years, the properties of magnesium alloys, especially the strength properties at ambient and elevated temperatures, have been significantly improved. It was mainly done by using for alloying different rare earth metals. Rare earth metals form a large group of elements including those of lanthanum series and also yttrium and scandium. All of them occupy IIIA subgroup of the periodic table of elements and, therefore, have a number of similar features. Nevertheless, the effect of each rare earth metal on the strength properties of magnesium may be quite different. Some rare earth metals have already been used as alloying elements in magnesium alloys for a long time, for example cerium and lanthanum. The influence of some others has been studied sufficiently, and magnesium alloys including them are being assessed by industry now. The third part of rare earth metals is still studied as possible additives to magnesium alloys. The opportunities of rare earths to improve the strength properties and other important characteristics of magnesium alloys are not yet fully exhausted. So it may be expected that new commercial magnesium alloys with these elements will be suggested for industry.

Magnesium alloys with rare earth metals are also quite interesting as the subject of scientific investigations. With these alloys, it is quite convenient to follow how the gradual change of the rare earth atom constitution with increase in atomic number results in the gradual change of the intermetallic compound formation, the solubility in magnesium solid solution, and other characteristics of the physico-chemical interaction between metals. The alloys have specific structures and specific behaviours during heat treatment and plastic deformation. These also change regularly as the rare earth atomic number increases. By studying these alloys, new features of the phase transformations, recovery and recrystallization in the hexagonal crystal lattice may be recognised. Investigations of magnesium alloys with the rare earth metals contribute to knowledge about the structure of metal alloys and its influence on the mechanical and other properties of materials.

Magnesium alloys with rare earth metals were thoroughly investigated in Russia. Most of the investigations were carried out in Baikov Institute of Metallurgy, Moscow. The Russian studies included phase diagrams, solid solution decomposition, influence of individual rare earth metals on the properties of magnesium, and other features of the alloys. Besides, a number of the commercial magnesium alloys with rare earth metals were developed and assessed in industry. Most of these investigations, as well as the investigations carried out in other countries up to about 1980, were described by the author in the book *L. L. Rokhlin "Magnesium Alloys Containing Rare Earth Metals"* issued by Nauka Publishers in 1980. Unfortunately, it is written in Russian and is practically unknown abroad. Since the above mentioned book was published, a lot of work has been carried out in the field of magnesium alloys with rare earth metals, both in Russia and in other countries too. It was stimulated by the lasting interest in these alloys from the scientific and industry viewpoint. So, a great number of phase diagrams were additionally investigated. Electron microscopy studies of the different alloys resulted in the discovery of quite complicated phase transformations in the structure of the alloys during solid solution decomposition and their effect on the properties. Special effort was devoted to detailed studies of supersaturated solid solutions resistance to decomposition during isothermal annealing at different temperatures and reversion after ageing. The structure and some properties of the alloys after high speed solidification were determined. A number of new commercial magnesium alloys with rare earth metals were developed and their properties were studied in detail. This new book *"Magnesium Alloys Containing Rare Earth Metals"*, which is offered to readers, includes the main contents of the first issue, although with a certain shortening, but also covers descriptions of the new investigations of Mg–RE alloys conducted after the first issue was published. In addition, it contains descriptions of the investigations of the acoustic properties of magnesium alloys. These investigations were aimed to develop materials with low ultrasound attenuation, and some of the rare earth metals turned out to be very effective additives in this case. As in the first issue, the new book contains a great deal of information about the investigations of these alloys in Russia. Nevertheless, many investigations of magnesium alloys with rare earth metals conducted in other countries are also presented.

Introduction

The first commercial applications of magnesium alloys with rare earth metals were connected with development of light structural materials having high strength properties at elevated temperatures. The need for such materials arose in aircraft from significant increase of the engine power and the flight speeds of planes after World War II. At the beginning, rare earth metals were used for alloying magnesium alloys in the form of "mishmetal'. This was the alloy of different rare earth metals in proportions corresponding approximately to their ratio in the ores where the rare earth metals were presented commonly together. The main constituent part of mishmetal was cerium and this was the reason why mishmetal was often named incorrectly as cerium. In that time there was the widespread opinion that properties of different rare earth metals were quite close. Assuming this opinion it was supposed that effects of mishmetal and cerium on magnesium were practically the same. Nevertheless, the investigations [1, 2] where effects of mishmetal and cerium on magnesium were compared showed this view point to be incorrect. The effects of mishmetal and cerium on mechanical properties of magnesium established in [1, 2] turned out to be close, but different.

The important step in production and application of magnesium alloys with rare earth metals began after works of T.E. Leontis from Dow Chemical Co [3, 4]. Working with both cast and extruded magnesium alloys in various conditions T.E. Leontis established significant differences between four initial rare earth metals of the lanthanum series, lanthanum, cerium, praseodymium and neodymium in their effects on mechanical properties of magnesium. Magnesium alloys containing neodymium showed significantly higher strength at elevated temperatures as compared with magnesium alloys containing other rare earth metals used, including cerium and mishmetal. Magnesium alloys with neodymium surpassed also magnesium alloys with cerium in strength at room temperature. As a result, magnesium alloys containing neodymium attracted attention for development of commercial structural materials. Soon such commercial magnesium alloys were developed and began to be used in industry despite the higher cost of neodymium as compared with cerium and mishmetal. Commercial production and application of magnesium alloys containing neodymium required, of course, a corresponding increase of commercial neodymium production.

At about that time, new magnesium alloys containing thorium were studied and developed for applications at elevated temperatures [5–7]. The main advantage of magnesium alloys with thorium was the high creep resistance at elevated temperatures. At room temperature and a range above, these alloys had moderate strength properties. Magnesium alloys with thorium were produced and used in industry in spite of the natural radioactivity of thorium.

Some later increase of commercial production of individual rare earth metals stimulated investigation of magnesium alloys with yttrium and scandium. The investigations revealed the possibility of reaching in these alloys higher strength at elevated temperatures than magnesium alloys with neodymium [8, 9]. The alloys with yttrium showed also the highest strength properties at room temperature amidst known magnesium alloys. These discoveries stimulated continuation of the investigations on magnesium alloys with individual rare earth metals. So, the investigations of magnesium alloys with the rest of the rare earth metals were conducted. This group of rare earth metals included samarium, gadolinium, terbium, dysprosium, holmium, erbium, thulium, ytterbium and lutetium. Some of these elements showed quite high strengthening effect on magnesium and could be considered as useful additives for its alloys. The investigation revealed also strict regularities in constitution and properties of magnesium alloys with different rare earth metals.

Commercial production and application of magnesium alloys with separate rare earth metals depends, to a great extent, on volumes of rare earth commercial production and the price, which is significantly different for each of the rare earth metals. Because of low production and high cost of some rare earth metals, commercial usage of magnesium alloys with them remains possible now only in very limited quantities, although these alloys may have very attractive properties. Meanwhile, commercial production of individual rare earth metals tends to grow and the price of them tends to fall. These factors must promote the possibility to enlarge usage of magnesium alloys with separate rare earth metals, even with the most expensive of them in the shortest supply.

Chapter 1

Physico-chemical Interaction Between Magnesium and Rare Earth Metals

General Characteristics of Rare Earth Metals

The term "rare earth metals" is usually applied to the group of chemical elements with atomic numbers from 57 to 71. All of them occupy the only cell in the common Periodic Table of the Elements belonging to sixth period and IIIA subgroup of it. However, as a rule, only the first of them, lanthanum with atomic number 57, is shown in the cell. The others of the group are shown out of the general table as a row of chemical elements arranged in order of successive increase of their atomic numbers. According to the rules suggested by International Union of Pure and Applied Chemistry [10], two other elements of the IIIA subgroup of the Periodic Table are also recommended to be called rare earth metals. They are scandium and yttrium belonging to the fourth and fifth periods of the Periodic Table, respectively. Nevertheless, sometimes the term "rare earth metals" is used only for the elements of atomic numbers from 57 to 71 with yttrium and scandium being noted separately. Thorium is also an element of IIIA subgroup of the Periodic Table, but it was never considered to belong to the rare earth metals. All rare earth metals are listed in Table 1 where their symbols, atomic numbers and atomic masses are also presented.

Rare earth metals with atomic numbers from 57 to 71 are recommended also to be called "elements of the lanthanum series" [10]. The elements following lanthanum in the lanthanum series are recommended to be called "lanthanides" or "lanthanoids" [10]. Sometimes instead of the "lanthanum series" the term "lanthanum row" is used. The rare earth metals are divided into two main subgroups. One of them is the "cerium subgroup" which includes the rare earth elements successively from lanthanum (atomic number 57) up to europium (atomic number 71) at the middle of the lanthanum series. The other one is the "yttrium subgroup" and this includes the rest of the elements belonging to the lanthanum series and yttrium. As a rule, scandium is not included in any subgroup. Division of the rare earth metals into two subgroups conforms to the constitution of their atoms and is manifested by certain similarities and differences in properties of the individual rare earth metals and their presence in the ores. Sometimes the rare earth metals of the first half of the lanthanum series are named "light" rare earth metals and those of the second half of it are named "heavy" rare earth metals. The common abbreviation of the rare earth metals is "RE". Also, often the generalised symbol "Ln" is used for rare earth metals.

In the second half of the 20th century the rare earth metals, their alloys and compounds were studied thoroughly and many particular features of them were discovered. They are reviewed in a number of books. In this book rare earth metals and their properties are considered only briefly and from the viewpoint of possible use as alloying elements for magnesium alloys.

Although rare earth metals reveal many similar features, especially in chemical properties, the differences between them may be also quite significant. As a rule, there are certain regularities in the change of properties with successive increase of the rare earth

Table 1. Rare earth metals [10].

Period	Atomic number	Symbol	Name	Atomic mass [11]
4	21	Sc	Scandium	44.95591
5	39	Y	Yttrium	88.9059
6	57	La	Lanthanum	138.9055
6	58	Ce	Cerium	140.12
6	59	Pr	Praseodymium	140.9077
6	60	Nd	Neodymium	144.24
6	61	Pm	Promethium	(145)
6	62	Sm	Samarium	150.36
6	63	Eu	Europium	151.96
6	64	Gd	Gadolinium	157.25
6	65	Tb	Terbium	158.9254
6	66	Dy	Dysprosium	162.50
6	67	Ho	Holmium	164.9304
6	68	Er	Erbium	167.26
6	69	Tm	Thulium	168.9342
6	70	Yb	Ytterbium	173.04
6	71	Lu	Lutetium	174.967

metal atomic number. The regularities may be complicated, but, in general, they conform to the respective change of the atom constitution. The similarities and differences between chemical properties of rare earth metals are connected with the particular electron configuration of their atoms. Table 2 shows electronic shells of the isolated atoms of the rare earth metals [12]. The remarkable features of them are as follows. Scandium, yttrium and lanthanum are the common transition metals with similar constitution of the outer electronic shells. Each of them have two electrons in the outer *s* shell and one electron in the near to it partially-filled *d* shell. This fact causes the similarity of Sc, Y and La in their chemical properties. In the elements of lanthanum series after lanthanum the deep electron shell 4*f* begins to be filled gradually with increasing atomic number of the elements. The filling of the shell 4*f* is not, however, regular. Thus, cerium following lanthanum in the lanthanum series has at once two electrons in the 4*f* shell with disappearance of one electron in 5*d* shell. Increase of the atomic number results then in the successive increase of the electron number in 4*f* shell each by one up to 7 for europium. No change of the 4*f* shell occurs in the next element gadolinium, but one electron appears again in the 5*d* shell of it. Having the next atomic number terbium has the 4*f* shell with nine electrons. This electron number is two more as compared with that of gadolinium. Simultaneously in terbium the 5*d* shell becomes again without electrons. The subsequent filling of the electron 4*f* shell from terbium to lutetium proceeds in the same way with increasing atomic number as from cerium to gadolinium. The number of electrons in the 4*f* shell increases successively up to the limiting value 14 for ytterbium and in the next element lutetium only the 5*d* shell is filled additionally by one electron again. Thus, the change of the electron configuration of the lanthanum series atoms is periodic with two periods. Each of the periods corresponds to the first or second half of the 4*f* shell being filled by electrons. Periodic change in the

Table 2. Electron configuration of the isolated atoms of the rare earth metals [12].

Atomic number	Element	$n = 1$	$n = 2$		$n = 3$			$n = 4$				$n = 5$			$n = 6$
		1s	2s	2p	3s	3p	3d	4s	4p	4d	4f	5s	5p	5d	6s
21	Sc	2	2	6	2	6	1	2							
39	Y	2	2	6	2	6	10	2	6	1		2			
57	La	2	2	6	2	6	10	2	6	10		2	6	1	2
58	Ce	2	2	6	2	6	10	2	6	10	2	2	6		2
59	Pr	2	2	6	2	6	10	2	6	10	3	2	6		2
60	Nd	2	2	6	2	6	0	2	6	10	4	2	6		2
61	Pm	2	2	6	2	6	10	2	6	10	5	2	6		2
62	Sm	2	2	6	2	6	10	2	6	10	6	2	6		2
63	Eu	2	2	6	2	6	10	2	6	10	7	2	6		2
64	Gd	2	2	6	2	6	10	2	6	10	7	2	6	1	2
65	Tb	2	2	6	2	6	10	2	6	10	9	2	6		2
66	Dy	2	2	6	2	6	10	2	6	10	10	2	6		2
67	Ho	2	2	6	2	6	10	2	6	10	11	2	6		2
68	Er	2	2	6	2	6	10	2	6	10	12	2	6		2
69	Tm	2	2	6	2	6	10	2	6	10	13	2	6		2
70	Yb	2	2	6	2	6	10	2	6	10	14	2	6		2
71	Lu	2	2	6	2	6	10	2	6	10	14	2	6	1	2

electron configuration of the rare earth series atoms is explained by the 4*f* shell being most stable when it consists of 7 or 14 electrons (half-filled and completely filled). Division of the rare earth metals into cerium and yttrium subgroups corresponds to the electron 4*f* shell being filled periodically with the two periods.

The certain similarity between the chemical properties of the rare earth metals is explained by the proximity of the energy levels of the 4*f* and 5*d* shells in atoms [13]. On the other hand, the noted periodic change in the electron configuration of them corresponds to many examples when rare earth metals are compared with each other. Especially this feature is manifested in anomalous behaviour of europium and ytterbium amongst the rare earth metals. Europium and ytterbium are the elements of the rare earth series where the 4*f* electron shell is filled respectively by a half and fully and there is no electron in the 5*d* shell. Such electron configurations cause in many cases these elements to reveal significantly more different chemical and physical properties than the neighbouring rare earth metals in the lanthanum series. Anomalous behaviour of europium and ytterbium is connected with their ability to show valence of two, as compared with valence of three for other rare earth metals, except cerium showing often valence of four [14, 15]. However, europium and ytterbium also reveal valence of three as other rare earth metals in interaction with high electronegative elements, for example fluorine, chlorine and oxygen.

One of the most significant characteristics of the rare earth metal atoms is their size. It controls, to a great extent, the ability of the rare earth metals to form solid solutions and compounds with other elements. The atom size is commonly characterised by the atomic radius. Such a characterisation is based on some models which are considered in many fundamental books, for example [16]. One of the models suggests atoms to have the form of balls in a crystal lattice which contact each other if they are the nearest neighbours. In general, consideration of atoms in lattice as balls is justified, although it is only an approximation. Another assumption of the atomic radius conception is connected with a certain dependence of them on the crystal lattice type and the nature of other atoms being present in the lattice. Actually, this is the dependence of the atomic radius on the nature of the chemical bond in the lattice where the atom is a participant. In lattices with metallic bond the atomic radius depends on the co-ordination number (CNN). Differences between the radius values of the same atom in the lattices with different co-ordination numbers may be essential. For metals existing in several allotropic forms the atomic radius change with changing co-ordination number may be estimated immediately. Dependence of the atomic radius on the chemical bond and co-ordination number is the reason for another approximation in consideration of atomic radius of elements.

Certain approximations arise also from the calculation of the atomic radius. They are calculated from the distances between the nearest atoms in the lattice which are calculated from the lattice parameters. There are two characteristics of the metal atomic radius. One of them is obtained from the lattice parameters of pure metals and it is named "metallic" atomic radius. This is the most widespread characteristic of the atomic radii of metals in their alloys. As a rule, when the atomic radius of metal is mentioned then it means the metallic radius of the atom. In cubic metal lattices each atom is surrounded by the nearest atoms of the only kind remote at the same distance. It is not so in the crystal lattice of other symmetry. For example, in a hexagonal close-packed lattice with co-ordination number 12 there are two kinds of the nearest atoms for every one. Six of them lay in the same basis plane and six others lay in two neighbouring basis planes. Both distances between the nearest atoms are close, yet different. This fact is the reason for some approximation in calculation of the metallic radius of an atom assumed to be a ball in hexagonal close-packed lattice. A similar problem exists for calculation of the metallic atom radius from other lattices with lower than cubic symmetry. It is assumed to compare atomic radii of different metals for the same co-ordination number 12 making respective corrections for the co-ordination number if it is necessary. These corrections give additional approximations in determination of the metallic atomic radius.

Another characteristic of the metal atomic radius is called the "ionic" atomic radius. This value characterises the atom size in the ion form and is calculated from the compound lattices with ionic bond where metal atoms coexist with atoms of electronegative elements.

Despite certain assumptions and approximations, the atomic radius concept is quite effective to describe the metal atom size in alloys and compounds, especially when metals of close chemical properties are compared with each other. All the above mentioned features of the atomic radius estimations are significant also for rare earth metals and must be taken into consideration in their physico-chemical interaction with other elements.

Metallic atomic radii of the rare earth metals are presented in Table 3. They are derived from [15]. The metallic radii were calculated from the room temperature lattice parameters

Table 3. Metallic radii of rare earth metals and magnesium [15, 16].

Metal	Atomic number	Lattice at room temperature	Metallic radius, nm
Mg	12	hcp	0.1602
Sc	21	hcp	0.16406
Y	39	hcp	0.18012
La	57	dhcp	0.18791
Ce	58	fcc	0.18247
Pr	59	dhcp	0.18279
Nd	60	dhcp	0.18214
Pm	61	dhcp	0.1811
Sm	62	rh	0.18041
Eu	63	bcc	0.20418
Gd	64	hcp	0.18013
Tb	65	hcp	0.17833
Dy	66	hcp	0.17740
Ho	67	hcp	0.17661
Er	68	hcp	0.17566
Tm	69	hcp	0.17462
Yb	70	fcc	0.19392
Lu	71	hcp	0.17349

Abbreviations: hcp, hexagonal close-packed; dhcp, double hexagonal close-packed; fcc, face-centred cubic; bcc, body-centred cubic; rh, rhombohedral.

of pure rare earth metals. The corresponding lattice types are presented also in Table 3. For comparison in Table 3 the lattice and metallic radius of magnesium are presented which are derived from [16]. All atomic radii conform to co-ordination number (CNN) of 12. Most rare earth metals have at room temperature double hexagonal close-packed (dhcp) or simple hexagonal close-packed lattices (hcp). Both lattices have a co-ordination number of 12, but with two different distances between neighbouring atoms. The atomic radii for these rare earth metals were calculated from the average of the two distances. Four rare earth metals have room temperature lattices different from double hexagonal close-packed and hexagonal close-packed ones. They are cerium, samarium, europium and ytterbium. Cerium and ytterbium have face-centred cubic (fcc) lattice with co-ordination number of 12 and for each of them the metallic atomic radius was calculated immediately from the only nearest distance between atoms without any averaging. Europium has a body-centred cubic lattice with co-ordination number of 8. Its metallic atomic radius was calculated from the only nearest distance between atoms with correction to the co-ordination number of 12. The rhombohedral lattice of samarium consists of the same atom close-packed planes as the double hexagonal and simple hexagonal close-packed lattices, but with different, more complicated, alternation of these planes in volume. Co-ordination number of samarium lattice is also 12. Consequently, the metallic radius of samarium was calculated in the same way as those of elements with hexagonal close-packed lattices, from the average distances between two sorts of the nearest atoms.

The ionic atomic radii of the rare earth metals together with that of magnesium are presented in Table 4 according to compatible works [17, 18] where they were reviewed and assessed in detail. The ionic radius values were obtained by calculations based on the lattice parameters of the compounds of the rare earth metals with fluorine, chlorine and oxygen. All presented values of the ionic radii conform to the same rare earth metal valence of 3 and are corrected to the same co-ordination number of 6.

Table 4. Ionic radii of rare earth metals and magnesium.

Metal	Atomic number	Ion	Ionic radius, nm	Metal	Atomic number	Ion	Ionic radius, nm
Mg	12	Mg^{+2}	0.0720	Eu	63	Eu^{+3}	0.0950
Sc	21	Sc^{+3}	0.0730	Gd	64	Gd^{+3}	0.0938
Y	39	Y^{+3}	0.0892	Tb	65	Tb^{+3}	0.0923
La	57	La^{+3}	0.1061	Dy	66	Dy^{+3}	0.0908
Ce	58	Ce^{+3}	0.1034	Ho	67	Ho^{+3}	0.0894
Pr	59	Pr^{+3}	0.1013	Er	68	Er^{+3}	0.0881
Nd	60	Nd^{+3}	0.0995	Tm	69	Tm^{+3}	0.0869
Pm	61	Pm^{+3}	0.0979	Yb	70	Yb^{+3}	0.0858
Sm	62	Sm^{+3}	0.0964	Lu	71	Lu^{+3}	0.0848

For convenience of discussion, the metallic and ionic radii of rare earth metals are shown in Figure 1 as curves characterising their dependence on the atomic number. The main feature of them is decrease of the rare earth atomic radius with increasing atomic number for the elements of the lanthanum series. This regularity is especially evident for the ionic radii which decrease with increasing atomic number successively. Metallic atomic radii of the rare earth metals are significantly more than ionic ones. In general, they also decrease with increasing atomic number for the elements of the lanthanum series, but with several exceptions. The first of them is cerium whose metallic atomic radius is determined to be somewhat less than that of the next element of the lanthanum series, praseodymium. This small effect, however, may be explained by some uncertainty arising from the calculation of the cerium atomic radius using the parameters of its face-centred cubic lattice whereas the atomic radii of praseodymium and most other rare earth metals are calculated using parameters of the hexagonal close-packed lattices. This viewpoint is supported by the lattice parameters of the intermetallic compounds of different rare earth metals with one of such other metals as zinc and aluminium [19, 20]. Amongst these compounds having the same formula and crystal structure, the lattice parameters for cerium are less than lattice parameters for praseodymium. The same situation takes place also amongst magnesium compounds with cerium and praseodymium. These compounds will be considered later. Two other exceptions are europium and ytterbium. Their atomic radii are anomalously high and depart significantly from the general tendency of atomic radius change with increasing atomic number. The differences between atomic radii of europium and ytterbium and atomic radii of their neighbours in the lanthanum series are too high to explain by the cubic lattices of europium and ytterbium being unlike hexagonal lattices of most rare earth metals. In the opinion of [15] the significantly higher metallic atomic radii of europium and ytterbium are connected with valence of 2 for these rare earth metals in metallic state as compared with valence of 3 for all other rare earth metals.

The phenomenon of the decrease of atomic radius with increasing atomic number for the elements of the lanthanum series is known as "lanthanoid shortening". It amounts to

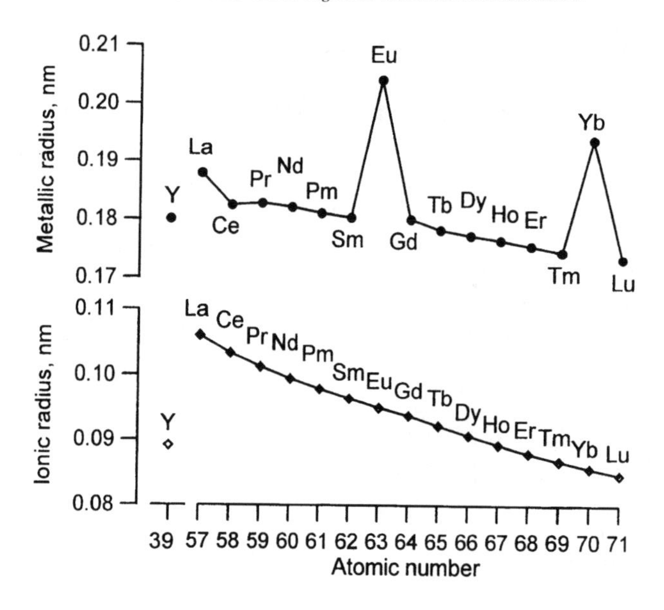

Figure 1. Atomic radius of the rare earth metals versus atomic number.

about 20% for ionic radii and about 8% for metallic radii. Lanthanoid shortening is explained by increase of attraction between electrons in the electron shells and nucleus with increasing nuclear charge [12].

At this point it is reasonable to note atomic radii of yttrium. Its ionic radius is quite close to that of holmium positioned in the middle of the second half of the lanthanum series (from gadolinium to lutetium). Meanwhile, the metallic radius of yttrium coincides practically with that of gadolinium which is the first element of the second part of the lanthanum series. Such a shift of the yttrium atomic radius as compared with the rare earth metals of the lanthanum series shows its different electron behaviour in interaction with other chemical elements. Anyway, both sorts of the yttrium atomic radius remain within the limits of those for the elements of the second half of the lanthanum series, and this fact justifies consideration of them as the united yttrium subgroup of the rare earth metals. Both metallic and ionic atomic radii of scandium are significantly less than the respective atomic radii of all other rare earth metals. Finally, the atomic radius of every rare earth metal is more than the atomic radius of magnesium. This is evident from Tables 3 and 4.

Along with atomic radius, an important factor for physico-chemical interaction of element is "electronegativity". Electronegativity characterises the ability of the valence electrons of an element to form chemical bonds with the valence electrons of another element. The highest electronegativity values are typical of the most chemically active nonmetals such as fluorine and oxygen. The least electronegativity values are determined for the most chemically active metals, caesium and rubidium. The greater the difference between electronegativity of the elements the more stable the compound formation between them. By contrast, small differences between electronegativities of the elements favour the formation of solutions between them. Electronegativity values calculated for the rare earth metals are presented in Table 5 [14]. For the elements of the lanthanum series, except europium and ytterbium, and also for yttrium they are close to each other and are within the limits 1.17–1.22. These values are close also to the electronegativity of magnesium of 1.2 calculated in the same way [21]. Europium and ytterbium fall out of the general tendency and show lower electronegativity values of about 1.0. Electronegativity of scandium, on the contrary, is more than that of most rare earth metals.

Table 5. Electronegativity of the rare earth metals [14].

Metal	Electronegativity	Metal	Electronegativity	Metal	Electronegativity
Sc	1.27	Pm	1.20	Ho	1.21
Y	1.20	Sm	1.18	Er	1.22
La	1.17	Eu	0.97	Tm	1.22
Ce	1.21	Gd	1.20	Yb	0.99
Pr	1.19	Tb	1.21	Lu	1.22
Nd	1.19	Dy	1.21		

All crystal structures of rare earth metals at atmospheric pressure are listed in Table 6 generally according to [15] with some additions from [11, 22]. As one can see from Table 6, some of the rare earth metals exist in several allotropic forms. These metals are Sc, Y, La, Ce, Pr, Nd, Pm, Sm, Gd, Tb, Dy, and Yb [11, 15]. The crystal structures of them at the highest temperatures are body-centred cubic. Body-centred cubic structures at high temperatures were discovered earlier also for Ho, Er, Tm and Lu [23, 24], but in recent publications they are rejected. Two metals, cerium and ytterbium, are known to show phase transformations during cooling below room temperature. Cerium has two allotropic forms below room temperatures, αCe and βCe. The first of them, αCe, forms after deep cooling from βCe or immediately from the room temperature form, γCe. Transformations proceed in a wide temperature range with hysteresis. Temperatures of the beginning and end of the transformations both during cooling and heating have not been established reliably. They are reported to be in the limits −248 – (−73)°C [22]. Rapid cooling hinders βCe →αCe transformation. αCe has the same face-centred cubic structure as γ Ce, but with significantly smaller parameter. βCe forms from γCe during cooling below about –15°C [22]. For ytterbium one allotropic form exists below room temperature, αYb. The phase transformation of ytterbium below room temperature shows hysteresis. Its temperature is reported to be −13°C during cooling and 3°C during heating [22]. In Table 6 the intermediate temperature of −3°C for this transformation is shown. There is also information about existence of orthorhombic forms of terbium and dysprosium below room temperature [11]. A number of specific crystal structures were established in rare earth metals at high pressures [11, 15].

In Table 7 some physical properties of rare earth metals are presented. One of the important physical characteristics of rare earth metals as alloying additives to magnesium is their density. Density depends upon atomic mass, atomic radius and crystal structure of the element. As all rare earth metals, except europium, have the closest packed lattices, their density is determined largely by atomic masses and atomic radii. In accordance with its small atomic mass scandium has the least density of 2.989 g/cm^3 amongst the rare earth metals. This value is only about 70% more than the density value of magnesium (1.741 g/cm^3 for the same temperature [25]). Yttrium has more atomic mass than scandium, but less atomic mass than the elements of the lanthanum series. Respectively, yttrium density of 4.469 g/cm^3 is intermediate between those of scandium and elements from lanthanum to lutetium. Within the lanthanum series, density of the rare earth metals generally successively increases with increasing atomic number as a result of the lanthanoid shortening and increase of atomic mass. Its increase from lanthanum to lutetium amounts to about 60%. Europium and ytterbium fall out of the general tendency having a smaller density than that of their neighbours in the lanthanum series in accordance with the anomalously higher atomic radii.

Table 6. Crystal structure of the rare earth metals [11, 15, 22].

Atomic number	Allotropic form	Temperature range, °C	Lattice	Pearson symbol	Space group	Prototype
21	αSc	<1337, RT	hcp	hP2	$P6_3/mmc$	Mg
	βSc	1337-m.p.	bcc	cI2	Im3m	W
39	αY	<1478< RT	hcp	hP2	$P6_3/mmc$	Mg
	βY	1478-m.p.	bcc	cI2	Im3m	W
57	αLa	<310, RT	dhcp	hP4	$P6_3/mmc$	αLa
	βLa	310–865	fcc	cF4	Fm3m	Cu
	γLa	865-m.p.	bcc	cI2	Im3m	W
58	αCe	below RT	fcc	cF4	Fm3m	Cu
	βCe	below RT	dhcp	hP4	$P6_3/mmc$	αLa
	γCe	<726, RT	fcc	cF4	Fm3	Cu
	δCe	726-m.p.	bcc	cI2	m	W
					Im3m	
59	αPr	<795, RT	dhcp	hP4	$P6_3/mmc$	αLa
	βPr	795-m.p.	bcc	cI2	Im3m	W
60	αNd	<863, RT	dhcp	hP4	$P6_3/mmc$	αLa
	βNd	863-m.p.	bcc	cI2	Im3m	W
61	αPm	<890, RT	dhcp	hP4	$P6_3/mmc$	αLa
	βPm	890-m.p.	bcc	cI2	Im3m	W
62	αSm	<734, RT	rh	hR3	R3m	αSm
	βSm	734–922	hcp	hP2	$P6_3/mmc$	Mg
	γSm	922-m.p.	bcc	cI2	Im3m	W
63	Eu	<m.p., RT	bcc	cI2	Im3m	W
64	αGd	<1235, RT	hcp	hP2	$P6_3/mmc$	Mg
	βGd	1235-m.p.	bcc	cI2	Im3m	W
65	αTb	<1289, RT	hcp	hP2	$P6_3/mmc$	Mg
	βTb	1289-m.p.	bcc	cI2	Im3m	W
66	αDy	<1381, RT	hcp	hP2	$P6_3/mmc$	Mg
	βDy	1381-m.p.	bcc	cI2	Im3m	W
67	Ho	<m.p., RT	hcp	hP2	$P6_3/mmc$	Mg
68	Er	<m.p., RT	hcp	hP2	$P6_3/mmc$	Mg
69	Tm	<m.p., RT	hcp	hP2	$P6_3/mmc$	Mg
70	αYb	<–3	hcp	hP2	Pr_3/mmc	Mg
	βYb	–3–795, RT	fcc	cF4	Fm3m	Cu
	γYb	795-m.p.	bcc	cI2	Im3m	W
71	Lu	<m.p., RT	hcp	hP2	$P6_3/mmc$	Mg

Abbreviations: fcc, face-centred cubic; bcc, body-centred cubic; hcp, hexagonal close-packed; dhcp, double hexagonal close-packed; rh, rhombohedral; RT means "exists at room temperature"; m.p., melting point.

Melting points of the rare earth metals of the lanthanum series change similarly to density with increasing atomic number, but with one exception for lanthanum. The melting point of lanthanum is higher than the melting point of the next element in the series, cerium. Beginning with cerium the melting points of the rare earth metals successively increase with increasing atomic number, except europium and ytterbium, whose melting points are lower than those of their neighbours in the series. Increase of melting points in the lanthanum series is significant and amounts to about 110% from cerium to lutetium. The melting point of yttrium is close to that of the elements in the middle of the yttrium subgroup of the lanthanum series. In this case the closest element is erbium having a melting point of 1545°C as compared with 1522°C for yttrium. Melting point of scandium is higher than that of yttrium.

Table 7. Some physical properties of the rare earth metals.

Metal	Density, g/cm^3 [15]	Melting point, °C [15]	Heat capacity, cal/g.-at· °C [24]	Heat of melting, kcal/g.-at [24]	Shear modulus, GPa [24]	Young's modulus, GPa [24]
Sc	2.989	1541	6.09	3.369	31.3	79.4
Y	4.469	1522	6.34	2.732	25.8	64.8
La	6.146	918	6.25	1.482	14.9	38.0
Ce	6.770	798	6.47	1.238	12.0	30.0
Pr	6.773	931	6.45	1.652	13.5	32.6
Nd	7.008	1021	6.55	1.705	14.5	38.0
Pm	7.264	1042	(6.50)	(1.94)	16.6	42.2
Sm	7.520	1074	6.80	2.061	12.7	34.1
Eu	5.244	822	6.48	2.202	5.9	15.2
Gd	7.901	1313	6.56	2.403	22.3	56.2
Tb	8.230	1365	6.81	2.583	22.9	57.5
Dy	8.551	1412	6.58	2.68	25.4	63.2
Ho	8.795	1474	6.50	2.91	26.7	67.1
Er	9.066	1529	6.72	4.76	29.6	73.4
Tm	9.321	1545	6.46	4.02	30.4	75.5
Yb	6.966	819	6.16	1.830	7.0	17.9
Lu	9.841	1663	6.46	4.55	33.8	84.4

The latent heat of fusion and modulus of elasticity change, in general, following the same relationship with increasing atomic number of the rare earth metals as melting point. In the lanthanum series the values of these characteristics tend to increase with increasing atomic number, and for lanthanum, as with the melting point, they are somewhat higher than those of cerium. Europium and ytterbium, as a rule, have lower values than their neighbours in the lanthanum series. The latent heat of fusion and modulus of elasticity of yttrium and scandium are also within the limits of those for the elements of the yttrium subgroup of the lanthanum series. All these characteristics are connected with interatomic bonds in the lattice of the metals and their change with increasing atomic number of the rare earth metals reflects regular change of the interatomic bonds. However, there are also some deviations from the above mentioned regularity. The modulus of elasticity of samarium is lower than that of promethium and even neodymium, and the latent heat of fusion of europium is intermediate between those of promethium and gadolinium. The deviations may be explained by the complicated nature of the physical properties, so that their dependence upon interatomic bonds in lattices may be different. All rare earth metals have approximately the same specific heat.

The rare earth metals are chemically reactive elements. They form various chemical compounds, including hydrides, fluorides, chlorides, oxides, sulphides, nitrides, carbides, silicides, salts of organic and inorganic acids, and complex compounds [14, 17, 24, 26–35]. Some of the non-metals, namely, oxygen, hydrogen, nitrogen, carbon, halogens can be present in rare earth metals of commercial purity. Contamination of the rare earth metals by these elements can happen during processes of their preparation. As the rare earth metals are quite chemically reactive, the removal of non-metallic impurities from them is commonly subject to various difficulties [24, 36].

Lanthanum, cerium, praseodymium, neodymium and europium are not stable in air at room temperature and quickly become oxidised. The rest of the rare earth metals are stable at least in dry air and may be stored in such conditions for a long time. According to [15], there is a correlation between the oxidation resistance and metallic radius of the rare earth metals. The higher the metallic radius of the rare earth metals, the less is their oxidation resistance in air. Ytterbium is the lone exception to the rule. Despite the large metallic radius (second in order after that of europium) its reactivity with the atmosphere is intermediate between those of samarium and gadolinium. Scandium is the most resistant in air corresponding to its least metallic radius of the rare earth metals. The oxidation resistance of rare earth metals in air is controlled by firmness of the oxide films on their surfaces.

Physico-chemical interaction of rare earth metals with other metals is very varied. Nevertheless, some common features can be identified. The rare earth metals interact with each other forming continuous solid solutions if both of them have the same crystal structures at corresponding temperatures. If both rare earth metals have not the same crystal structures they form limited solid solutions. The solid solutions are commonly quite extended. Intermediate phases in the rare earth binary systems occur only in some of them when each of the metals belong to different subgroups. Commonly there is only one intermediate phase in the system. It has one of the typical crystal structures for rare earth metals and significant homogeneity range [11, 14, 24, 37–41].

There are metals, for example vanadium, niobium and tantalum, which are immiscible with rare earth metals in the liquid state. On the other hand, many metals, for example aluminium, zinc, indium, cobalt and many others form alloys with the rare earth metals within the whole concentration range. One of the remarkable features of the interaction between the rare earth metals and other metals is formation of many intermetallic compounds in some systems. For example, in the system Ce–Zn nine intermetallic compounds of different compositions are formed. Both in the system Au–Yb and in the system Co–Ho eight compounds are formed [41–45]. Certain chemical similarities of rare earth metals suggested possible similarity of their binary phase diagrams with the same metal. This viewpoint was confirmed in [46]. In this investigation the binary phase diagrams of the rare earth metals published by that time were analysed. The binary phase diagrams included those of the three first elements of the lanthanum series, lanthanum, cerium and praseodymium, with silver, aluminium, gold, copper, magnesium and some other metals. Analysis showed that the phase diagrams of different rare earth metals with the same other metal were actually of the same general view with similar compounds and equilibria. The difference between them consisted mainly of the temperatures and concentrations of the critical points. However, later investigations showed this viewpoint not to be true when systems with more rare earth metals are considered. The views of the phase diagrams of separate rare earth metals with the same metal may be different, especially if the rare earth metals belong to different subgroups. As a rule, the similarity of the phase diagrams of the separate rare earth metals becomes less if one of the rare earth metals is europium or ytterbium.

Analysis of rare earth binary systems shows that the rare earth metals tend to form intermetallic compounds of the same formula and crystal structure with the same other metal. Such a row of isomorphous compounds exists commonly for a certain number of rare earth metals arranged successively in the lanthanum series beginning and ending at certain rare earth elements. As a rule, the rows of the isomorphous compounds are interrupted for europium and ytterbium. Meanwhile, the rare earth metals of different subgroups form often different intermetallic compounds with the same other metal. Yttrium forms commonly the same compounds as the heavy rare earth metals of the lanthanum series. As compared with yttrium, scandium shows more differences with rare earth metals of the lanthanum series in the intermetallic compound formation. However, isomorphous compounds amongst scandium and the rest of the rare earth metals also often occur.

The lattice parameters of the isomorphous compounds successively change with changing the rare earth metal atomic number. In compounds with metallic bonds, change of the lattice parameters shows corresponding change of the metallic radius of the rare earth metals. Nevertheless, the sequence of the lattice parameters values may differ from the sequence of the above mentioned metallic radii determined from the lattice parameters of the pure rare earth metals. Commonly a discrepancy takes place for cerium as compared with praseodymium as was noted earlier. In isomorphous compounds with ionic bonds formed by the rare earth metals with oxygen, chlorine and other non-metals the lattice parameters change shows a corresponding change of the ionic radius of the rare earth metals. The lattice parameters of the compounds with yttrium are commonly within the limits of the lattice parameter ranges of the same isomorphous compounds of the heavy rare earth metals. The lattice parameters of the scandium compounds fall out of the lattice parameters of the isomorphous compounds of other rare earth metals as a result of its significantly smaller metallic and ionic radii [33, 41, 47, 48].

Rare earth metals occur in the earth's crust in the form of many minerals, but only some of them are of commercial importance. They are monozite, bastnasite, xenotime, loparite and some others. Monozite is the most usual and relatively quite rich in rare earth metals. As a rule, the rare earth metals of the lanthanum series and yttrium are present always in minerals together. Nevertheless, quantities of individual rare earth metals in them are different. The minerals may be enriched by the elements of the cerium or yttrium subgroup, but ratios between contents of the elements in one of the subgroups, especially amongst the neighbouring ones, tend to remain the same. For example, monazite contains mainly rare earth metals of the cerium subgroup, although some metals of the yttrium subgroup are also always present. The typical chemical composition of the enriched monazite concentrate from Idaho, USA is as follows [26]: La_2O_3 – 17.0%, Ce_2O_3 – 29.9%, Pr_2O_3 – 3.9%, Nd_2O_3 – 11.0%, Sm_2O_3 – 1.3%, Eu_2O_3 – 0.001%, Gd_2O_3 – 0.5%, Y_2O_3 – 0.9%, Yb_2O_3 – 0.3%, ThO_2 – 3.5%, MgO – 0.1%, CaO – 0.3%, Al_2O_3 – 1.3%, Fe_2O_3 – 0.6%, TiO_2 – 0.2%, ZrO_2 – 0.1%, P_2O_3 – 28%, SiO_2 – 1.0%. This composition shows the general features of the rare earth distribution in the minerals. One of them is the relative contents of the rare earth metals. The highest content is typical of cerium. In order of content decrease, cerium is followed by lanthanum, neodymium, praseodymium, samarium and europium from the cerium subgroup. Amongst them cerium and lanthanum show significantly greater contents than the rest of the rare earth metals. On the other hand, europium shows significantly lower content than other rare earth metals of the cerium subgroup. Content of neodymium is commonly less than that of cerium, but more than contents of praseodymium and samarium. The composition of the monozite concentrate presented above shows also that the yttrium content is greater than other elements of the yttrium subgroup, gadolinium and ytterbium. The highest content of yttrium as compared with contents of all other rare earth metals of the yttrium subgroup is typical of minerals. Table 8 contains the results of chemical analysis of some other minerals [49]. They may be considered as minerals with high enough content of the rare earth metals of the yttrium subgroup and show typical ratios between those of them belonging to the lanthanum series. As Table 8 shows, amongst these rare earth metals the highest contents are typical

Table 8. Typical relative contents of the rare earth metals of the lanthanum series in some ores (%) [49].

La	Ce	Pr	Nd	Sm	Eu	Gd	Tb	Dy	Ho	Er	Tm	Yb	Lu
Fossil bone detrite. Total RE ~ 0.10%													
12.6	39.6	4.5	18.0	5.8	1.1	5.8	0.9	5.0	1.1	2.9	0.5	1.8	0.5
11.7	42.2	4.6	16.2	3.5	0.6	4.5	1.0	5.1	1.0	1.9	–	1.0	–
11.2	39.3	3.8	16.2	5.6	1.1	6.4	0.8	9.0	1.5	2.6	0.4	1.9	0.4
Apatite. Total RE ~ 0.4–0.8%													
20.4	42.8	5.8	20.5	3.4	0.3	3.4	0.3	1.6	0.2	0.9	<0.1	0.4	<0.1
Phosphorite. Total RE ~ 0.06–0.07%													
28.6	20.5	5.5	21.6	5.1	0.9	4.4	0.9	3.8	1.1	4.0	0.5	2.5	0.6
12.0	41.8	3.6	19.0	5.6	1.4	6.5	0.6	5.0	0.5	1.8	0.3	1.4	0.5
31.4	25.1	6.3	14.7	6.3	0.9	3.0	0.9	3.4	0.9	4.4	0.4	1.7	0.6
Mica-quartz albite. Total RE ~ 0.08%													
3.4	22.0	0.9	7.3	4.7	0.9	4.3	1.3	12.5	3.4	9.9	2.6	24.1	2.6
Yttrium bastnesite													
9.2	22.1	4.2	11.7	6.0	–	6.4	2.4	12.3	3.5	10.0	1.6	9.1	1.5
4.8	19.1	2.9	14.3	3.8	–	9.5	1.0	11.9	2.9	11.8	1.3	16.7	–
Uranite													
1.6	3.2	3.2	3.1	3.0	3.0	11.6	2.9	22.5	8.3	10.9	5.4	16.0	5.3

of gadolinium, dysprosium and erbium and the lowest contents are typical of thulium and lutetium. The significant feature of rare earth metals of both subgroups is the reduced contents of elements with odd atomic numbers, as compared with neighbouring elements with even atomic numbers. As a rule, scandium is present in the joint rare earth minerals in

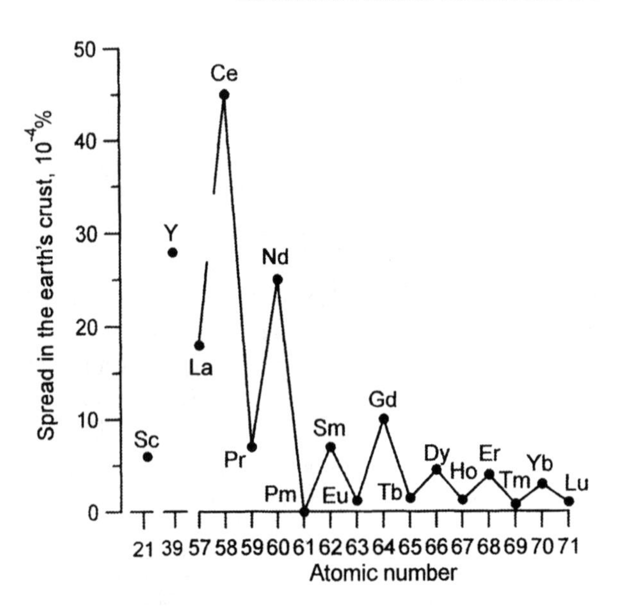

Figure 2. Spread in the earth's crust of rare earth metals.

insignificant amounts and tends to form its own minerals. This fact is connected with the substantially smaller atomic radius of scandium as compared with all other rare earth metals [13].

In general, the relative contents of rare earth metals in minerals conform to their average content (spread) in the earth's crust. There are several estimations of this quantity. The results obtained according to one of them are presented in Figure 2 [50]. Conformity between the above mentioned features of the rare earth metal distributions in the ores and their spread in the earth's crust is evident. The used estimation (by A.P. Vinogradov method) is common in Russia, but results of it agree, in general, with results of other estimations considered in [51]. Figure 2 shows clearly the reduced spread in the earth's crust for elements of the lanthanum series having odd atomic numbers as compared with those having even atomic numbers. One of the elements having odd atomic number 61, promethium, is practically absent in the earth's crust. The relative spread of rare earth metals in the earth's crust is of geochemical nature and the relative contents of the individual rare earth metals in ores reflect this phenomenon [49, 51].

Preparation of rare earth metals from ores includes several successive processes. Commonly, they are extraction of the rare earth metals jointly in the form of united chemical compounds, separation of the individual rare earth metals in the form of chemical compounds, reduction of the compounds to the individual rare earth metals, and refining and purification of the metals obtained [15, 24, 36, 52–57]. Compounds for reduction are commonly fluorides or chlorides of the rare earth metals. They are reduced commonly with calcium and rarely with magnesium. Reduction of the heavy rare earth metals and yttrium is also carried out with lanthanum. After reduction the rare earth metals are commonly contaminated by gas impurities. The most common method to remove them is remelting in vacuum. Metals of the highest purity are obtained using distillation, zone melting and other well-known methods of deep purification. In rare earth metals of commercial purity the main impurities are other rare earth metals, metals used for reduction (calcium, lanthanum and others), and metals of crucibles where the rare earth metals are obtained during reduction or remelted for refining, for example, tantalum and molybdenum. Commercial rare earth metals can contain also some gas impurities retained after refining. By the 1970s all separate rare earth metals had been produced already as commercial materials. In Russia most of them had been delivered in the 1960s at various levels of purity with total contents of the main element in the limits 93.80–99.7 mass % with main impurities being other rare earth metals of 4.8–0.10 mass %, respectively [58]. Since the 1970s the purity of commercial rare earth metals produced in Russia has been significantly increased so that the usual contents of the main element in them has amounted to 97.7–99.85 mass %, with the main impurities remaining other rare earth metals [59].

The cost of separate rare earth metals is of great importance for their use as alloying additives in magnesium alloys. The cost depends upon many factors including the kinds of raw materials, technology used, purity of metals obtained, production and consumption volumes, market situation and others. Consequently, the cost of the rare earth metals may change over a wide range. However, some features of it are quite invariable. They are significant and deserve to be underlined. The cost of separate rare earth metals is different, but each of them, even the cheapest one, is very much more expensive than pure magnesium. So, the rare earth alloying elements contribute significantly to the cost of magnesium alloys containing them. The other important feature concerns the cost ratios between different rare earth metals. They reflect, in general, the ratios of their contents in the ores and, respectively, the ratios of the rare earth spread in the earth crust. The rare earth metal is the cheaper, the more it is spread in the earth's crust. This rule is especially strict if metals of the same purity and other identical conditions are compared. In accordance with it, cerium turns out to be the cheapest rare earth metal. Lanthanum is commonly the second after cerium in order of the cost increase, but the difference between them is not so high. Neodymium is always significantly more expensive than cerium and lanthanum, but cheaper than praseodymium, samarium and yttrium. Yttrium is the cheapest metal amongst the rare earth metals of the yttrium subgroup. Gadolinium, dysprosium, erbium and ytterbium having even atomic numbers are cheaper than other heavy rare earth metals having odd atomic numbers. The most expensive rare earth metals are thulium, lutetium and scandium. The same cost ratios are typical also of the rare earth oxides [60, 61].

The high cost of rare earth metals may be reduced if they are used jointly as alloys of some of them. Separation of the rare earth metals is an expensive technological operation and missing or shortening this process it is possible to decrease the final cost of the product. Usage of such obtained "mixtures" of rare earth metals is typical of alloying magnesium alloys. The most known of them is mishmetal. At first its composition corresponded approximately to the contents of the individual rare earth metals in the most widespread ore, monozite, and included mainly cerium, lanthanum, neodymium and praseodymium. An example of it is presented in [3] as follows (in mass %): La 22.6, Ce 50.6, Pr 6.4, Nd 18.2, Fe 0.59, Si 0.16, Cr 0.03, other impurities 1.42. Content of other rare earth metals was estimated to be not more than 1–2%. It was included in the shown La obtained by difference between total rare earth and sum of the Ce, Pr, Nd contents. The main peculiarity of this mishmetal was sufficiently high content of neodymium and praseodymium. Increased usage of neodymium in industry caused separation of it from other rare earth metals contained in ores. Simultaneously some praseodymium was separated, too. As a result, the more common mishmetal began to be produced with lower neodymium and praseodymium contents. The main rare earth metal in it remained cerium with common content in limits 50–75 mass % and the main other element was lanthanum [58]. Another known "mixture" of different rare earth metals used for magnesium alloys is named didymium in the Western World. It is an alloy of the rare earth metals consisting essentially of neodymium (not less than 70 mass %) with other rare earth metal being substantially praseodymium [62]. Typical contents are about 85 mass % Nd and 15 mass % Pr [63]. Used in Russia for commercial magnesium alloys "technical neodymium" consists mainly of Nd and Pr approximately in the same proportions as in didymium. In commercial magnesium alloys containing yttrium, certain heavy rare earth metals are present [62, 64, 65]. So, in production of these alloys yttrium may be added together with other elements of the yttrium subgroup without deep separation from them.

Binary Phase Diagrams of Magnesium with Individual Rare Earth Metals

The attractive properties of magnesium alloys containing various rare earth metals are closely connected with respective phase diagrams. Significantly different behaviour of separate rare earth metals in magnesium alloys is caused to a great extent by certain differences in the phase diagram constitutions. These differences were recognised only after a great number of investigations.

Investigations of the binary phase diagrams of magnesium with individual rare earth metals were started in early 20th century and went on actually up to the present. Some significant investigations of the phase diagrams were carried out only recently and, therefore, might not be still included in most known reference books. The first investigations were conducted on systems with cerium and two other elements of the cerium subgroup, lanthanum and praseodymium [46, 66–71]. Investigations of the phase diagrams with other rare earth metals were undertaken later. Amongst them great attention was paid to the systems Mg–Nd and Mg–Y which were of the most importance for commercial alloys [72–76]. Along with these the phase diagrams of the systems Mg–Gd [77], Mg–Yb [78], Mg–Eu [79] and Mg–Sc [80, 81] were studied, as well. In [82, 83] some Mg–RE phase diagrams were investigated on the RE sides. The latter investigations showed significant solid solubility of magnesium in the rare earth metals and eutectoid decomposition of the high temperature phases with the body-centred cubic crystal structure in these parts of the systems. The phase diagıams of magnesium with other rare earth metals were studied over the whole concentration range mainly only in late 1980s and early 1990s by R. Ferro with co-workers [84–88]. In Baikov Institute of Metallurgy the Mg–RE phase diagrams were studied on the magnesium sides in detail [89–108]. Many efforts were made also to investigate the crystal structure of many binary compounds formed in the Mg–RE systems [109–137]. The Mg–RE phase diagrams were revised and evaluated in many reviews [106, 138–144].

Very many works were devoted to the studies of Mg–RE phase diagrams which are now almost fully covered. Nevertheless, some details of the phase diagrams need to be clarified. In this review the binary phase diagrams of magnesium with individual rare earth metals are presented in two forms. The first form includes representation of the phase diagram in the whole concentration range with the most reliable and full version of them being adopted. All rare earth metals, except scandium, differ significantly from magnesium by atomic mass. On the other hand, there are many compounds in the majority of the systems. Because of this fact it is reasonable to draw the whole phase diagram in atomic percentages. Phase diagrams presented in atomic percentages are more convenient also for comparison with each other. However, for magnesium alloys the Mg-rich parts of the Mg–RE phase diagrams (up to the closest compounds) are the most important. These have been studied in more works and there were more contradictions in them. Taking into account the importance of the Mg-rich parts of the Mg–RE phase diagrams they are considered especially and in detail. These parts of the binary phase diagrams are presented in mass %. Such representation are more convenient in this case because the considered parts of the phase diagrams become more extended along the concentration axis and, moreover, for the assessment of alloys for possible commercial applications.

The Mg-rich parts of the Mg–RE phase diagrams were thoroughly investigated in Baikov Institute of Metallurgy. These investigations were carried out keeping, in general, the same conditions of the sample preparation and the same experimental techniques. The starting materials had to be of sufficiently high purity. Melting of the alloys was performed under flux consisting of about 50% LiCl + 50% KCl preventing them from burning, contamination and losses of the rare earth metals. Composition of all prepared alloys was checked by chemical analysis. Thermal analysis of the alloys was carried out also under the same flux with the same acting. Cooling rate during thermal analysis was within the limits 2–3 or 4–5°C/min. Great attention was paid to the solubility of rare earth metals in solid magnesium. It was determined using samples of the alloys annealed at different temperatures aiming to reach an equilibrium state with subsequent quenching in cold water to fix such a state. Possible losses of the rare earth metals in the alloys were taken into account and prevented. The structure of the quenched alloys was analysed by common light microscopy, by X-ray diffraction determination of the Mg solid solution lattice parameters (or interplane distances) and by measurements of electrical resistivity of the samples. The last method turned out to be especially convenient for exact determination of small values of the solid solubility in magnesium for the rare earth metals of the cerium subgroup and comparison of them. In this case the electrical resistivity method for solid solubility determination was significantly more exact and reliable than the X-ray diffraction method based on measurements of the Mg solid solution parameters [145]. Advantages of the electrical resistivity method resulted from the significantly higher change of electrical resistivity as magnesium dissolved small contents of the rare earth metals as compared with the change of the Mg lattice parameters. Microscopy method was very approximate for determination of the solid solution boundaries when the solubility was small because of the difficulty in distinguishing with confidence the initial appearance of the second phase in the structure, especially, after low temperature annealings resulting in precipitation of the second phase in the form of quite small particles. The resistivity method could be also used partially for determination of the solidus line along the solid solution field. After annealing at temperatures higher than the solidus and eutectic temperatures the samples retained their form and could be used for resistivity measurements. In this case the resistivity/concentration curves manifested the same character, and kinks on them corresponded to the solid solution limits as after annealings without incipient melting (below eutectic temperature).

However, for large solubility values observed in the alloys with rare earth metals of the yttrium subgroup at high temperatures, the electrical resistivity method turned out to be less effective for solubility determination than the X-ray lattice parameter measurements and microscopy observation. In this case the kinks on the resistivity/concentration curves corresponding to the solubility values could not be established precisely because resistivities of Mg solid solution and the second phase became comparable. Anyway, three methods were used in solid solution determinations and they supplemented each other. The results of each method of the solid solubility determination could be compared and the most reliable data of them could be chosen. Some of the binary Mg–RE phase diagrams constructed in Baikov Institute of Metallurgy were confirmed in the following investigations of the corresponding ternary phase diagrams. The author considers the Mg-rich parts of the Mg–RE phase diagrams constructed in Baikov Institute of Metallurgy to be the most reliable and gives descriptions of them based mainly on these data.

As a rule, in every Mg–RE system, a number of binary compounds are formed. For convenience, they are listed in the tables showing their formulas and crystal structures. However, the real composition ranges of the compounds can deviate from their formulas. This fact is connected with ascription of the formulas based on the types of the compound crystal structures, but not the real compositions of them. Really, the lattice of the compounds can contain some excess of any component atoms as compared with stoichiometry. The crystal structures of the compounds are characterised in the tables by the Pearson symbol and prototype. The Pearson symbol includes the main peculiarities of the crystal structure. The first letter shows its symmetry (c – cubic, h – hexagonal, t – tetragonal, o – orthorhombic, m – monoclinic, a – triclinic or anorthic). The second capital letter means the lattice character (F – face-centred, I – body-centred, P – primitive, R – rhombohedral, C – base-centred). The number in the Pearson symbol corresponds to the amount of the atoms in the lattice cell. The lattice parameters of the compounds could be determined many times yielding close, but somewhat different values. Only some of them are included in the tables. The chosen values of parameters are considered to be most reliable and convenient for comparison with the lattice parameters of the same type of compounds in other systems of magnesium with different rare earth metals.

Mg–La Phase Diagram

The Mg–La phase diagram for the whole concentration range is shown in Figure 3. It is drawn in general according to the reviews [139, 147] with change of Mg side according to our data [93, 106] and some insignificant amendments. On the lanthanum side (100–60 at. % La) the Mg–La phase diagram corresponds to the last investigation [148]. The main characteristics of the Mg–La phase diagram as a whole are formation of at least four binary compounds, $Mg_{17}La_2$, Mg_3La, Mg_2La, MgLa, with one of them (Mg_3La) being congruent melting one and two eutectic equilibria on the Mg and La sides. There are also two eutectoid decompositions, $MgLa_2 \rightarrow Mg_3La + MgLa$ and $\gamma La \rightarrow MgLa + \beta La$. The congruent melting compound Mg_3La is shown by dashed line on the Mg-rich side supposing some homogeneity range of it [70, 147]. Existence of this compound is confirmed in a number of works [46, 112, 133, 147]. Other binary compounds are formed from liquid by peritectic reactions. The compound $Mg_{12}La$ was observed in an investigation [149]. It is shown in the phase diagram by the dashed lines, because its homogeneity range is not determined exactly. On the contrary, in [115, 117] in equilibrium with magnesium solid solution the next phase $Mg_{17}La_2$ was observed. Composition of the $Mg_{17}La_2$ corresponds also more than $Mg_{12}La$ to the composition of the closest to Mg phase in the system (Mg_9La)

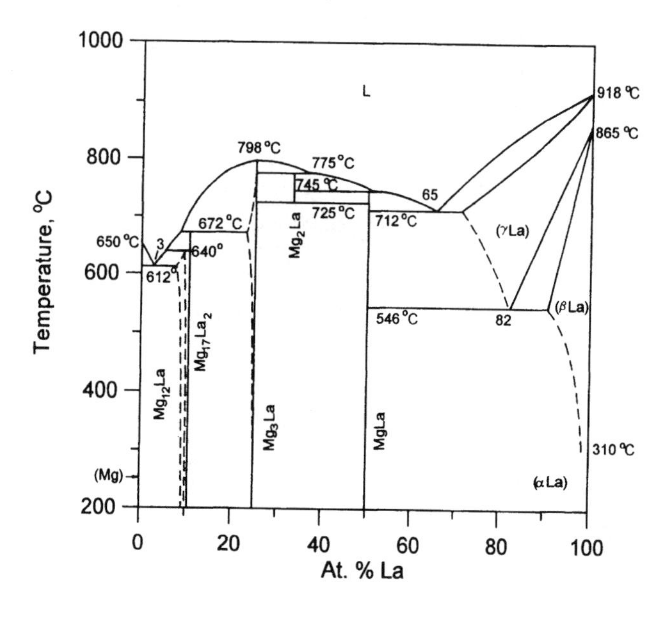

Figure 3. Mg–La phase diagram.

estimated in the earlier investigation of the Mg–La phase diagram [67]. Formation of the phase $Mg_{12}La$ was also reported in [150], but it was considered to be a metastable one. Peritectic reaction of the $Mg_{12}La$ formation was estimated by [147] approximately to be at near 650°C based on the analogy with the Mg–Ce system. However, in the latest

Table 9. Crystal structure of the Mg–La binary compounds.

Formula/ Temperature range, °C	Homogeneity range, at. % La	Symbol Pearson/ Prototype	Lattice parameters, nm	Comments
$Mg_{12}La$ <640	~7.7	oI338 $Mg_{12}Ce(II)$	$a=b=1.0320-1.0373$ $c = 7.724-7.756$	[147, 149, 175]
$Mg_{17}La_2$ <672	~10.5	hP38 $Ni_{17}Th_2$		[139, 147]
			$a = 1.036, c = 1.024$	[115]
			$a = 1.038, c = 1.027$	[117]
Mg_3La <798	? to 25	cF16 BiF_3		[139,147]
			$a = 0.7509$	[112]
			$a = 0.7467$	[133]
Mg_2La 775–725	~33.3	cF24 Cu_2Mg		[139, 147]
			$a = 0.8787$	[112]
			$a = 0.88007$	[162]
MgLa <745	~50	cP2 CsCl		[139, 147]
			$a = 0.3963$	[112]
			$a = 0.3970$	[125]

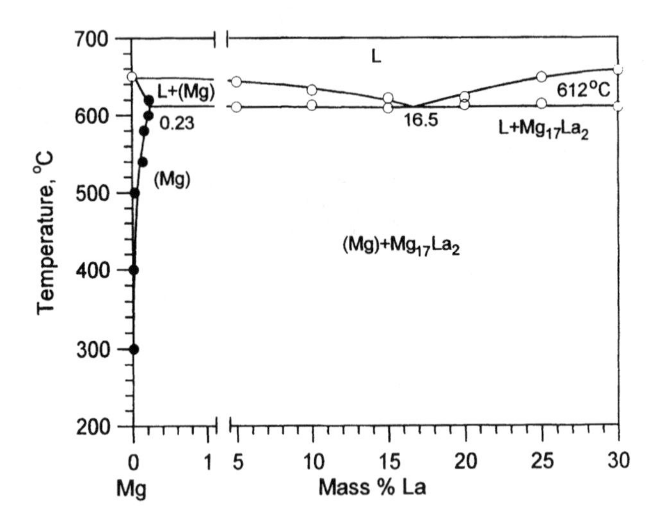

Figure 4. Mg-rich side of the Mg–La phase diagram.

investigation of the Mg–La phase diagram [175] existence of the compound $Mg_{12}La$ and its crystal structure was confirmed. The peritectic reaction of the $Mg_{12}La$ formation was established in [175] to be at 640°C in good accord with 650°C estimated in [147]. Existence of the other compounds $Mg_{17}La_2$, Mg_2La and MgLa is established quite reliably in many works [110–113, 115, 117, 125, 131, 147, 148].

The crystal structure of the binary compounds is presented in Table 9. The composition ranges of the Mg–La compounds were not determined specially and are shown in Table 9 assuming them to be near the stoichiometry ones. The composition range 7.14–8.3 at. % La shown in review [147] for $Mg_{12}La$ can not be considered to be true, because it is based only on compositions of three alloys where the phase was observed by [149] and the change of the lattice parameters of the phase was determined in them. The lattice parameter values determined in [149] are, however, doubtful, because they decreased with increasing the lanthanum content whereas the atomic radius of lanthanum is more than that of magnesium.

The phase diagram Mg–La on the Mg side constructed in our investigations is shown in Figure 4 [93, 106]. The eutectic temperature and eutectic point concentration were determined to be 612°C and 16.5 mass % La (3.34 at. % La), respectively, as compared with 613°C with 11.9 mass % La (2.31 at. % La) reported in the reviews [139, 147]. In other investigations for temperature and concentration of the eutectic point the following values were obtained, respectively: 610°C and 11.2 mass % La (2.16 at. % La) [71], 616°C and 12.3 mass % La (2.40 at. % La) [72], 611°C and 15.5 mass % La (3.11 at. % La) [150]. In our investigation the phase $Mg_{12}La$ (at about 32.3 mass % La) was not observed. Although the composition of the phase being in equilibrium with magnesium solid solution was not determined specially, the microstructure observations led to the conclusion that it is near 40 mass %, which corresponds approximately to the $Mg_{17}La_2$ composition. The solubility of La in solid magnesium was determined using electrical resistivity measurements. The results are shown in Figure 5. Despite the visible scattering connected with small scale resistivity change, the values obtained show evident decrease of the resistivity and,

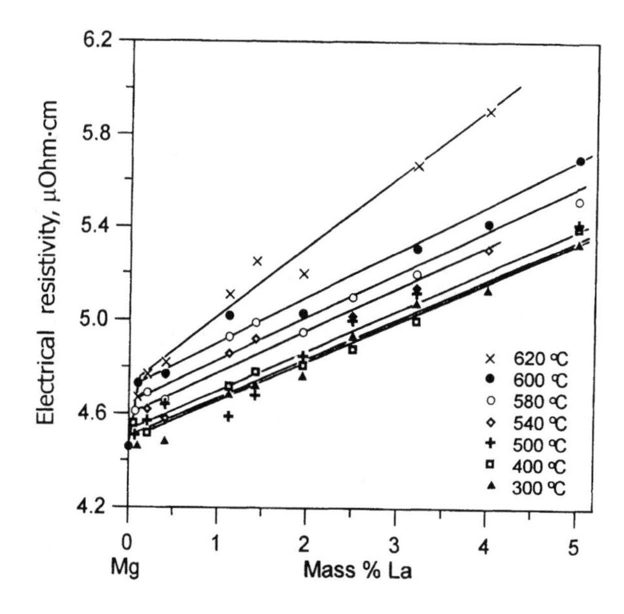

Figure 5. Electrical resistivity of the Mg–La alloys quenched from different temperatures.

consequently, existence and decrease of La solubility in solid magnesium with lowering temperature. The kinks on the resistivity curves are distinctly manifested and correspond to quite small solubility values. At separate temperatures the solubility of La in solid magnesium was determined to be: 600°C – 0.22 mass % (0.039 at. %), 580°C – 0.16 mass % (0.029 at. %), 540°C – 0.14 mass % (0.024 at. %), 500°C – 0.03 mass % (0.005 at. %), 400°C – 0.01 mass % (0.002 at. %), 300°C – 0.01 mass % (0.002 at. %). Actually the lines corresponding to the resistivity change with increasing La concentrations after quenching from 400 and 300°C coincided with each other. At eutectic temperature 612°C the maximum solubility value obtained by extrapolation of the values established for the separate temperatures amounted to 0.23 mass % La (0.042 at. % La). The solubility values of lanthanum in solid magnesium determined in the previous investigations differed significantly from each other and could not be considered to be exact enough. They were 3.0 mass % La at 609°C established by observation of resistivity change during continuous heating and X-ray method [71], 0.2 mass % La at 545°C established by microscopy observation [151], 0.8 mass % La at 616°C (eutectic temperature) established in [72] by X-ray method. In the review [147] the results of some other investigations of La solubility in solid magnesium are presented, but they are, in general, within the limits of the above mentioned values. Figure 5 presents also the results of resistivity measurements of Mg–La alloys annealed at a temperature of 620°C which is above eutectic temperature. The kink on this resistivity curve was used for determination of the solidus line along the Mg solid solution field in the phase diagram shown in Figure 4. In general, the results of our investigations on the Mg–La phase diagram agree with the results of previous investigations and the reviews [139, 147].

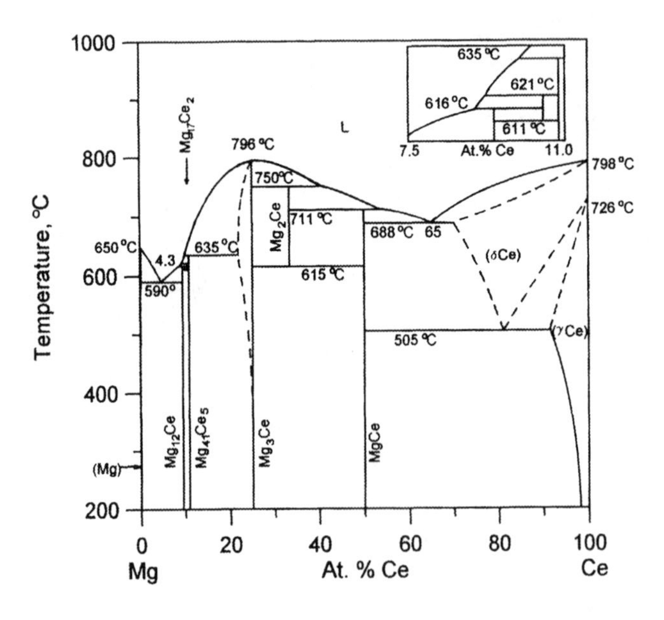

Figure 6. Mg–Ce phase diagram.

Mg–Ce Phase Diagram

The general view of the Mg–Ce phase diagram is shown in Figure 6. It is reproduced from the reviews [139, 152] with correction of the Mg-rich side according to our investigations [89, 90] and some other insignificant amendments. The Mg–Ce phase diagram as a whole is characterised by the existence of six intermetallic compounds with formulas $Mg_{12}Ce$, $Mg_{17}Ce_2$, $Mg_{41}Ce_5$, Mg_3Ce, Mg_2Ce and MgCe. Only one of them, Mg_3Ce is a congruently melting compound. All other compounds are formed from liquid by peritectic reactions. In the system there are two eutectic transformations on Mg and Ce sides and two eutectoid transformations. One of the eutectoid transformations is decomposition of the compound Mg_2Ce, and the second one is decomposition of the solid solution of the high temperature modification of Ce. The binary compounds Mg_3Ce, Mg_2Ce and MgCe are well established. Their existence is confirmed in a number of works [109–113]. For Mg_3Ce the homogeneity range was established with the solubility of magnesium in it decreasing with lowering temperature. The change of the magnesium solubility in Mg_3Ce was not determined exactly and is shown, therefore, by the dashed line.

In the area of three other binary compounds, $Mg_{12}Ce$, $Mg_{17}Ce_2$ and $Mg_{41}Ce_5$, there are some uncertainties and it is reasonable to consider them in detail. In the earlier investigation of the Mg–Ce phase diagram [66] the phase established in equilibrium with magnesium solid solution was described by the formula Mg_9Ce. The formula was derived only from estimation of the phase composition. A number of investigations were carried out then where the crystal structure of the compound was studied by the X-ray diffraction method. Only one of them recognised the formula Mg_9Ce and ascribed a certain crystal lattice to it [153]. All other investigators rejected this formula because it did not correspond to the phase crystal structure. In earlier investigations two phases in the Mg-rich Mg–Ce alloys were established by X-ray diffraction. Their crystal structures corresponded to the formulas $Mg_{12}Ce$ [122, 123] and $Mg_{17}Ce_2$ [115, 117, 120]. These investigations were not accompanied by thermal analysis and microscopy observations. However, [122] managed to perform chemical analysis of the phase they studied by the X-ray method and obtained its composition to be near, but Ce-richer, as compared with the established formula $Mg_{12}Ce$ (from $Mg_{11.45}Ce$ to $Mg_{11.96}Ce$). In a later investigation [127], thermal analysis and microscopy methods were used, and three binary compounds were established in the concentration area near that corresponding to the formula Mg_9Ce pointed out by [66] for the intermediate phase closest to Mg. These three compounds were described by the formulas, $Mg_{12}Ce$, $Mg_{17}Ce_2$ and $Mg_{8.25}Ce$. The corresponding part of the phase diagram is included in the general view of the Mg–Ce phase diagram in Figure 6 with corrections of the $Mg_{12}Ce$ phase composition according to our investigation. It is shown also in the insert of Figure 6 on an enlarged scale. All three compounds are formed following three successive peritectic reactions. One of the compounds, $Mg_{17}Ce_2$, decomposes eutectoidly into two others and exists as an equilibrium phase only within a narrow temperature range. According to [127], this compound, however, is stable and can be retained easily after cooling down to room temperature. Existence of the phase $Mg_{12}Ce$ was also confirmed in [149].

In [124] the crystal structure of the compound $Mg_{12}Ce$ was determined repeatedly and two allotropic forms of it were established. The first form was of the same tetragonal type

Table 10. Crystal structure of the Mg–Ce binary compounds.

Formula/ Temperature range, °C	Homogeneity range, at. % Ce	Symbol Pearson/ Prototype	Lattice parameters, nm	Comments
$Mg_{12}Ce(I)$ <616	~9.4	tI26 $Mn_{12}Th$	$a = 1.033$, $c = 0.5964$	[124, 127, 152, 154]
$Mg_{12}Ce(II)$?	~9.4	oI338 $Mg_{12}Ce(II)$	$a = b = 1.033$ $c = 7.75$	[124, 127, 152, 154]
$Mg_{17}Ce_2$ 621–611	~10.53	hP38 $Ni_{17}Th_2$		[115, 127, 152]
			$a = 1.035, c = 1.026$	[115]
			$a = 1.033, c = 1.025$	[130]
$Mg_{41}Ce_5$ <635	10.81	tI92 $Mg_{41}Ce_5$		[152]
			$a = 1.454, c = 1.028$	[127]
			$a = 1.478, c = 1.043$	[129]
Mg_3Ce <796	~21.7–25	cF16 BiF_3		[152]
			$a = 0.7428$	[112]
			$a = 0.7424$	[133]
Mg_2Ce 750–615	33.33	cF24 Cu_2Mg		[152]
			$a = 0.8733$	[112]
			$a = 0.87326$	[162]
MgCe <711	50	cP2 CsCl		[152]
			$a = 0.3898$	[110]
			$a = 0.3912$	[112]

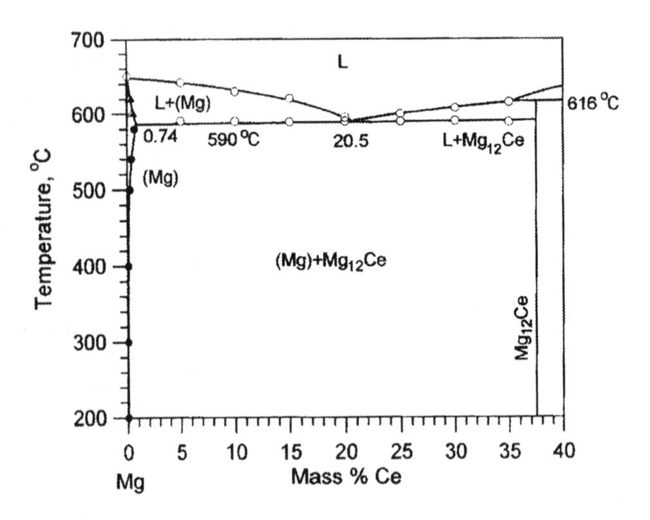

Figure 7. Mg-rich part of the Mg–Ce phase diagram.

as reported in [115, 117, 120] and the second form was of the orthorhombic type derived from the first one. The second allotropic form of $Mg_{12}Ce$ was considered to be a long ordering modification of its first allotropic form, and a possible schematic addition to the phase diagram was proposed [127, 152]. However, neither temperature of the ordering/disordering transition was established. [124] studied the phase crystals extracted immediately from the slowly cooled hypereutectic Mg–Ce alloy and performed chemical analysis of the crystals. It showed composition of the crystals to be $Mg_{11.59-11.71}Ce$, somewhat richer in Ce as compared with stoichiometry of the compound. In [130] the crystal structure of the $Mg_{17}Ce_2$ compound was studied repeatedly. Its lattice type reported in [115, 117, 120] was confirmed. Meanwhile, analysing the lattice parameters of the $Mg_{17}Ce_2$ compound the authors [130] concluded it was enriched in Mg as compared with the stoichiometric composition and, therefore, must be described by the formula $Mg_{10.3}Ce$. This formula was used for the compound in the reviews [139, 152], but it can not be considered to be convincing enough yet, in that it is not confirmed by chemical analysis or metallography. For the third compound $Mg_{8.25}Ce$ the stoichiometric formula $Mg_{42}Ce_5$ was initially assumed based on the study of its crystal structure [129]. It was replaced then by a more exact one $Mg_{41}Ce_5$ [132].

The crystal structures of the Mg–Ce binary compounds are described in Table 10. Composition of the phase $Mg_{12}Ce$ is shown in it according to our data [154] which will be described later. The crystal structure of $Mg_{12}Ce$(II) is shown in the form presented in [124, 152]. It can be considered as a particular case of the orthorhombic lattice. The homogeneity range of Mg_3Ce is shown according to the data on the maximum solubility of Mg in the compound reported in [152]. Compositions of other compounds are shown in Table 10 according to their stoichiometry formulas. At least for $Mg_{17}Ce_2$ the composition shown is quite approximate as far as a certain excess of Mg in it is suggested [130, 132, 139, 152]. Only two typical values of the lattice parameters for each compound are chosen to be presented in Table 10.

Figure 7 shows the Mg-rich part of the Mg–Ce phase diagram constructed experimentally in Baikov Institute of Metallurgy [89, 90]. Temperature and composition of

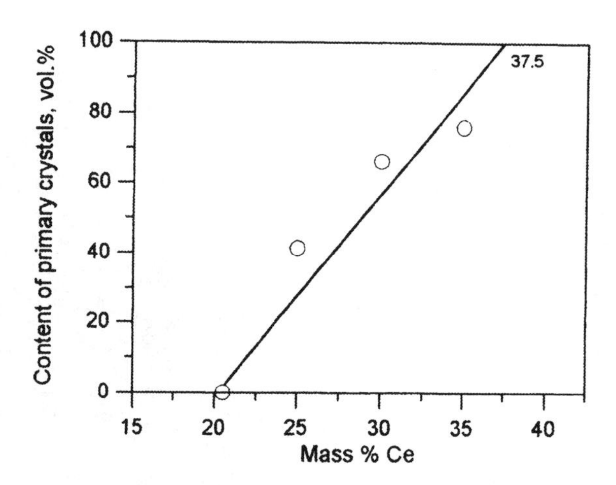

Figure 8. Content of the Ce-rich primary crystals versus Ce concentration in the structure of the Mg-rich Mg–Ce alloys.

the eutectic point, 590°C and 20.5 mass % Ce (4.28 at. % Ce), respectively, established in this investigation, turned out to be in good agreement with results of other works. The other established values characterising the eutectic point were reported to be 585°C and about 24 mass % Ce (5.20 at. % Ce) [66], 590°C and 21 mass % Ce (4.41 at. % Ce) [69], 593°C and 17.3 mass % Ce (3.50 at. % Ce) [72], 593°C and 20.2 mass % Ce (4.21 at. % Ce) [127], and 591°C and 20.5 mass % Ce (4.28 at. % Ce) [93]. In the review [152], the average values 592°C and 20.6 mass. % Ce were assumed for it. However, composition of the compound being in equilibrium with magnesium solid solution, according to our data, differs significantly from those reported by other investigators [122, 124] and assumed in review [152]. In [122, 124] composition of the compound $Mg_{12}Ce$ closest to Mg was determined to be in limits 32.9–33.5 mass % Ce (7.84–8.02 at. % Ce), whereas in our investigation the alloy Mg–35 mass % Ce higher in Ce had a structure after thermal analysis consisting only of the eutectic and the primary crystals of the compound. Figure 8 shows results of quantitative metallography of Mg–Ce alloy structures [154]. In quantitative metallography the volume content of the primary crystals of the compound was determined in the structure of the hypereutectic alloys obtained after thermal analysis. Correlation between the volume content of the primary crystals and composition of the alloys is manifested by the curve in Figure 8. Extrapolation of the curve to the volume content of 100% gives composition of the phase to be about 37.5 mass % Ce corresponding to the formula $Mg_{9.60}Ce$ (9.43 at. % Ce). These data support the assumption that the compound $Mg_{12}Ce$ is significantly enriched in Ce as compared with its stoichiometry. Consequently, composition corresponding to the formula $Mg_{10.3}Ce$ can not be ascribed to the next phase $Mg_{17}Ce_2$ as was done in [130, 152]. We believe our data on the composition closest to Mg phase to be quite reliable because they are immediately connected with results of thermal analysis and microstructure observations unlike data [122, 124]. Nevertheless, the existing contradiction on this phase composition requires additional investigations.

Solubility of Ce in solid magnesium was established in our investigations mainly by the electrical resistivity method. The results of the resistivity measurements are presented in Figure 9. They show the resistivity curves for every quenching temperature with distinct

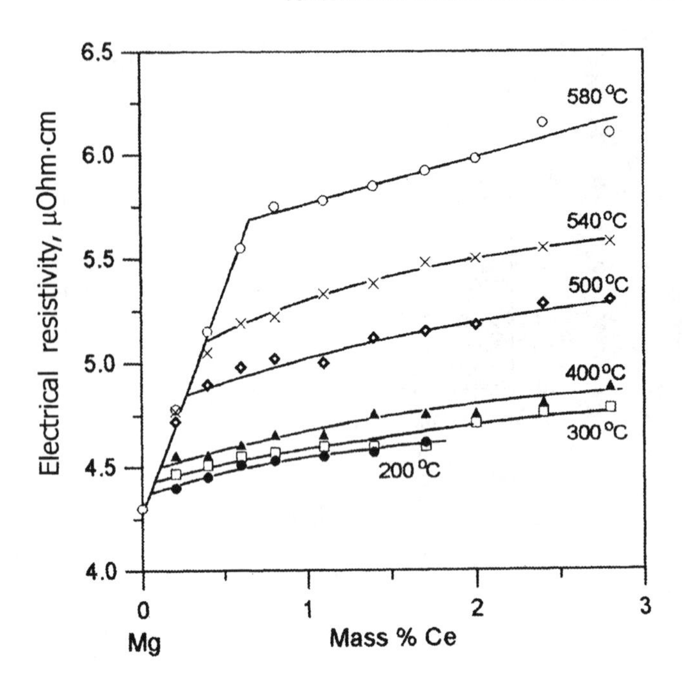

Figure 9. Electrical resistivity of the Mg–Ce alloys quenched from different temperatures.

kinks corresponding to the solubility values. As compared with the Mg–La alloys, the resistivity ascent in the Mg solid solution area for the Mg–Ce alloys is more, and consequently there are more distances between the resistivity curves for separate temperatures in the two-phase area. This fact suggests more solubility of Ce in solid magnesium as compared with La. The solid solubility of Ce in solid magnesium was established to be at 590°C – 0.74 mass % (0.13 at. %), at 580°C – 0.66 mass % (0.13 at. %), at 540°C – 0.38 mass % (0.066 at. %), at 500°C – 0.26 mass % (0.045 at. %), at 400°C – 0.08 mass % (0.014 at. %), at 300°C – 0.06 mass % (0.010 at. %), and at 200°C – 0.04 mass % (0.007 at. %). The solubility value at eutectic temperature 590°C was obtained by the extrapolation of the other solubility values determined immediately from the resistivity curves. The solubility of Ce in solid magnesium was determined earlier by many investigators and the values obtained by them were in quite a wide range. At the highest of the temperatures used they amounted to 2.2 mass % at 590°C (electrical resistivity measurements during heating) [71], 0.3–0.4 mass % at 565°C (microscopy observations) [151], 0.5 mass % at 593°C (X-ray method) [72], 1.6 mass % at 590°C (microscopy observations) [69], 0.8 mass % at 577°C (quantitative metallography) [155], 0.85 mass % (unknown method) [156]. Our solubility values are inside the limits of those reported by other investigators, but as Figure 9 shows, they could be determined quite exactly and within wide temperature range.

The solidus line along the Mg solid solution field was determined in our investigation by microscopy observations of the incipient melting after annealing at temperatures above eutectic one. Results for two edge alloys are shown in part of the Mg–Ce phase diagram (Figure 7).

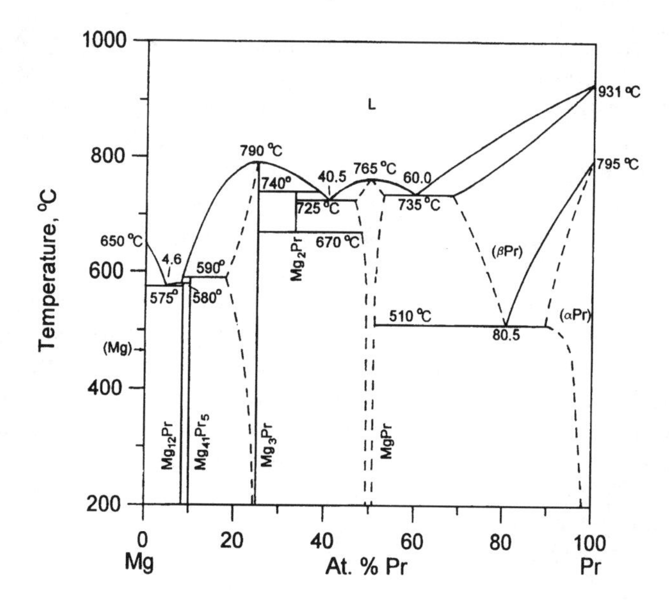

Figure 10. Mg–Pr phase diagram.

Mg–Pr Phase Diagram

Figure 10 shows the general view of the Mg–Pr phase diagram. It is drawn based on the reviews [139, 158] corrected and supplemented by the data of the Baikov Institute of Metallurgy [92, 94, 106] in the part from Mg to the compound closest to Mg and by the data from

Table 11. Crystal structure of the Mg–Pr binary compounds.

Formula/ Temperature range, °C	Homogeneity range, at.% Pr	Symbol Pearson/ Prototype	Lattice parameters, nm	Comments
$Mg_{12}Pr$ <580	at least >8.2 and <9.9	tI26 $Mn_{12}Th$	$a = 1.034$, $c = 0.598$	[92, 119, 136, 159, 160, 161]
$Mg_{41}Pr_5$ <590	~10	tI92 $Mg_{41}Ce_5$	$a = 0.1475(2)$ $c = 0.1441(1)$	[136, 158]
Mg_3Pr <790	?–25	cF16 BiF_3		[159, 160]
			$a = 0.7430$	[112]
			$a = 0.7411$	[133]
Mg_2Pr 740–670	33.33	cF24 Cu_2Mg		[159, 160]
			$a = 0.8689$	[112]
			$a = 0.86957$	[162]
MgPr <765	50	cP2 CsCl		[159, 160]
			0.3885	[112]
			0.3898	[125]

[159, 160] in the part from the closest to Mg compound to Pr. Five binary compounds were found in the Mg–Pr system. They are described by the formulas $Mg_{12}Pr$, $Mg_{41}Pr_5$, Mg_3Pr, Mg_2Pr and MgPr derived from their crystal structures. According to the last investigation [159], two compounds, Mg_3Pr and MgPr, are congruently melting ones. Other compounds are formed from the liquid phase by the peritectic reactions. There are three eutectic transformations and two eutectoid transformations in the system. One of the eutectoid transformations is decomposition of the compound Mg_2Pr into Mg_3Pr and MgPr, and another of them is decomposition of the (βPr) solid solution into MgPr and the (αPr) solid solution. For the compounds $Mg_{41}Pr_5$, Mg_3Pr and MgPr, the existence of some homogeneity ranges is assumed. Crystal structures of the compounds are presented in Table 11.

Figure 11 demonstrates the Mg-rich area of the Mg–Pr phase diagram constructed in our investigation [92, 94]. The phase diagram was studied by thermal analysis, microscopy observations and electrical resistivity measurements. Results of the thermal analysis are shown in the plot. Temperature of the eutectic equilibrium in the Mg-rich part of the system was established to be 575°C [92, 94] as compared with 583°C [46], 573°C [72], 570°C [150], and 560°C [159, 160]. Concentration of the eutectic point was established to be 22 mass % Pr (4.7 at. % Pr) [92, 94] as compared with 9.1 mass % Pr (1.7 at. % Pr) [72], 27 mass % Pr (6 at. % Pr) [150], and 23 mass % Pr (5 at. % Pr) [159, 160]. In general, our results on temperature and concentration of the eutectic point are at the middle of the values obtained by other investigators.

There are some discrepancies on the compound closest to Mg with the assumed formula $Mg_{12}Pr$ between our results and the data by other authors. This compound was reported first in [119] after X-ray diffraction investigation of one alloy of the composition corresponding to 7.7 at. % Pr (32.6 mass % Pr). In the investigation [119] the type of the compound crystal structure and its parameters were determined. The authors of [119] assumed the compound to be in equilibrium with Mg solid solution, but they did not report how much Mg solid solution could be present in the samples studied. This fact led to the conclusion that there was a possible error in their determination of the compound composition. In our investigation [92] carried out by thermal analysis and microscopy observations, the Mg–Pr alloys containing up to 34 mass % Pr (8.2 at. % Pr) remained hypereutectic. Thus, according to our data [92], the compound closest to Mg contains more Pr than that corresponding to the formula $Mg_{12}Pr$. Taking into account our microscopy observations, the Pr content in the compound is at least more than 8.2 at. % conforming to the formula $Mg_{11.2}Pr$. The upper possible limit of the Pr content in the compound was not determined. However, it could be assumed to be less than 9.9 at. % Pr. The alloy of such a composition consisted actually of the next compound $Mg_{41}Pr_5$ only [136]. These estimations are included in Table 11.

Existence of $Mg_{12}Pr$ was assumed then in [139, 150, 158–161]. According to [159, 160] it is formed by peritectic reaction $L+Mg_{41}Pr_5 \rightarrow Mg_{12}Pr$ at 565°C (by 5°C more than temperature of the eutectic reaction $L \rightarrow (Mg)+Mg_{12}Pr$). Temperature 565°C [159, 160] of the peritectic reaction $L+Mg_{41}Pr_5 \rightarrow Mg_{12}Pr$ contradicts our investigation [92] where temperature of the eutectic reaction $L \rightarrow (Mg)+Mg_{12}Pr$) was established to be 575°C. Therefore, it was reasonable to assume temperature of the peritectic reaction $L+Mg_{41}Pr_5 \rightarrow Mg_{12}Pr$ to be 580°C retaining the same excess over the eutectic temperature. The same temperature 580°C was also reported for the peritectic reaction $L+Mg_{41}Pr_5 \rightarrow Mg_{12}Pr$ by [161]. Assumption of these reaction temperatures required to be corrected in the same way the temperature 590°C for the next invariant reaction

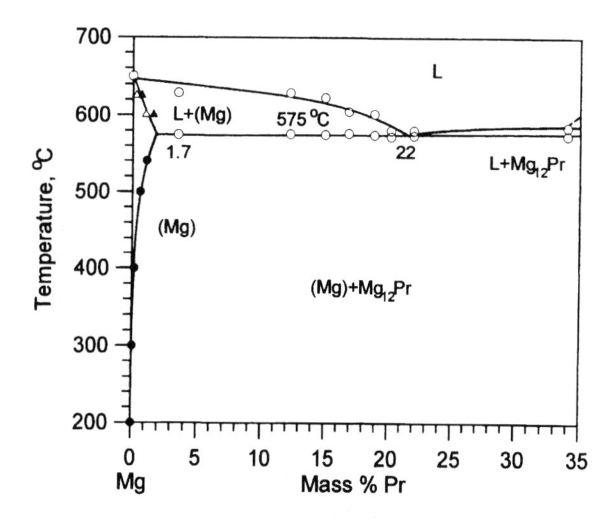

Figure 11. Mg-rich side of the Mg–Pr phase diagram.

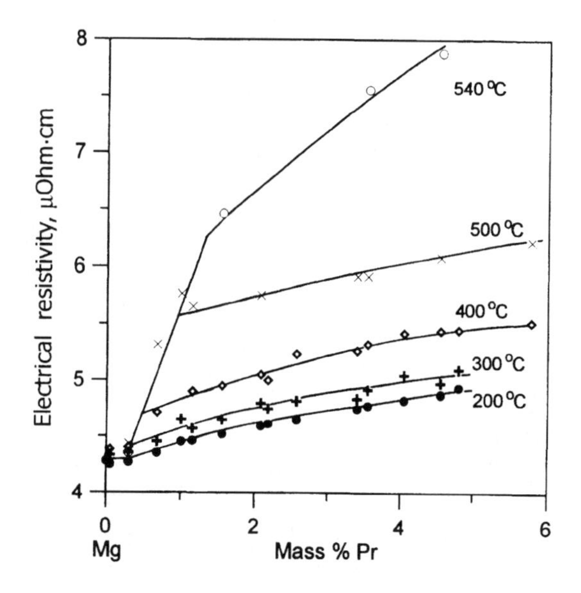

Figure 12. Electrical resistivity of the Mg–Pr alloys quenched from different temperatures.

L+Mg_3Pr → $Mg_{41}Pr_5$ as far as temperature 575°C reported by [159, 160] for it was not compatible with the assumed 575°C and 580°C for the reactions L → (Mg)+$Mg_{12}Pr$ and L+$Mg_{41}Pr_5$ → $Mg_{12}Pr$, respectively.

Solubility of praseodymium in solid magnesium was investigated by microscopy observations and measurements of electrical resistivity [92, 94, 106]. The solubility values obtained by both methods were compatible, although the latter one was considered to be more exact. The results of the electrical resistivity measurements are presented in Figure 12. They are typical with one peculiarity showing absence of the resistivity increase

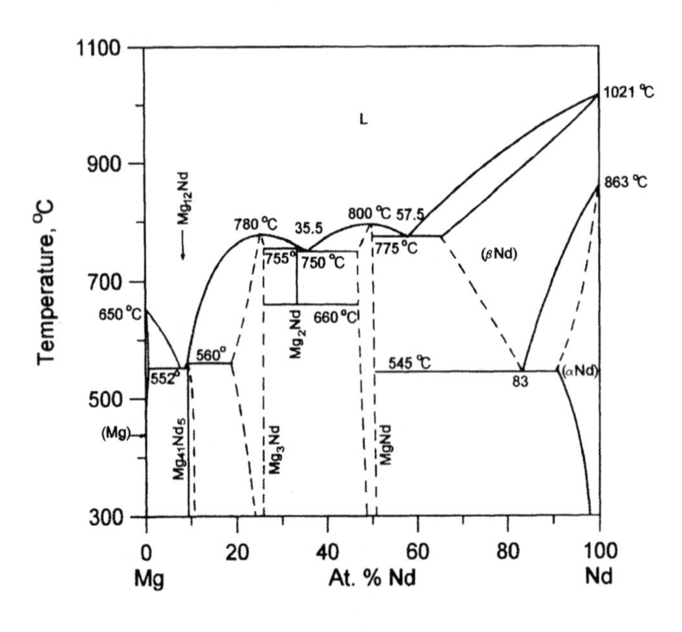

Figure 13. Mg–Nd phase diagram.

at the small initial concentrations of praseodymium (up to 0.3 mass %). Absence of the resistivity increase at the small Pr concentrations was explained by the interaction of it with small quantities of hydrogen which could be caught by the alloys accidentally and connect Pr into a hydride. The high sensibility of the electrical resistivity for the solid solution concentration enabled this effect to be revealed. The true solubility values were calculated by subtraction of 0.3 mass % Pr from the values corresponding to the main kinks on the resistivity/concentration curves shown in Figure 12 for the different quenching temperatures. They amounted to 540°C – 1.09 mass % Pr (0.19 at. % Pr), at 500°C – 0.6 mass % Pr (0.10 at. % Pr), at 400°C – 0.2 mass % Pr (0.034 at. % Pr), at 300°C – 0.05 mass % Pr (0.009 at. % Pr) and at 200°C ~0.01 mass % Pr (~0.002 at. % Pr). All of them are shown in Figure 11. At eutectic temperature 575°C the solubility value of 1.7 mass % Pr (0.31 at. % Pr) was obtained by extrapolation of the values calculated for different temperatures. The solidus line outlining Mg solid solution field in the phase diagram (Figure 11) is drawn using results of the microscopy observations after annealing at different temperatures. The open triangles in the plot correspond to the structure without any signs of partial melting during annealing and the dark triangles correspond to the structure with the first signs of partial melting.

Mg–Nd Phase Diagram

Figure 13 shows the general view of the Mg–Nd phase diagram. It is drawn following mainly the last study of it by [163] with correction only of the Mg-rich part according to the investigations by the Baikov Institute of Metallurgy [89, 90, 106]. The phase diagram is characterised by existence of at least four stable binary compounds, $Mg_{41}Nd_5$, Mg_3Nd, Mg_2Nd and MgNd. Two of them, Mg_3Nd and MgNd, are congruently melting ones, and the two others, $Mg_{41}Nd_5$ and Mg_2Nd, are incongruently melting compounds. The fifth binary compound $Mg_{12}Nd$, which had also occurred in Mg-rich Mg–Nd alloys, was recognised in [163] to be a metastable phase. Nevertheless, $Mg_{12}Nd$ is also shown in the phase diagram in Figure 13. The other main features of the Mg–Nd phase diagram are the existence of three eutectic equilibria and two eutectoid equilibria. The eutectoid equilibria correspond to decomposition of either Mg_2Nd or the solid solution of the Nd high temperature allotropic form (βNd) during cooling. Two compounds, Mg_3Nd and MgNd, were suggested to have some homogeneity ranges from the lattice parameters measurements [163], but their boundaries were not established. Therefore, these homogeneity ranges are shown in the phase diagram in Figure 13 by the dashed lines only.

Compositions and crystal structures of the compounds are described in Table 12. It contains also description of the compound $Mg_{12}Nd$, although it was concluded in [163] to be a metastable phase. $Mg_{12}Nd$ was discovered by analysing the X-ray diffraction pattern

Table 12. Crystal structure of the Mg–Nd binary compounds.

Formula/ Temperature range, °C	Homogeneity range, at. % Nd	Symbol Pearson/ Prototype	Lattice parameters, nm	Comments
$Mg_{12}Nd$ ~552	~8.31	tI26 $Mn_{12}Th$		Metastable, [89, 106, 115, 163]
			$a = 1.030, c = 0.593$	[115]
			$a = 1.0307, c = 0.5947$	[163]
$Mg_{41}Nd_5$ <560	~9.36	tI92 $Mg_{41}Ce_5$		[89, 136, 163]
			$a = 1.476, c = 1.039$	[136]
			$a = 1.4741, c = 0396$	[163]
Mg_3Nd <780	~25	cF16 BiF_3		[133, 163]
			$a = 0.7397$	[133]
			$a = 0.7399–0.7413$	[163]
Mg_2Nd ~755–660	33.3	cF24 Cu_2Mg		[112, 162, 163]
			$a = 0.8662$	[112]
			$a = 0.86744$	[162]
			$a = 0.8671$	[163]
MgNd <800	~50	cP2 CsCl		[112, 163]
			$a = 0.3867$	[112]
			$a = 0.3881$	[125]
			$a = 0.3865–0.3869$	[163]

from the Mg–Nd alloy containing 10.5 at. % Nd [115]. Its formula was assumed then according to the type of the lattice established without taking into account the real composition. In the next investigations [122, 123] the closest to Mg compound was electrochemically extracted from the two-phase Mg–Nd alloys containing 1.1 and 3.71 at. % Nd. The extracted crystals were studied by X-ray diffraction and chemical analysis. The type of crystal structure of the compound of the formula $Mg_{12}Nd$ was confirmed, but its composition turned out to be poorer in Mg as compared with the formula up to $Mg_{9.94}Nd$.

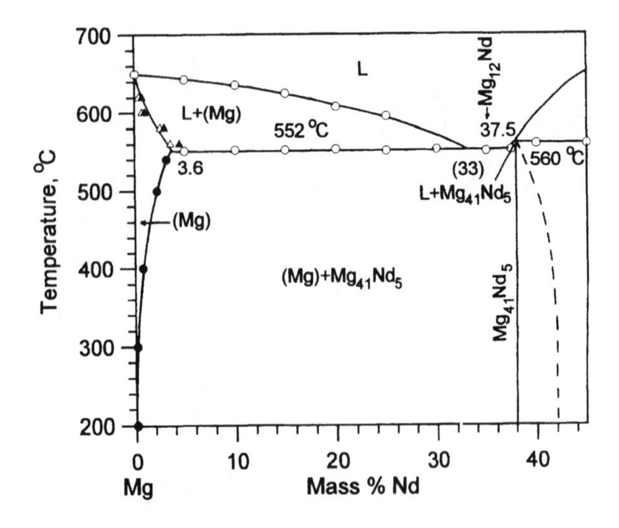

Figure 14. Mg-rich side of the Mg–Nd phase diagram.

In [163] the phase $Mg_{12}Nd$ was recognised to be metastable because it was observed only in the alloys quenched from liquid state.

Figure 14 presents results of our investigation of the Mg-rich part of the Mg–Nd phase diagram [89, 90, 106]. Thermal analysis with a cooling rate of $2°C \cdot min^{-1}$ showed eutectic equilibrium at temperature 552°C with eutectic point at about 33 mass % Nd (7.66 at. % Nd). In other investigations the following eutectic temperatures and the eutectic point compositions were reported: 548°C and 25–28 mass % Nd (5.31–6.15 at. % Nd) [73], 546°C and 9.3 mass % Nd (1.7 at. % Nd) [72], and 545°C and 32.5 mass % Nd (7.5 at. % Nd) [163]. In general, characteristics of the eutectic equilibrium shown in Figure 14 [89, 90, 106] agree with the results of the other works [72, 73, 163], except for the too low value of 9.3 mass % Nd (1.7 at. % Nd) for the eutectic point reported by [72]. We consider our data on the temperature and eutectic point to be the most reliable, because they were obtained using more alloys, high purity of magnesium and neodymium and quite slow cooling during thermal analysis.

The eutectic point of 33 mass % Nd (7.66 at. % Nd) [89, 106] turned out to be very near the stoichiometric composition of $Mg_{12}Nd$ (7.69 at. % Nd) suggesting that the compound was enriched by Nd. In the structure of the hypoeutectic alloys obtained after thermal analysis the typical eutectic constituent was not observed [89, 106]. It consisted of Mg solid solution and coarse continuous crystals of an intermetallic phase only. Quantitative metallographic analysis (Figure 15) showed the phase composition to be in limits 35–40 mass % Nd (8.32–10.1 at. % Nd). These values correspond to more Nd than the stoichiometric composition of the compound $Mg_{12}Nd$ (7.69 at. % Nd). Taking into consideration the results of the investigations [115, 122, 123], the observed intermetallic phase in the Mg–Nd alloys which solidified during thermal analysis was identified by us as $Mg_{12}Nd$ enriched in Nd [106]. Behaviour of the observed phase was, however, unusual. In separate places continuous crystals of $Mg_{12}Nd$ showed fine plate-like inner constitution suggesting eutectoid decomposition. Such a possibility was confirmed by the annealing of the alloys resulting in a cheese-type structure of the crystals which is typical of a globular

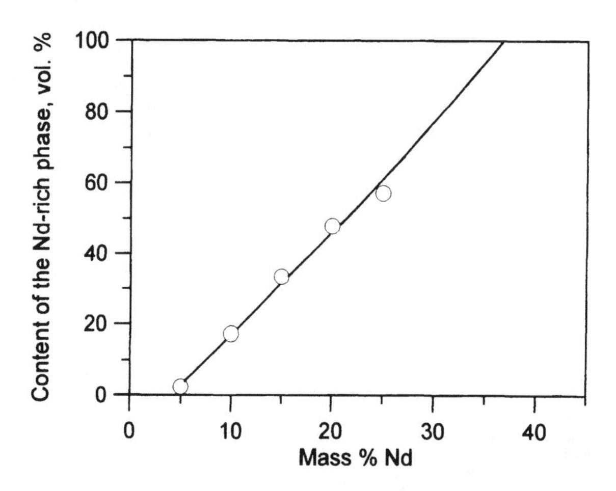

Figure 15. Content of the Nd-rich phase versus Nd concentration in the structure of the Mg-rich Mg–Nd alloys.

eutectoid. On the other hand, at 37.5 mass % Nd (9.18 at. % Nd), being within the limits of the possible $Mg_{12}Nd$ composition, the structure of the alloy changed abruptly and immediately after thermal analysis consisted completely of typical fine eutectic. The significantly different view of the eutectic suggested it consisted of Mg solid solution together with the next binary compound $Mg_{41}Nd_5$. Based on thermal analysis and these microscopy observations the part of the phase diagram between 33 and 40 mass % Nd was proposed [106], where the phase $Mg_{12}Nd$ was presented as a congruent melting compound at about 35 mass % Nd (8.31 at. % Nd) with two adjoining eutectic equilibria, at 33 and 37.5 mass % Nd. Both eutectic equilibria took place at temperatures close to each other. The phase $Mg_{12}Nd$ was proposed to exist within a quite narrow temperature range with the melting and decomposition temperatures being also close to those of both eutectics. Consideration of the compound $Mg_{12}Nd$ as a metastable phase by [163] agrees with our data [89, 90, 106] of the ability of this phase to decompose into Mg solid solution at a temperature near its formation temperature during crystallisation. On the other hand, we observed this phase after quite slow cooling rate (2°C$\cdot$min^{-1}) unlike [163] where $Mg_{12}Nd$ was observed only after quenching from the liquid state. In [115, 122, 123] where $Mg_{12}Nd$ was revealed by X-ray diffraction the details of the alloy crystallisation were not reported, but most likely these authors did not use quenching from the liquid state, either. In summary, recognition of $Mg_{12}Nd$ as a metastable phase [163] may be considered to be justified, but the phase is easily formed in Mg–Nd alloys during crystallisation and the possibility of its presence in cast structures must be always taken into account. In this case the eutectic point at 33 mass % Nd may be ascribed to the three-phase equilibrium with metastable phase, $L \rightleftharpoons (Mg)+Mg_{12}Nd$, and that at 37.5 mass % Nd may be ascribed to the three-phase equilibrium with the stable phase, $L \rightleftharpoons (Mg)+Mg_{41}Nd_5$ (Figure 14). The part of the Mg–Nd phase diagram between 30 and 40 mass % Nd needs, however, additional study.

The phase $Mg_{41}Nd_5$, according to our data [89, 106], is formed by peritectic reaction $L+Mg_3Nd \rightarrow Mg_{41}Nd_5$ at 560°C in complete accordance with [163]. The real composition of $Mg_{41}Nd_5$ is estimated in [136] to be about Mg–38 mass % Nd ($Mg_{9.68}Nd$) on Mg side.

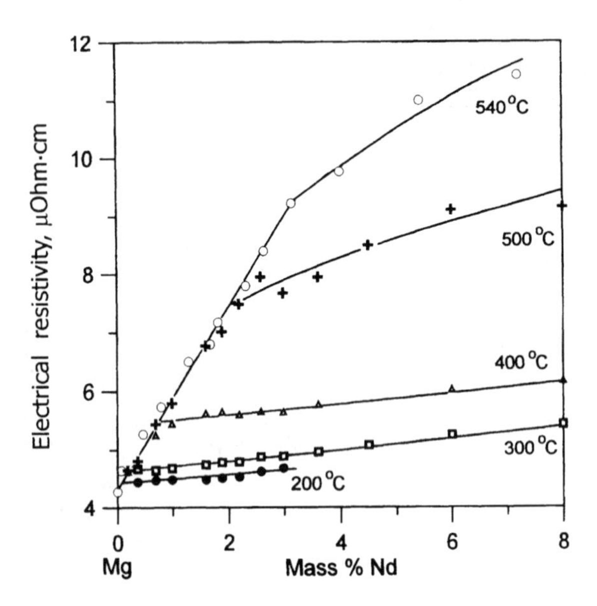

Figure 16. Electrical resistivity of the Mg–Nd alloys quenched from different temperatures.

In [163] a homogeneity range for $Mg_{41}Nd_5$ was suggested, but its limits were not determined.

Solubility of neodymium in solid magnesium was investigated by microscopy observations and electrical resistivity measurements [89, 90]. The results of the latter method were considered to be more exact, especially at low temperatures. They are shown in Figure 16. At the highest temperature 540°C the kink on the resistivity/concentration curve was not manifested clearly enough and its position needed to be checked by microscopy observations of the sample structures. The solubility values established for different temperatures were: at 540°C – 3.2 mass % Nd (0.55 at. % Nd), at 500°C – 2.2 mass % Nd (0.38 at. % Nd), at 400°C – 0.70 mass % Nd (0.12 at. % Nd), at 300°C – 0.16 mass % Nd (0.027 at. % Nd), and at 200°C – 0.08 mass % Nd (0.013 at. % Nd). These values are shown in the phase diagram in Figure 14. Their extrapolation to the eutectic temperature 552°C gave the maximum solubility of Nd in solid Mg to be 3.6 mass % (0.63 at. %). The solidus line in Figure 14 is drawn along the points showing results of the structure observations after quenching of the samples from the temperatures above the eutectic. The open triangles on the plot show absence of incipient melting in the structure of the samples, and the solid triangles show its presence.

Mg–Pm Phase Diagram

The author does not know of any investigation in which the Mg–Pm phase diagram has been studied experimentally. Nevertheless, the strict regularities in the Mg–RE phase diagram enable one to propose the Mg–Pm one with a high probability. These regularities were considered in [106, 112,128, 133, 164] and will be described in the next paragraph. Following these regularities, the Mg–Pm phase diagram should be, in general, intermediate

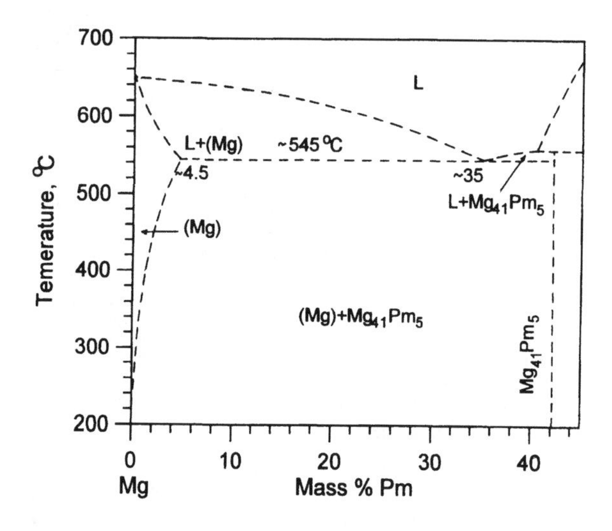

Figure 17. Proposed Mg-rich side of the Mg–Pm phase diagram.

between those of the Mg–Nd and Mg–Sm systems. The proposed Mg-rich part of the Mg–Pm phase diagram is shown in Figure 17. It is of the eutectic type with temperature and eutectic point concentration being about 545°C and about 35 mass % Pm (8.3 at. % Pm), respectively. The maximum solubility of promethium in solid magnesium is estimated to be about 4.5 mass % (0.78 at. %). The compound closest to Mg is proposed to be $Mg_{41}Pm_5$ in the Mg–Pm system, taking into consideration the tendency to stability decrease and disappearance of the $Mg_{12}Ln$ type compound in systems with Ce, Nd and Sm, and the existence of compounds of the $Mg_{41}Ln_5$ type in all of them.

Three other compounds can be assumed in the Mg–Pm system following its analogy with the systems Mg–Nd and Mg–Sm. They are Mg_3Pm of the cubic BiF_3 type (cF16) with $a = \sim 0.740$ nm, Mg_2Pm of the cubic Cu_2Mg type (cF24) with $a = \sim 0.865$ nm and MgPm of the cubic CsCl type (cP2) with $a = \sim 0.385$ nm. The assumed $Mg_{41}Pm_5$ is of the tetragonal $Mg_{41}Ce_5$ type (tI92) with $a = \sim 1.476$ nm, $c = \sim 1.035$ nm.

Mg–Sm Phase Diagram

The whole Mg–Sm phase diagram is presented in Figure 18. It is derived from the latest work [84] with some corrections of the Mg-rich part according to the data of Baikov Institute of Metallurgy [96, 105]. The same Mg–Sm phase diagram was reproduced, in general, in the reference book [139]. The main characteristics of the system are the existence of five binary compounds with two of them being congruently melting ones. In accordance with this, there are three eutectic equilibria where the congruently melting compounds, the Mg- and Sm-base solid solutions and the liquid phase are participants. There is also a eutectoid equilibrium connected with the existence of several allotropic forms of samarium. Four of the compounds are similar to those observed in the system Mg–Nd. They are $Mg_{41}Sm_5$, Mg_3Sm, Mg_2Sm, MgSm. The compound Mg_5Sm belongs to the type which does not occur in the systems of magnesium with other rare earth metals of the cerium subgroup, but it occurs in the Mg–Gd system. On the other hand, the compound

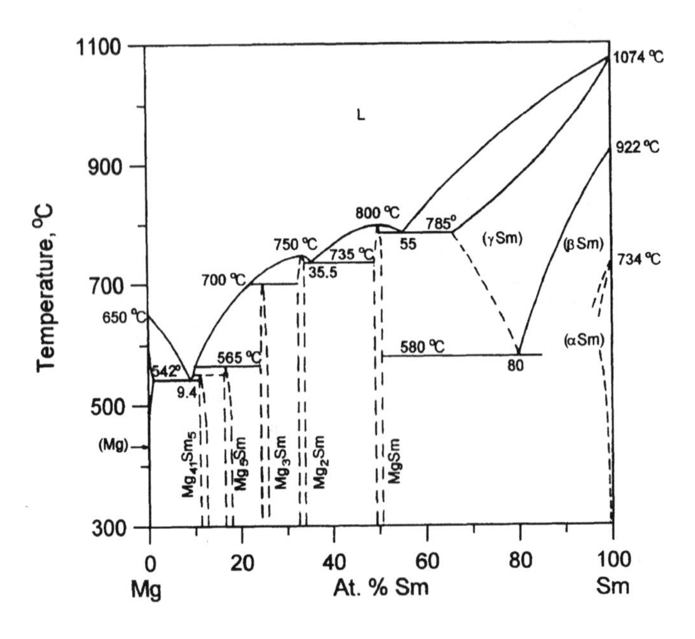

Figure 18. Mg–Sm phase diagram.

of the $Mg_{12}Ln$ type had not been observed in the Mg–Sm system. The congruently melting compounds are Mg_2Sm and MgSm. Other compounds are incongruently melting ones. For all the compounds, small homogeneity ranges were proposed. For the compounds Mg_3Sm, Mg_2Sm and MgSm the existence of the homogeneity ranges was supported by the difference between the lattice parameters determined on the Mg-rich and Sm-rich sides of

Table 13. Crystal structure of the Mg–Sm binary compounds.

Formula/ Temperature range, °C	Homogeneity range, at.% Sm	Symbol Pearson/ Prototype	Lattice parameters, nm	Comments
$Mg_{41}Sm_5$ <~550	~10.87	tI92 $Mg_{41}Ce_5$	a = 1.480, c = 1.036	[84, 139]
			a = 1.477, c = 1.032	[134]
Mg_5 Sm <565	~16.67	cF448 $Cd_{45}Sm_{11}$	a = 2.246	[84, 139, 135]
Mg_3Sm <700	~25	cF16 BiF_3	a = 0.7346–0.7371	[84, 139]
			a = 0.7360	[133]
Mg_2Sm <750	~33.3	cF24 Cu_2Mg	a = 0.8622–0.8639	[84, 139]
			a = 0.86289	[162]
MgSm <800	~50	cP2 CsCl	a = 0.3832–0.3848	[84, 139]
			a = 0.3845	[125]
			a = 0.3810	[112]

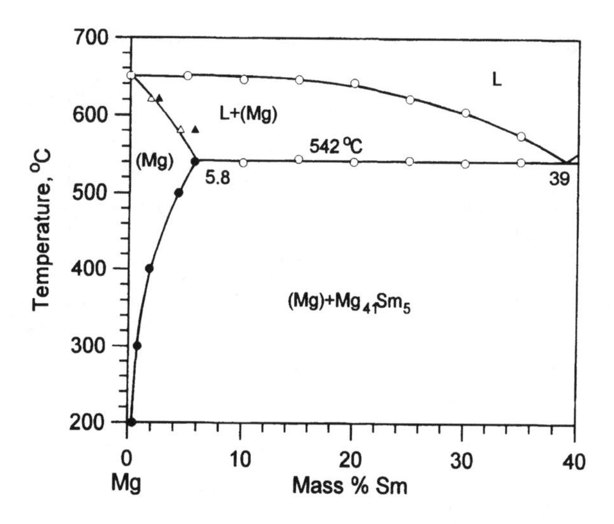

Figure 19. Mg-rich side of the Mg–Sm phase diagram.

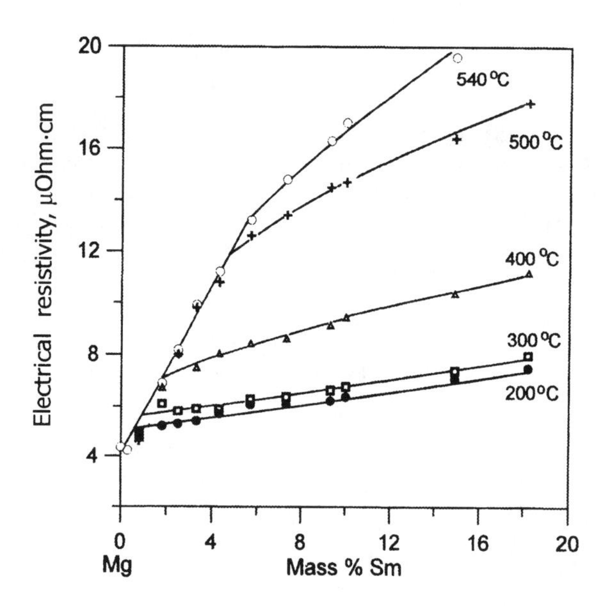

Figure 20. Electrical resistivity of the Mg–Sm alloys quenched from different temperatures.

the phases [84]. The crystal structures of the Mg–Sm compounds are listed in Table 13. For the compound Mg_5Sm, the prototype Mg_5Gd was assumed in [139], but it is replaced more correctly in Table 13 by the prototype $Cd_{45}Sm_{11}$ assumed for Mg_5Gd [135].

The Mg-rich part of the phase diagram with results of the experiments is presented in Figure 19 [96, 105, 134]. The compound closest to Mg was identified as $Mg_{41}Sm_5$ [84, 134]. Its composition was assumed to be about $Mg_{8.2}Sm$ corresponding to 43 mass % Sm (10.87 at. % Sm) [84]. The earlier composition $Mg_{6.2}Sm$ [105] was quite approximate and

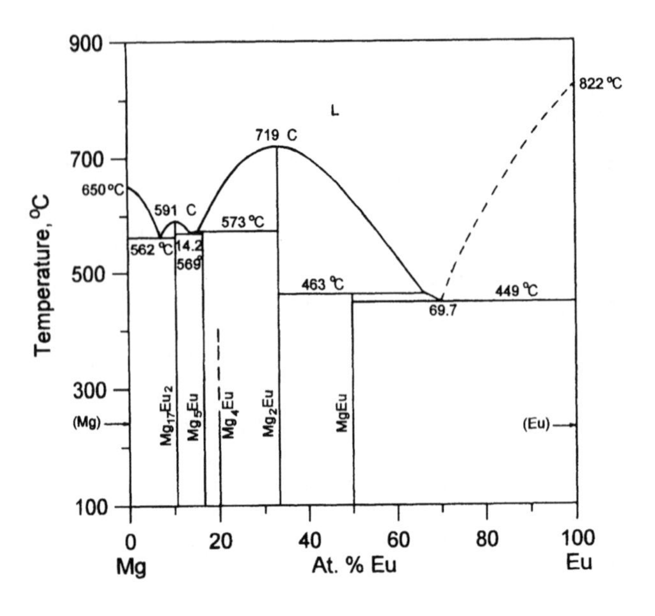

Figure 21. Mg–Eu phase diagram.

can be rejected. The eutectic temperature and concentration were determined to be 542°C and 39 mass % Sm (9.37 at. % Sm), respectively [105]. In [84] the eutectic temperature was determined to be 530°C during cooling and 540°C during heating with eutectic concentration of 35 mass % Sm (8.0 at. % Sm). In [165] the eutectic temperature was determined to be 532°C. In general, all results of the eutectic temperature and concentration

Table 14. Crystal structure of the Mg–Eu binary compounds.

Formula/ Temperature range, °C	Homogeneity range, at. % Eu	Symbol Pearson/ Prototype	Lattice parameters, nm	Comments
$Mg_{17}Eu_2$ <591	10.53	hP38 $Ni_{17}Th_2$	$a = 1.049$, $c = 1.033$	[121, 138]
Mg_5Eu <573	16.67	hP42 $EuMg_5$	$a = 1.0395$, $c = 1.074$	[138, 141]
Mg_4Eu ?	20	hP90 Mg_4Sr	$a = 1.0416$, $c = 2.8051$	[141]
			$a = 1.042$, $c = 2.805$	[137]
Mg_2Eu <719	33.33	hP12 $MgZn_2$	$a = 0.6379$, $c = 1.0308$	[121, 138]
MgEu <463	50	cP2 CsCl	$a = 0.4102$	[121, 138]

determinations are in reasonable agreement. Solubility of Sm in solid magnesium was determined by microscopy observations and the electrical resistivity method. Both methods gave consistent values of solubility at high temperatures of 540 and 500°C. At low

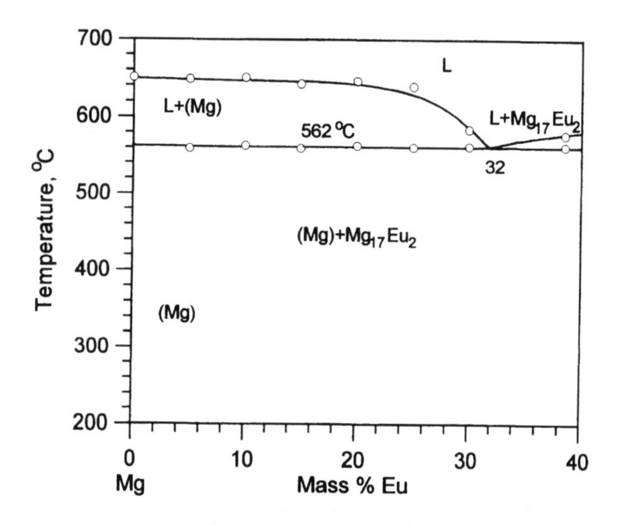

Figure 22. Mg-rich side of the Mg–Eu phase diagram.

temperatures 400, 300 and 200°C, the resistivity method gave lower solubility values than microscopy observations. In this case the data obtained by the resistivity method were considered to be more exact. The results of the resistivity measurements confirming the solubility values obtained and their decrease with lowering temperature are shown in Figure 20. At separate temperatures the solubility values were: 540°C – 5.7 mass % Sm (5.7 at. % Sm), 500°C – 4.3 mass % Sm (0.72 at. % Sm), 400°C – 1.8 mass % Sm (0.30 at. % Sm), 300°C – 0.8 mass % Sm (0.13 at. % Sm), and 200°C – 0.4 mass % Sm (0.063 at. % Sm). The maximum solubility value of 5.8 mass % Sm (0.99 at. % Sm) at 542°C was determined by extrapolation of the values at separate temperatures. They are shown in Figure 19. The solidus line along the Mg solid solution field was determined by microscopy observations of incipient melting in structure of the alloys. Their results showing absence of melting (open triangles) and existence of melting (solid triangles) are also inserted in Figure 19. The open circles on the plot show results of thermal analysis.

Mg–Eu Phase Diagram

The Mg–Eu phase diagram in the full concentration range is shown in Figure 21. It is drawn following the review [138] with correction of the Mg-rich side according to the investigation by the Baikov Institute of Metallurgy [106]. Also, one more phase Mg_4Eu is shown according to [137] as an assumption. The review [138] is based mainly on the work [79]. There are in the system two congruently formed compounds, $Mg_{17}Eu_2$ and Mg_2Eu, and two compounds, Mg_5Eu and MgEu, which are formed by peritectic reactions. The reaction type of the Mg_4Eu formation from the liquid phase was not reported in [137]. Between Mg-base, Eu-base solid solutions and the congruently formed compounds three eutectic equilibria take place. Characteristics of the compounds in the Mg–Eu system are presented in Table 14.

The Mg-rich part of the Mg–Eu phase diagram [106] is shown in Figure 22. It is of the eutectic type with the eutectic point at 32 mass % Eu (7.0 at. % Eu) and eutectic

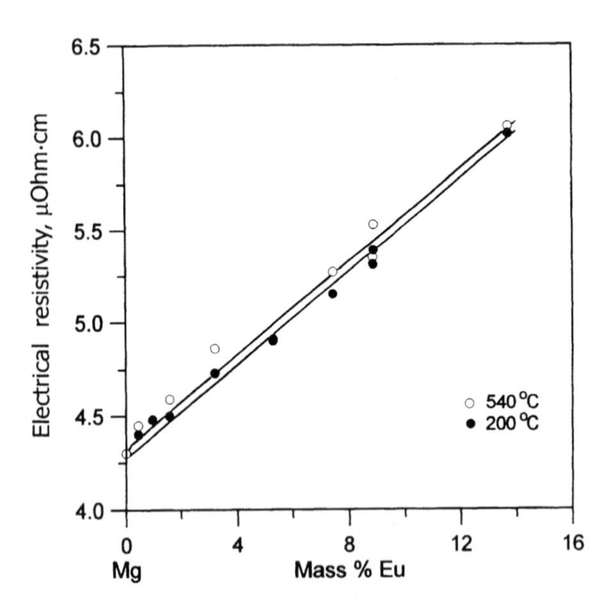

Figure 23. Electrical resistivity of the Mg–Eu alloys quenched from different temperatures.

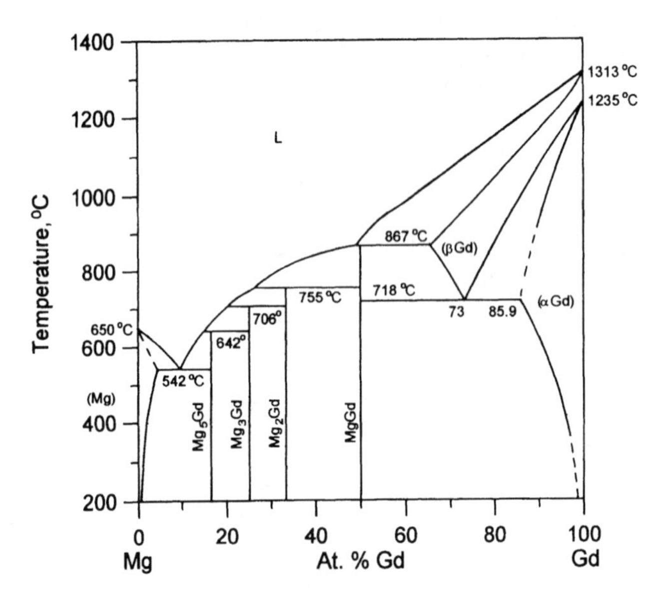

Figure 24. Mg–Gd phase diagram.

temperature 562°C. These data are in reasonable agreement with 29 mass % Eu (6 at.% Eu) and 571°C [79]. The electrical resistivity method showed the actual absence of the Eu solubility in solid magnesium [106]. The resistivity values of the alloys did not actually change with changing the quenching temperature, or at least their change was close to the measurement error. Nevertheless, comparison of the resistivity of the alloys annealed at the highest and lowest temperatures, 540 and 200°C, respectively, revealed some difference between them at the average (Figure 23). Estimation of this resistivity difference gave the maximum solubility of Eu in solid magnesium as about 0.03 mass % ($4.8 \cdot 10^{-5}$ at. %). The phase being in equilibrium with magnesium solid solution is assumed to be $Mg_{17}Eu_2$ according to [138].

Mg–Gd Phase Diagram

The Mg–Gd phase diagram for the full concentration range is shown in Figure 24. It has been redrawn following mainly the last evaluated version [144] which has taken into account investigations [99, 148, 166]. Meanwhile, the Mg-rich part of the phase diagram [144] was to an extent corrected according to the last investigation of the Baikov Institute of Metallurgy [108]. The phase diagram shows the existence of four binary compounds, Mg_5Gd, Mg_3Gd, Mg_2Gd and MgGd. All of them are peritectically formed. There is one eutectic equilibrium on the Mg side, and one eutectoid equilibrium connected with two allotropic forms of gadolinium. The crystal structure of the binary compounds are described in Table 15. Unlike other Mg–RE compounds, the formula Mg_5Gd shows the

Table 15. Crystal structure of the Mg–Gd binary compounds.

Formula/ Temperature range, °C	Homogeneity range, at. % Gd	Symbol Pearson/ Prototype	Lattice parameters, nm	Comments
Mg_5Gd <642	16.5	cF448 $Cd_{45}Sm_{11}$	$a = 2.2344$	[108, 135, 141]
Mg_3Gd <706	25	cF16 BiF_3	$a = 0.7326$ $a = 0.7324$	[133, 144] [144, 148]
Mg_2Gd <755	33.33	cF24 Cu_2Mg	$a = 0.85762$ $a = 0.8575$	[144, 162] [144, 148]
MgGd <867	50	cP2 CsCl	$a = 0.3820$ $a = 0.3818$	[144, 148] [125, 144]

real composition of the compound, but not its stoichiometry corresponding to the type of the crystal structure [135]. In our investigation the real composition of the compound was established to be Mg_6Gd by using quantitative metallography [107]. Both formulas, Mg_5Gd and Mg_6Gd, may be considered to be in reasonable agreement, but the first of them is more exact, being derived without extrapolation. The stoichiometric formula of the compound is $Mg_{45}Gd_{11}$ [135].

The Mg-rich side of the Mg–Gd phase diagram is shown in Figure 25. The liquidus lines and temperatures of the invariant equilibria in it are taken from [[108] with thermal analysis arrests shown by open circles. Solubility of Gd in solid magnesium is taken from [99] as in the evaluated version [144]. The assumed eutectic temperature 542°C [108] is comparable with that of 544°C [99], 545 [77] and 548°C [144, 148]. The assumed eutectic concentration 40.4 mass % Gd (9.48 at. % Gd) [108] is close to that of 38.4 mass % Gd (8.8 at. % Gd) [144, 148], but significantly more than that of ~28 mass % Gd (5.7 at. % Gd) [77]. The intermetallic phase closest to Mg is assumed to be Mg_5Gd [135, 148]. Its

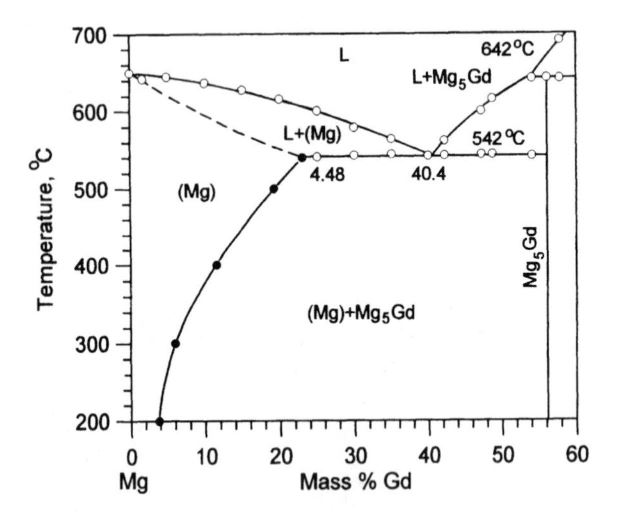

Figure 25. Mg-rich side of the Mg–Gd phase diagram.

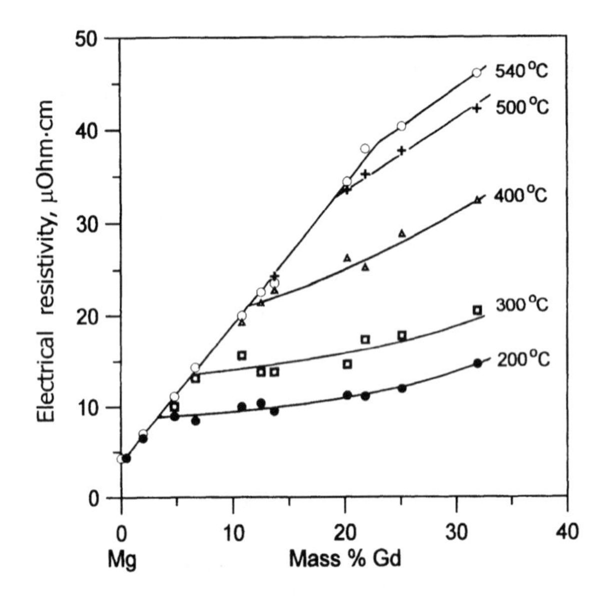

Figure 26. Electrical resistivity of the Mg–Gd alloys quenched from different temperatures.

formation temperature 642°C [108] by the peritectic reaction is in reasonable agreement with 640 [77] and 658°C [148] and coincides with that of [144].

The solubility values at separate temperatures shown in Figure 25 (dark circles) were obtained in [99] using the electrical resistivity method confirmed by microscopy observations. They are at 540°C – 23.0 mass % Gd (4.41 at. % Gd), 500°C – 19.2 mass % Gd (3.54 at. % Gd), 400°C – 11.5 mass % Gd (1.97 at. % Gd), 300°C – 6.0 mass % Gd (0.98 at. % Gd), and 200°C – 3.8 mass % Gd (0.61 at. % Gd). Extrapolation of the

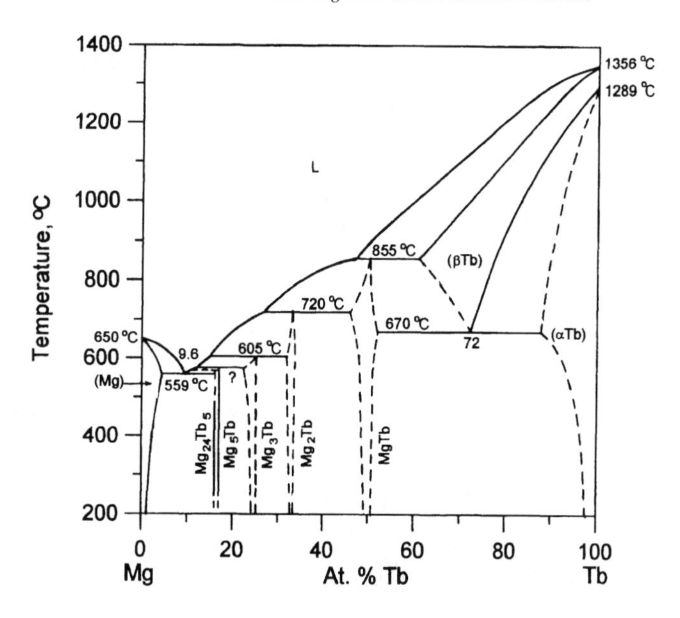

Figure 27. Mg–Tb phase diagrams.

solubility curve to the eutectic temperature 542°C gives a maximum solubility value of 23.3 mass % Gd (4.48 at. % Gd). The results of the resistivity measurements [99, 106] used for solubility determination are presented in Figure 26. They show certain limitations of the resistivity method. The kinks on the resistivity curves at high quenching temperatures, 540 and 500°C, appear to be quite weak as a result of the high resistivity of the rich Mg-base solid solution which becomes comparable with the resistivity of the second phase. So, the kink positions need to be confirmed by another analytical method. As was mentioned above, in this case microscopy observations were used. The solubility values [99] were in general confirmed by later investigation [108] where only microscopy observations were used (at 500°C – 18. 8 mass % Gd, at 400°C – 10.4 mass % Gd and at 300°C – 6.5 mass % Gd). They are significantly more than those established in the earliest work [77] (~2.5 mass % Gd at the eutectic temperature 545°C). The solidus line was not determined reliably and, therefore, it is drawn tentatively.

Mg–Tb Phase Diagram

The whole Mg–Tb phase diagram is shown in Figure 27. It is redrawn from the last investigation [88] with correction of the Mg-rich part in accordance with the investigation of Baikov Institute of Metallurgy [102]. The main features of the Mg–Tb phase diagram are five binary compounds being formed peritectically from the liquid phase, one eutectic transformation on the Mg side and one eutectoid transformation on the Tb side connected with allotropic transformation of solid terbium. The work [88] established experimentally many characteristics of the Mg–Tb system which were not determined earlier and were presented in the assessed Mg–Tb phase diagram in the reference book [139] as tentative. In addition to the earlier investigations [118, 120, 125, 128, 133, 162, 167] a new binary compound χ_1 was found in [88]. Its crystal structure was of the Mg_5Gd type. Other binary compounds in the system are $Mg_{24}Tb_5$ (χ_2), Mg_3Tb, Mg_2Tb and MgTb. Characteristics of the compounds are described in Table 16. X-ray investigations [88] showed some

Table 16. Crystal structure of the Mg–Tb binary compounds.

Formula/ Temperature range, °C	Homogeneity range, at.% Tb	Symbol Pearson/ Prototype	Lattice parameters, nm	Comments
χ_2, $Mg_{24}Tb_5$ at least <559*	~17	cI58 αMn	$a = 1.1283$	[88]
			$a = 1.1283$	[167]
χ_1, Mg_5Tb at least <559**	~17	cF448 $Cd_{45}Sm_{11}$	$a = 2.232$	[88, 141]
Mg_3Tb <605	~25	cF16 BiF_3	$a = 0.7300$	at ~22 at. % Mg [88]
			$a = 0.7296$	[133]
Mg_2Tb <720	~32–33.5	hP12 $MgZn_2$	$a = 0.6042$, $c = 0.9776$	at ~32 at. % Mg [88]
			$a = 0.60512$, $c = 0.98074$	[162]
MgTb <855	~48–54	cP2 CsCl	$a = 0.3775–0.3787$	[88]
			$a = 0.3796$	[125]

*The upper temperature of the compound existence is not known. Most likely it is between 605°C (that of the Mg_3Tb formation) and 559°C (eutectic temperature).
**The upper temperature of the compound existence is not known. Most likely it is higher than that of χ_2 and 605°C (that of the Mg_3Tb formation).

homogeneity ranges for the compounds Mg_3Tb, Mg_2Tb and MgTb, but their limits were not determined. These compounds, therefore, are shown in the phase diagram as the narrow fields outlined by the dashed lines. Both compounds, χ_1 (Mg_5Tb) and χ_2 ($Mg_{24}Tb_5$), were observed at close Tb content (~17 at.%). Nevertheless, microprobe analysis showed a difference of ~1–1.5 at. % between them. The authors admitted some shift of the $Mg_{24}Tb_5$ composition from stoichiometry corresponding to 17.24 at. % Tb. Obviously, some shift from the χ_1 stoichiometry ($Mg_{45}Tb_{11}$) may also be assumed. In accordance with the accepted eutectic temperature from [102] the peritectic temperatures of the χ_1 and χ_2 formation are shown deliberately in Figure 27 to be somewhat more than those established in [88]. The peritectic temperature of the χ_1 formation was established in [88] to be 555°C. Temperature of the χ_2 formation was not determined in [88] although it was supposed to be lower than that of the χ_1 formation.

Figure 28 displays the Mg-rich part of the Mg–Tb phase diagram redrawn according to our investigations [102, 106]. The phase closest to Mg is assumed to be $Mg_{24}Tb_5$. The liquidus lines and eutectic horizontal were drawn based on the thermal analysis arrests during cooling (open circles). The established eutectic temperature and eutectic concentration are 559°C and 41 mass % Tb (9.61 at. % Tb), respectively. The eutectic concentration obtained is in good agreement with 40.7 mass % Tb (9.5 at. % Tb) established in [88]. However, the eutectic temperature 559°C [102] is certainly higher than 530°C established in [88]. The solidus line between (Mg) and L+(Mg) fields was drawn based on

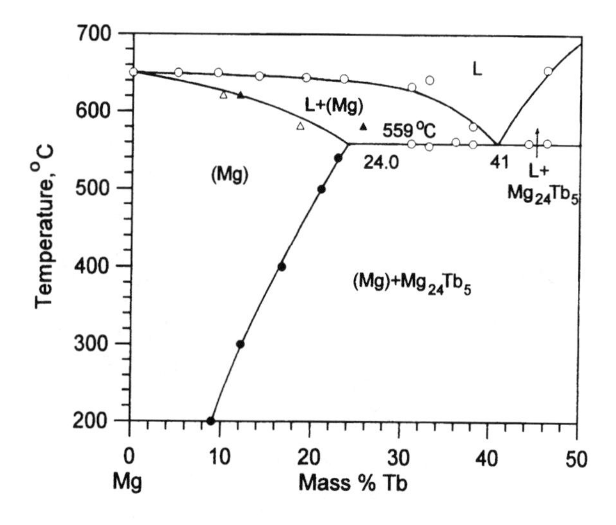

Figure 28. Mg-rich side of the Mg–Tb phase diagram.

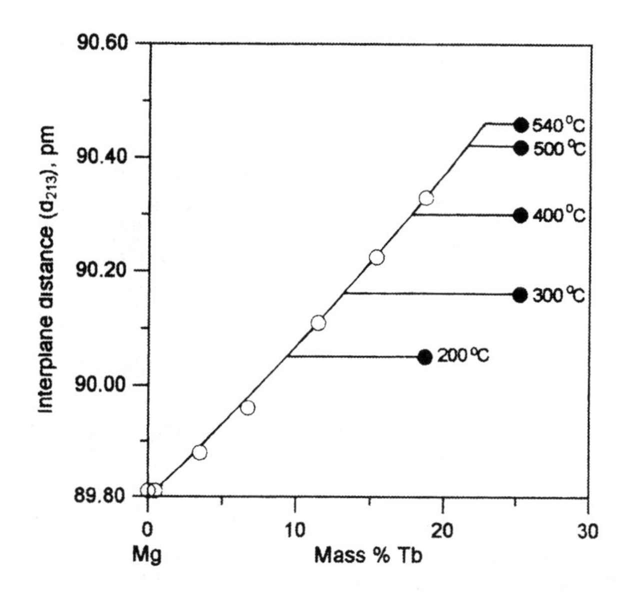

Figure 29. Interplane distances (d_{213}) of the Mg–Tb alloys. ○ – one-phase samples, ● – two-phase samples quenched from different temperatures.

microscopy observations of the incipient melting during annealing at different temperatures. The main results are shown on the phase diagram by open triangles (absence of melting) and dark triangles (beginning of melting). The solubility of terbium in solid magnesium was determined combining microscopy observations, X-ray determination of the interplane distances in the Mg solid solution lattice and the electrical resistivity method. In this case the electrical resistivity method could not be considered to be the most reliable, because high resistivity values of the solid solution and its high concentrations increased

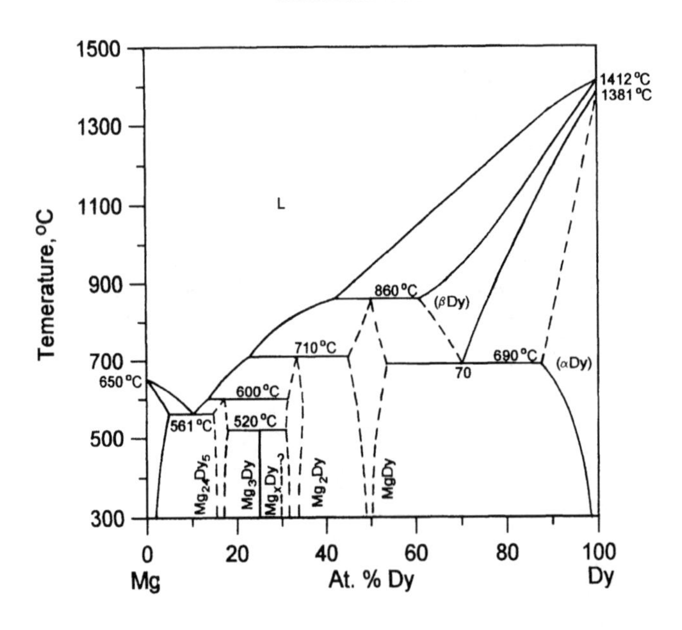

Figure 30. Mg–Dy phase diagram.

significantly the experimental error. On the other hand, the X-ray method became more effective. Figure 29 presents results of the measured interplane distances in the Mg solid solution lattice which were used for determination of the solid solubility in the Mg–Tb system. In the investigation, quite intense reflections from the lattice planes (213) were used. The open circles in Figure 29 correspond to the d_{213} values determined with one-phase samples and show the d_{213} increase with increasing Tb content in Mg solid solution. The dark circles in Figure 29 correspond to the d_{213} values determined with two-phase samples quenched after annealing at different temperatures shown. Absence or existence of the second phase in the structure of the samples were established by the previous microscopy observations. In general, the solubility values obtained by the X-ray method agreed with results of the microscopy observations and electrical resistivity method. For separate temperatures the solubility of Tb in solid Mg was established to be as follows: at 540°C – 22.8 mass % (4.45 at. %), at 500°C – 21.0 mass % (3.91 at. %), at 400°C – 16.7 mass % (2.99 at. %), at 300°C – 12.2 mass % (2.08 at. %), and at 200°C (1.49 at. %). These values are shown in Figure 28 (dark circles). Extrapolating the solubility curve up to the eutectic temperature 559°C gave the maximum solubility of 24.0 mass % Tb (4.57 at.% Tb).

Mg–Dy Phase Diagram

Figure 30 displays the Mg–Dy phase diagram in the whole concentration range. It is redrawn from [85] with correction of the Mg-rich part according to investigation of Baikov Institute of Metallurgy [97, 106]. The main features of the Mg–Dy phase diagram as a whole are formation of five binary compounds, one eutectic equilibrium on the Mg side and one eutectoid equilibrium connected with the existence of two allotropic forms of dysprosium. Three compounds, $Mg_{24}Dy_5$, Mg_2Dy and MgDy are formed from the liquid phase peritectically. The compound Mg_3Dy is formed from $Mg_{24}Dy_5$ and Mg_2Dy by the peritectoid reaction. In the phase diagram [85] the phase Mg_xDy is shown as supposed where x is more than 2. The phase with the same crystal structure (partial disordered type NaTl) was observed also in [128]. Its formation mechanism and exact composition were not established. Existence of the compound Mg_xDy is doubtful because, in addition to the noticed uncertainties, a phase of such type was not observed in the systems of magnesium with the neighbouring rare earth metals, terbium and holmium. The crystal structure of the compounds is described in Table 17. X-ray determination of the lattice parameters showed

Table 17. Crystal structure of the Mg–Dy binary compounds.

Formula/ Temperature range, °C	Homogeneity range, at. % Dy	Symbol Pearson/ Prototype	Lattice parameters, nm	Comments
$Mg_{24}Dy_5$ <600	~15–18	cI58 αMn	a = 1.1225–1.1264	[85]
			a = 1.1246	[167]
Mg_3Dy <520	~25	cF16 BiF_3	a = 0.7296	[85]
			a = 0.7267	[128]
Mg_xDy ?	~30	cF16 partially disordered NaTl	a = 0.742	[85]
			a = 0.740	[128]
Mg_2Dy <710	~33.3	hP12 $MgZn_2$	a = 0.6014–0.6029 c = 0.9740–0.9767	[85]
			a = 0.60330 c = 0.97786	[162]
MgDy <860	~45–54	cP2 CsCl	a = 0.3740–0.3776	[85]
			a = 0.3784	[125]

some homogeneity ranges for the compounds $Mg_{24}Dy_5$, Mg_2Dy and MgDy, but their limits were not established. Therefore, the compounds $Mg_{24}Dy_5$, Mg_2Dy and MgDy are shown in the phase diagram as homogeneity fields outlined by the dashed lines. The limits of the lattice parameter values for the compounds $Mg_{24}Dy_5$, Mg_2Dy and MgDy are presented in Table 17 according to [85] corresponding to those determined for the Mg or Dy side of the homogeneity ranges.

Figure 31 displays the Mg-rich part of the Mg–Dy phase diagram according to [97, 106]. The eutectic temperature and the eutectic concentration were established by thermal analysis (open circles) and microscopy observations to be of 561°C and of 46 mass % Dy (11.30 at. Dy), respectively. These values may be considered to be in agreement with 560°C and 41.2 mass % Dy (9.5 at. % Dy) established in [85]. The solubility of Dy in solid Mg was determined by microscopy observations, the electrical resistivity method and X-ray measurements of the interplane distance in the Mg solid solution lattice using samples of the alloys quenched after annealings at different temperatures. None of the methods can be considered as preferable, but by combining them the most reliable values can be obtained.

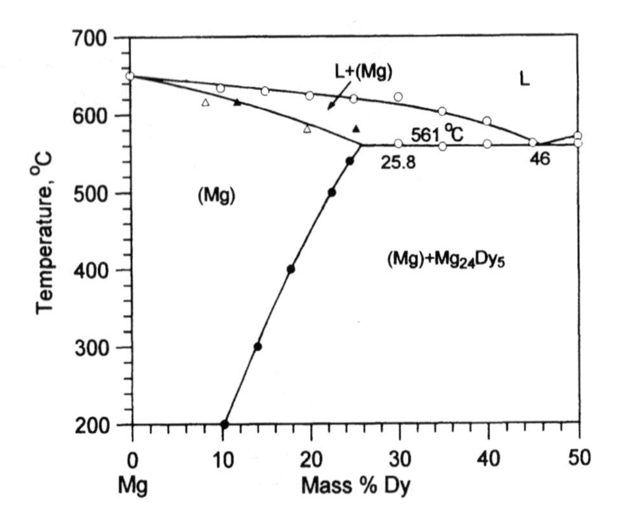

Figure 31. Mg-rich side of the Mg–Dy phase diagram.

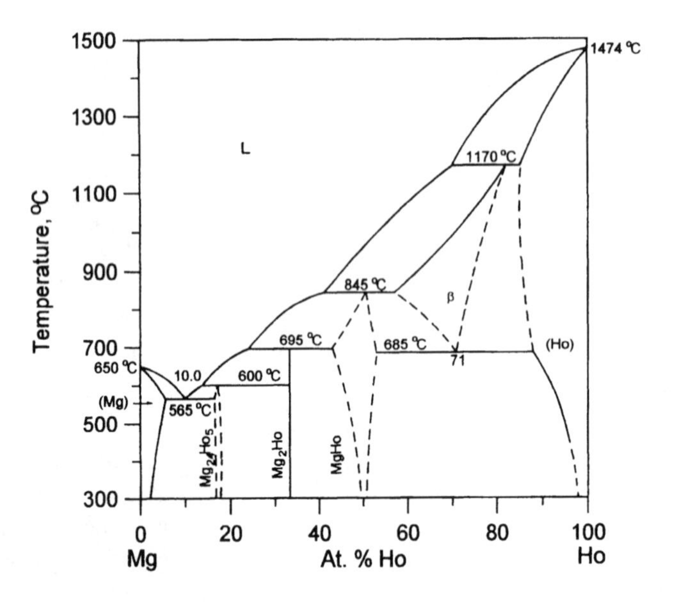

Figure 32. Mg–Ho phase diagram.

The solubility values of Dy in solid Mg were obtained at separate temperatures as follows: at 540°C – 24.5 mass % (4.65 at. %), at 500°C – 22.5 mass % (4.18 at. %), at 400°C – 17.8 mass % (3.13 at. %), at 300°C – 14 mass % (2.37 at. %), and at 200°C – 10.2 mass % (1.67 at. %). They are shown in Figure 31 by dark circles. Extrapolation of the solubility values at separate temperatures up to the eutectic temperature 561°C gave the maximum solubility of Dy in solid Mg to be 25.8 mass % (4.83 at. %).

The solidus line between the (Mg) and L+(Mg) fields was drawn taking into account results of microscopy observations of the samples quenched after annealing at temperatures above the eutectic temperature. Absence or existence of incipient melting in the structure of the sample indicated disposition of the alloy, respectively, in the (Mg) or L+(Mg) area at the annealing temperature. The corresponding results of microscopy observations for the edge samples are shown in Figure 31 by open triangles (absence of incipient melting) and dark triangles (existence of incipient melting).

Mg–Ho Phase Diagram

Mg–Ho phase diagram in the whole concentration range is shown in Figure 32. It is redrawn following the latest and most complete investigation [87] with correction of the Mg-rich side according to the data of the Baikov Institute of Metallurgy [103, 106]. There are three binary compounds in the Mg–Ho system with formulas $Mg_{24}Ho_5$, Mg_2Ho and MgHo. In addition to them the intermediate phase β with body-centred cubic lattice is formed. All these phases are formed from the liquid phase following peritectic reactions. One invariant equilibrium in the system is of eutectic type and this takes place on the Mg side. Phase β decomposes eutectoidly during cooling into Ho solid solution and MgHo. X-ray measurements of the lattice parameters showed certain homogeneity ranges for the compounds $Mg_{24}Ho_5$ and MgHo, but their limits were not established exactly. These phases are shown, therefore, in the phase diagram by narrow fields with dashed boundaries. The homogeneity range for the phase β was also supported by the results of thermal analysis. The crystal structures of the intermediate phases in the Mg–Ho system are described in Table 18. In [141] the phase β is described by the formula $MgHo_3$, but this formula is not connected with the phase crystal structure.

Table 18. Crystal structure of the Mg–Ho intermediate phases.

Formula/ Temperature range, °C	Homogeneity range, at. % Ho	Symbol Pearson/ Prototype	Lattice parameters, nm	Comments
$Mg_{24}Ho_5$ <600	16.5–~17.8	cI58 αMn	$a = 1.1235–1.1245$	[87]
			$a = 1.1233$	[167]
Mg_2Ho <695	33.3	hP12 $MgZn_2$	$a = 0.6011$, $c = 0.9737$	[87]
			$a = 0.60203$ $c = 0.97642$	[162]
MgHo <845	43–53	cP2 CsCl	$a = 0.3716–0.3785$	[87]
			$a = 0.3770$	[125]
β 1170–685	57–82	cI2 W	$a = 0.382$	at 68 at. % Ho [87]

The Mg-rich part of the Mg–Ho phase diagram according to our investigations [103, 106] is shown in Figure 33. It is of eutectic type with eutectic temperature of 565°C and concentration at the eutectic point of 43 mass. % Ho (10.0 at. % Ho). These values agree, in general, with those obtained in the later investigation [86], which are 555°C (575°C during heating) and the same 43 mass % Ho (10.0 at. % Ho). The arrests during thermal analysis used for the determination of the liquidus lines and the eutectic point disposition

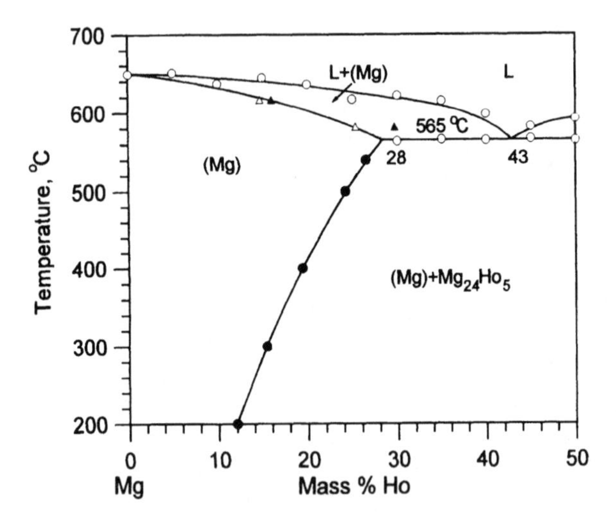

Figure 33. Mg-rich side of the Mg–Ho phase diagram.

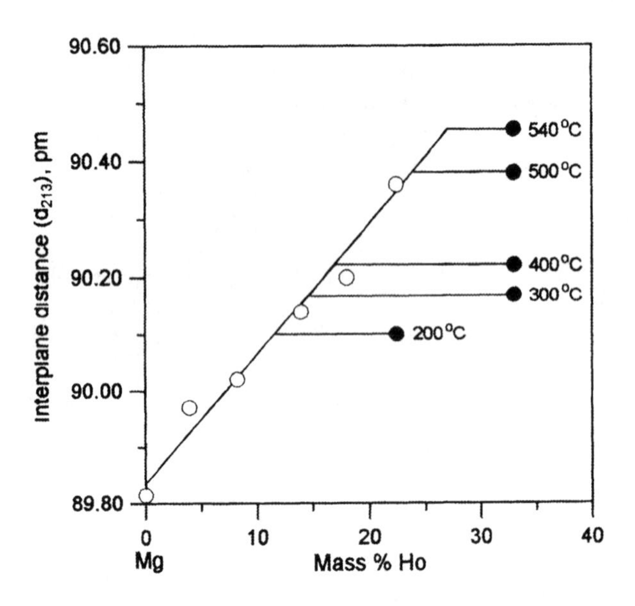

Figure 34. Interplane distances (d_{213}) of the Mg–Ho alloys. ○ – one-phase samples, ● – two-phase samples quenched from different temperatures.

[103, 106] are shown by the open circles. The phase closest to Mg is assumed to be $Mg_{24}Ho_5$. The solubility of Ho in solid Mg was determined combining microscopy observations and X-ray measurements of the interplane distances of the Mg solid solution lattice. Both methods supplemented each other making results of the solubility values obtained more reliable. The results of solubility determination at various temperatures are at 540°C – 26.5 mass % (5.05 at. % Ho), at 500°C – 24.2 mass % Ho (4.50 at. % Ho), at 400°C – 19.4 mass % Ho (3.43 at. % Ho), at 300°C – 15.4 mass % Ho (2.55 at. % Ho),

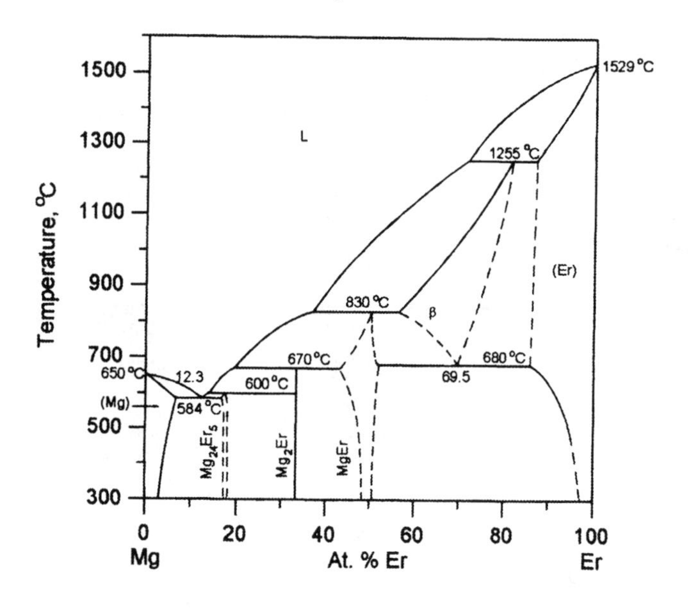

Figure 35. Mg–Er phase diagram.

and at 200°C – 12.0 mass % Ho (1.92 at. % Ho). These values are shown in Figure 33 by dark circles. Extrapolation of the solubility values at separate temperatures up to 565°C gave a maximum solubility of Ho in solid Mg of 28 mass % (5.44 at. %). The solidus line between the (Mg) and L+(Mg) fields was drawn using results of the microscopy observations of the alloys quenched after annealings at temperatures above temperature of the eutectic equilibrium. The open triangles in Figure 33 show the edge samples without signs of melting during annealing, and the dark triangles show the edge samples where partial melting during annealing took place. Figure 34 demonstrates the use of the X-ray method for the solubility determination of Ho in solid Mg.

Mg–Er Phase Diagram

The Mg–Er phase diagram in the concentration range from 0 to 100 at. % Er is presented in Figure 35. The phase diagram is redrawn from [86] with correction of the Mg-rich side according to the investigation of Baikov Institute of Metallurgy [104, 106]. The phase diagram shows the existence of four intermediate phases, $Mg_{24}Er_5$, Mg_2Er, MgEr and β. In [141] the formula $MgEr_3$ is ascribed to phase β, but it shows approximately the β composition only. All intermediate phases are formed peritectically. Apart from the peritectic invariant equilibria, there are in the system a eutectic equilibrium on the Mg-rich side and a eutectoid equilibrium conforming to the β phase decomposition into Er solid solution and the phase MgEr. The crystal structure of the intermediate phases is described in Table 19.

The Mg-rich part of the Mg–Er phase diagram [104, 106] is shown in Figure 36. Open circles on the plot are results of thermal analysis used for determination of the eutectic temperature 584°C and the eutectic concentration 49 mass % Er (12.3 at. % Er). These values are in reasonable agreement with 570°C and 44.7 mass % Er (10.5 at. % Er), respectively, obtained in [86]. The solubility of Er in solid Mg was determined combining mainly microscopy observation and X-ray method based on determination of interplane distances in the Mg solid solution lattice. Both methods supplemented each other and gave consistent results. The electrical resistivity method was used for the solubility

Table 19. Crystal structure of the Mg–Er intermediate phases.

Formula/ Temperature range, °C	Homogeneity range, at.% Er	Symbol Pearson/ Prototype	Lattice parameters, nm	Comments
$Mg_{24}Er_5$ <600	16.5–~18	cI58 αMn	$a = 1.1214–1.1255$	[86]
			$a = 1.1224$	[167]
Mg_2Er <670	33.3	hP12 $MgZn_2$	$a = 0.6003$ $c = 0.9710$	[86]
			$a = 0.60036$ $c = 0.97356$	[162]
MgEr <830	43–52	cP2 CsCl	$a = 0.3722–0.3761$	[86]
			$a = 0.3756$	[125]
β 1255–680	53–82	cI2 W	$a = 0.381$	at 68 at. % Er [86]

determination, too. However, it was quite approximate because of the high resistivity values and, as a result, increased uncertainty in disposition of the kinks on resistivity/composition curves corresponding to the solubility values. The solubility of Er in solid Mg was established to be at 540°C – 30.5 mass % (5.99 at. %), at 500°C – 28.3 mass % (5.42 at. %), at 400°C – 23.0 mass % (4.15 at. %), and at 300°C – 18.5 mass % (3.17 at. %). These values are shown in Figure 36 by dark circles. The maximum solubility of 32.7 mass % Er (6.59 at. % Er) at the eutectic temperature 584°C was established by extrapolation of the solubility values at separate temperatures. The solidus line between the (Mg) and L+(Mg) fields in the phase diagram is drawn using microscopy observations of incipient melting in the alloys quenched after annealing at one temperature 615°C between melting point of magnesium and eutectic temperature. The results of the microscopy observations for the edge alloys where melting was not observed (open triangle) and melting was observed (dark triangles) are shown in Figure 36.

Mg–Tm Phase Diagram

The Mg–Tm phase diagram in the concentration range from 0 to 100 at. % Tm is shown in Figure 37. It is reproduced with insignificant amendments from the reference book [139]. The Mg-rich side of the phase diagram coincides actually with that constructed in the investigation of Baikov Institute of Metallurgy [100, 106]. Three binary compounds, $Mg_{24}Tm_5$, Mg_2Tm and MgTm were well established in the system without investigation of the phase diagram [23, 125, 128, 162, 167]. The phase β was established in investigation of Tm-rich part of the phase diagram [23]. At first this phase was assumed to be the solid solution of the high temperature body-centred cubic modification of Tm. However, according to the last data this allotropic form of Tm does not exists [15]. The phase β

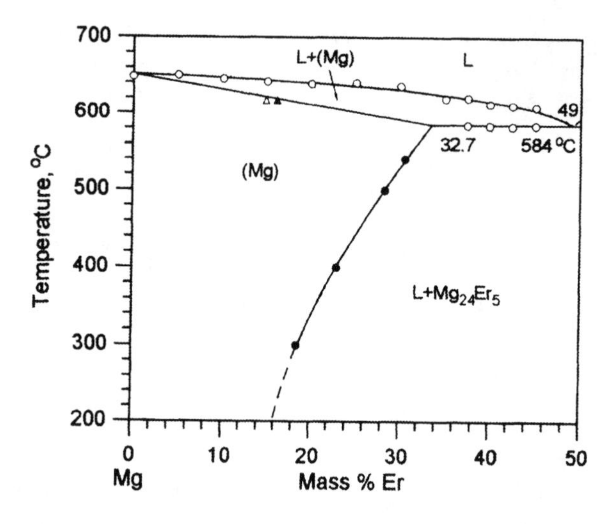

Figure 36. Mg-rich side of the Mg–Er phase diagram.

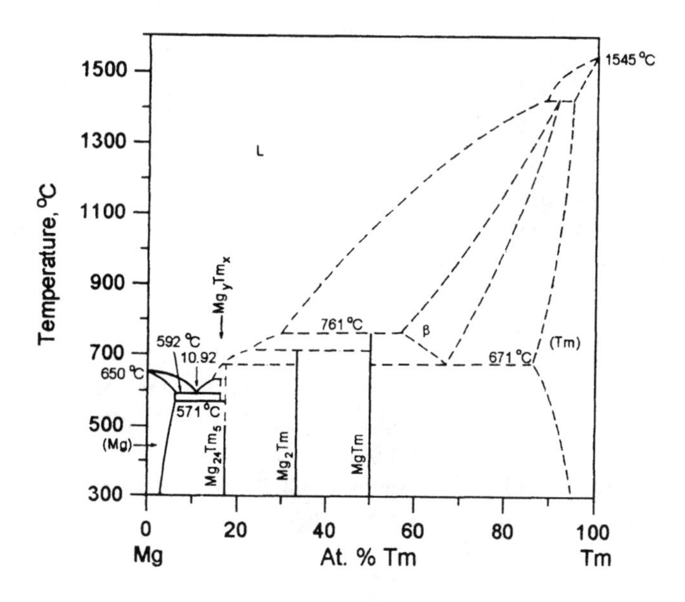

Figure 37. Mg–Tm phase diagram.

decomposes eutectoidally into MgTm and Tm solid solution. In addition to them, the phase of unknown composition, Mg_yTm_x, is assumed based on the investigation of the Mg-rich part of the phase diagram [100, 106]. This phase decomposes eutectoidally into Mg solid solution and $Mg_{24}Tm_5$. In [23] the peritectic character of the compound MgTm formation and temperature of this reaction was established. In the Mg-rich side eutectic equilibrium occurs. The most part of the Mg–Tm phase diagram is supposed in the assessment [139] "on extrapolation from the Mg–Gd system". Except the unknown compound Mg_yTm_x the general view of the suggested Mg–Tm phase diagram [139] is similar to that of the Mg–Ho and Mg–Er systems. The homogeneity ranges of the compounds Mg_yTm_x, $Mg_{24}Tm_5$, Mg_2Tm and MgTm were not established. Crystal structure of the intermediate phases in the Mg–Tm system is described in Table 20.

Table 20. Crystal structure of the Mg–Tm intermediate phases.

Formula/ Temperature range, °C	Homogeneity range, at. % Tm	Symbol Pearson/ Prototype	Lattice parameters, nm	Comments
Mg_yTm_x ?–571	>12.14 <17.24	?	?	[100, 139]
$Mg_{24}Tm_5$?	17.24	cI58 αMn	a = 1.1208	[139, 167]
Mg_2Tm ?	33.33	hP12 $MgZn_2$	a = 0.59868 c = 0.97031	[162]
			a = 0.597 c = 0.974	[128]
MgTm <761	50	cP2 CsCl	a = 0.3749	[23]
			a = 0.3744	[125]
β ?–671	at least ~70–80	cI2 W	a = ~0.382 a = ~0.384	at ~70 at. % Tm at ~79 at. % Tm [23]

The Mg-rich part of the Mg–Tm phase diagram [100, 106] is shown in Figure 38. Thermal analysis (open circles) showed two invariant equilibria in the system. One of them was recognised to be the eutectic at temperature 592°C and liquid concentration of 46 mass % Tm (10.9 at. % Tm). The other equilibrium at 571°C was assumed to be of eutectoid type. This suggestion was supported by the microstructure of the studied alloys after thermal analysis where the Tm-rich crystals in the eutectics formed had an inner disperse constitution which resembled eutectoid decomposition. Taking into consideration [167] where the existence of the compound $Mg_{24}Tm_5$ at room temperature was established, it was assumed in [100, 106] that during the eutectic transformation at 592°C the liquid phase decomposed into Mg solid solution and the binary Mg–Tm compound of an unknown composition and then during cooling this unknown Mg–Tm compound decomposed eutectoidally at 571°C into Mg solid solution and $Mg_{24}Tm_5$. In assessment [139] the unknown Mg–Tm compound was marked then by Mg_yTm_x with $y/x > 24/5$. Its composition was not established, but should be between the edge alloy with 49 mass % Tm (12.14 at. % Tm) studied in [100] and 59.2 mass % Tm (17.24 at. % Tm) corresponding to $Mg_{24}Tm_5$. This fact is noted in Table 20.

The solidus line between the fields (Mg) and L+(Mg) in the Mg-rich part of the phase diagram was established by microstructure observations of the alloys annealed at 615°C. The open triangle shown in Figure 38 corresponds to the edge sample where the signs of the melting during annealing were not observed. The dark triangle corresponds to the alloy where first signs of the melting during annealing was observed. The solubility of Tm in solid Mg was established only by observation of the alloy microstructure after quenching samples from different temperatures. The dark circles in Figure 38 correspond to the solubility values obtained. They are at 571°C – 31.8 mass % (6.29 at. % Tm), at 540°C –

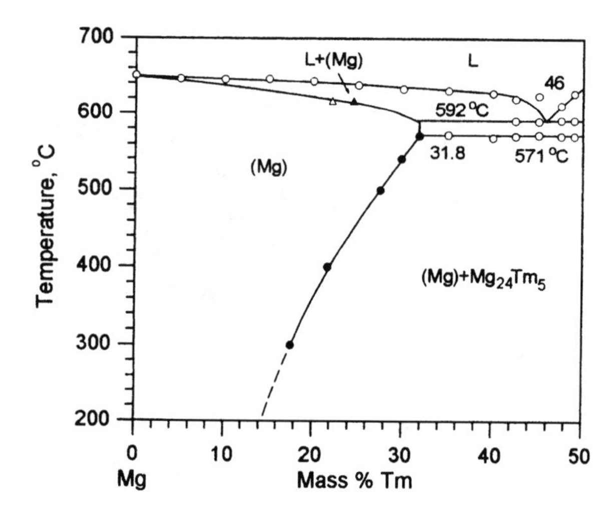

Figure 38. Mg-rich side of the Mg–Tm phase diagram.

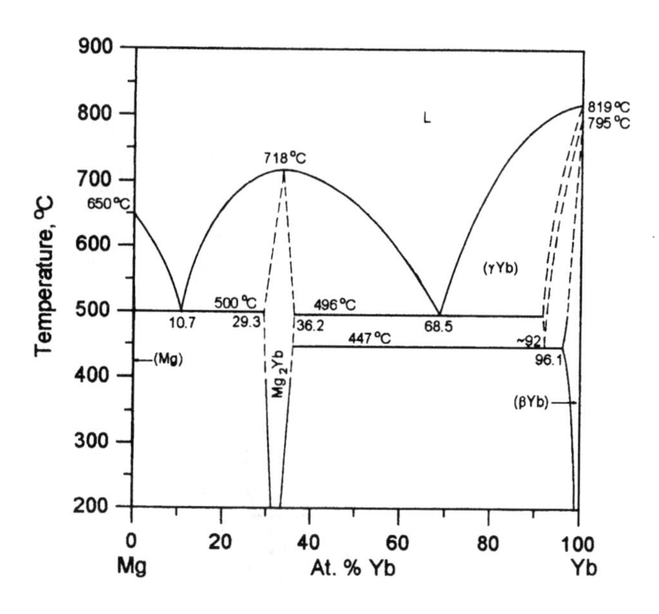

Figure 39. Mg–Yb phase diagram.

29.8 mass % Tm (5.76 at. % Tm), at 500°C – 27.5 mass % Tm (5.17 at. % Tm), at 400°C – 21.7 mass % Tm (3.83 at. % Tm), and at 300°C – 17.6 mass % Tm (2.98 at. % Tm).

Mg–Yb Phase Diagram

Mg–Yb phase diagram is shown in Figure 39. It is redrawn from the reference book [139] with correction of the Mg-rich side following the investigation of Baikov Institute of

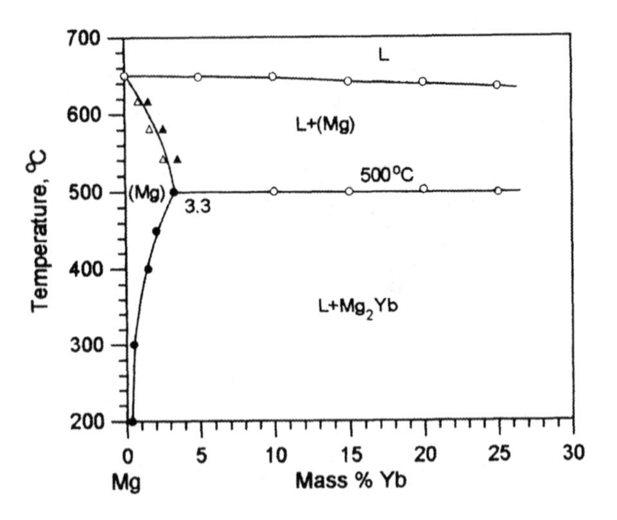

Figure 40. Mg-rich side of the Mg–Yb phase diagram.

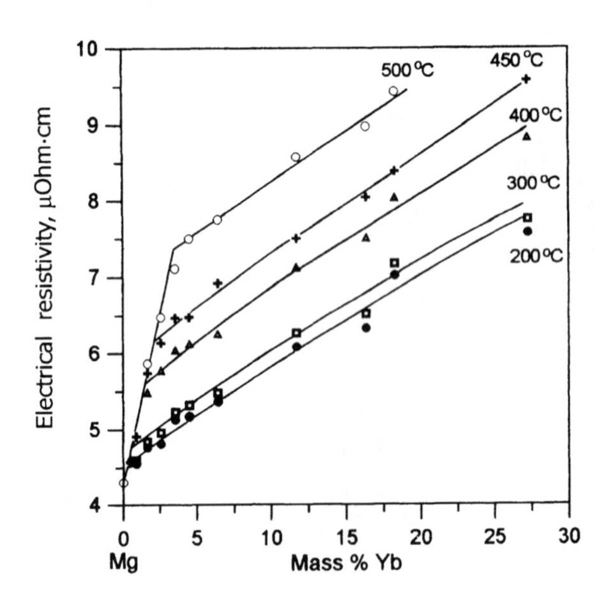

Figure 41. Electrical resistivity of the Mg–Yb alloys quenched from different temperatures.

Metallurgy [98, 106]. Presented in the reference book [139] the Mg–Yb phase diagram actually reproduces the results of the only investigation [78]. The Mg–Yb system is characterised by formation of only one congruently melting compound Mg_2Yb. Accordingly, there are in the system two eutectic equilibria on both sides of the compound. There is also a eutectoid equilibrium connected with the existence of the high temperature allotropic form of ytterbium. Mg_2Yb exists from melting point 718°C down to room temperature. Its homogeneity range is in the limits ~29.3 to ~36.2 at. % Yb. Mg_2Yb is the hexagonal Laves

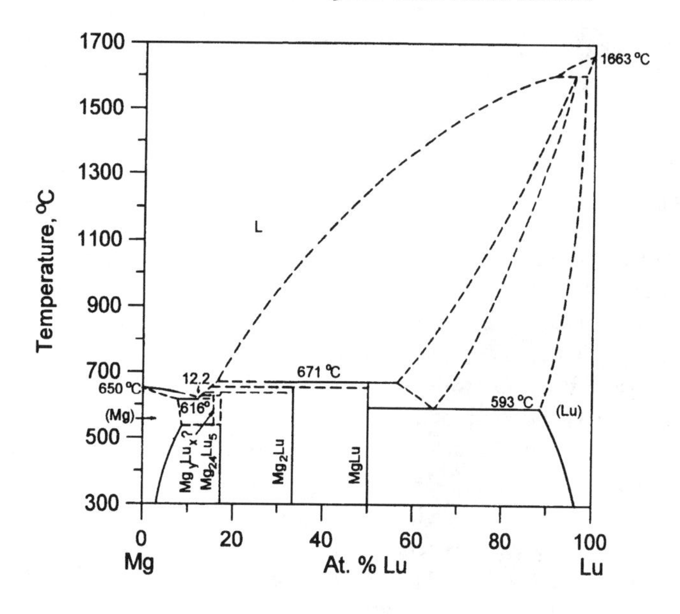

Figure 42. Mg–Lu phase diagram.

phase of the $MgZn_2$ type (Pearson symbol is hP12). The lattice parameters of Mg_2Yb were determined to be a = 0.62182–0.62362 nm, c = 1.00403–1.01028 nm for 200°C [78] and a = 0.62235 nm, c = 1.00850 nm [162].

The Mg-rich side of the Mg–Yb phase diagram studied in our investigation [98, 106] is presented in Figure 40. It was limited by the Yb concentrations significantly less than by the eutectic point which was not determined. The eutectic temperature was established to be 500°C in good agreement with 509°C of [78]. The main attention was paid in [98, 106] to the boundaries of Mg solid solution including the solubility of Yb in solid Mg and the solidus line between (Mg) and L+(Mg) fields. The solidus line was drawn using results of the microscopy observations of the initial melting in the structure of the alloys quenched after annealing at different temperatures. The solid solubility of Yb in Mg was determined by microscopy observations of the alloy structure and by the electrical resistivity method. Both methods gave compatible results, but the solubility values obtained by the resistivity method were assumed to be more exact. The results of the resistivity measurements are presented in Figure 41. They confirm unequivocally the existence of the specific solubility of Yb in solid Mg and its decrease with lowering temperature. The solubility of Yb in solid Mg is significantly less yet than those of other rare earth metals of the yttrium subgroup. The established solubility values of Yb are at 500°C – 3.3 mass % (0.48 at. %), at 450°C – 2.1 mass % (0.31 at. %), at 400°C – 1.5 mass % (0.21 at. %), at 300°C – 0.5 mass % (0.071 at. %), and at 200°C – 0.35 mass % (0.049 at. %) [98, 106]. These values are less than those obtained in [78] using the X-ray method (at 500°C –8.0 mass % Yb (1.2 at. % Yb), at 400°C – 4.8 mass % Yb (0.70 at. % Yb), and at 200°C – 2.5 mass % Yb (0.35 at. % Yb)). We believe the data [98, 106] to be more reliable than those obtained in [78] because at small solubility the electrical resistivity method is more sensitive and exact than the X-ray method. As in the previous descriptions of the phase diagrams, the symbols in Figure 40 show results of the experiments which are arrests during thermal analysis (open circles), absence and presence of melting signs in the structure of the edge alloys (open and dark triangles, respectively), and the solid solubility values established at separate temperatures (dark circles).

Mg–Lu Phase Diagram

Figure 42 shows Mg–Lu phase diagram reproduced from the reference book [139] with insignificant amendments. The assessed phase diagram takes into account the firmly established existence of the compounds $Mg_{24}Lu_5$, Mg_2Lu, MgLu and the phase β with body-centred cubic lattice without ordering [23, 83, 128, 167]. Formation of the compound MgLu is established to be according to the peritectic reaction [23]. In that work, the phase β was considered to be the solid solution of the high temperature allotropic form of Lu [23], but as the existence of this allotropic form of Lu is now refuted, β is assumed to be an independent phase. The phase β decomposes eutectoidly into MgLu and Lu-base solid solution. The solubility of Mg in solid Lu is assumed according to [83]. The Mg-rich side of the Mg–Lu phase diagram was constructed in [139] taking into account the study by the Baikov Institute of Metallurgy [101]. It is characterised by the extended Mg solid solution and eutectic equilibrium. Moreover, equilibrium in the solid state is suggested. In [139] the equilibrium in the solid state is supposed to be the eutectoid decomposition of the unknown solid phase Mg_yLu_x with $y/x > 24/5$ stoichiometry (~17 at. % Lu). Meanwhile, [139] believe that the existence of this phase and its stability range remains to be established. The other details of the Mg–Lu phase diagram were proposed in [139] assuming its similarity with that of the Mg–Gd system. Table 21 contains characteristics of the intermediate

Table 21. Crystal structure of the Mg–Lu intermediate phases.

Formula/ Temperature range, °C	Homogeneity range, at.% Lu	Symbol Pearson/ Prototype	Lattice parameters, nm	Comments
Mg_yLu_x ?–538	>12.20 <17.24	?	?	[101, 139]
$Mg_{24}Lu_5$?	17.24	cI58 αMn	$a = 1.1185$	[139, 167]
Mg_2Lu ?	33.33	hP12 $MgZn_2$	$a = 0.596$ $c = 0.971$	[128]
MgLu <671	50	cP2 CsCl	$a = 0.3727$	[23]
			$a = 0.3700$	[128]
β ?–593	at least ~70–80	cI2 W	$a = \sim 0.379$ $a = \sim 0.382$	at ~70 at. % Lu at ~80 at. % Lu [23]

phases occurring in the Mg–Lu system. The limits of the supposed phase Mg_yLu_x in it are shown taking into account the phase position between the established eutectic point (12.20 at. % Lu) and $Mg_{24}Lu_5$ (17.24 at. % Lu).

The Mg-rich side of the Mg–Lu phase diagram studied in Baikov Institute of Metallurgy [101] is presented in Figure 43. This part of the phase diagram was investigated

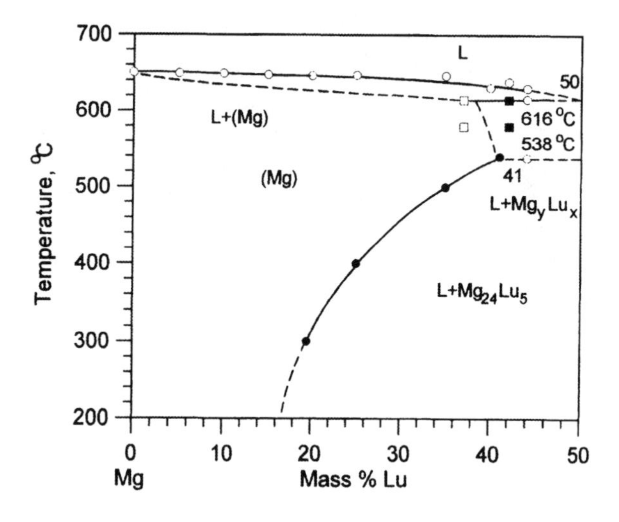

Figure 43. Mg-rich side of the Mg–Lu phase diagram.

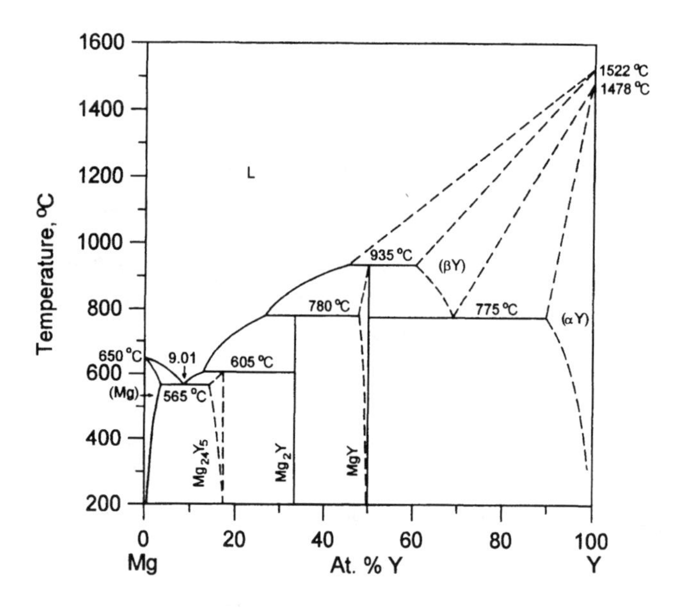

Figure 44. Mg–Y phase diagram.

by thermal analysis (open circles) and microscopy observations. Existence of the invariant eutectic reaction in the system was proved by results of thermal analysis and microstructure of the cast alloys. The second invariant reaction in the solid state was suggested by the additional arrests during thermal analysis. Extrapolation of the liquidus line enabled us to establish the eutectic point composition to be approximately 50 mass % Lu (12.2 at. % Lu) at 616°C. The solubility of Lu in solid Mg was determined only by microscopy observations. It was at 540°C – 41 mass % Lu (8.80 at. % Lu), at 500°C – 35 mass % Lu

(6.95 at. % Lu), at 400°C – 25 mass % Lu (4.76 at. % Lu), and at 300°C – 19.5 mass % Lu (3.27 at. % Lu). These values are shown in Figure 43 by dark circles, as in the previous descriptions of the phase diagrams. At 538°C (temperature of the invariant equilibrium in solid state) the solubility of Lu in solid Mg can be assumed to be the same 41 mass % (8.80 at. %) in the error limits as at 540°C. The solid solubility of Lu in magnesium at 580 and 615°C could be determined only with more uncertainty. According to the microscopy observations, it is between 37 mass % Lu (7.54 at. % Lu) and 42 mass % Lu (9.14 at. % Lu). These concentrations corresponded to the compositions of the edge alloys where the structure consisted of one phase (open squares) or two phases (dark squares), respectively, after annealing and quenching from the mentioned temperatures. The solidus line between the (Mg) and L+(Mg) fields could be drawn only approximately without any intermediate point determined experimentally.

Mg–Y Phase Diagram

The general view of the Mg–Y phase diagram is presented in Figure 44. It is an assessed version accepted from the reference book [139] with insignificant amendments made mainly to take into account the results of the Baikov Institute of Metallurgy on the Mg-rich side. The main features of the Mg–Y phase diagram are existence of three binary compounds, $Mg_{24}Y_5$, Mg_2Y, MgY, formed peritectically. There is also eutectic equilibrium on the Mg-rich side of the system and eutectoid equilibrium in the Y-rich side, connected with existence of the high temperature allotropic form of yttrium. For two compounds, $Mg_{24}Y_5$ and MgY, the tangible homogeneity ranges were established by X-ray determination of the lattice parameters [168]. At temperatures rather lower than temperatures of the corresponding peritectic reactions, the homogeneity ranges were determined to be within the limits 13–16 at. % Y for $Mg_{24}Y_5$ and 48–50 at. % Y for MgY. Thus, the homogeneity range of $Mg_{24}Y_5$ is shifted from the stoichiometric composition (17.24 at. % Y) to Mg side. The crystal structure characteristics of the compounds are given in Table 22.

Table 22. Crystal structure of the Mg–Y binary compounds.

Formula/ Temperature range, °C	Homogeneity range, at.% Y	Symbol Pearson/ Prototype	Lattice parameters, nm	Comments
$Mg_{24}Y_5$ <605	13–16	cI58 αMn	$a = 1.1251–1.1277$	at 525°C [168]
			$a = 1.1257$	[167]
Mg_2Y <780	33.3	hP12 $MgZn_2$	$a = 0.6037$ $c = 0.9752$	[168]
			$a = 0.60384$ $c = 0.98005$	[162]
MgY <935	48–50	cP2 CsCl	$a = 0.3782–0.3810$	at ~725°C [168]
			$a = 0.376$	[120]

The Mg-rich part of the Mg–Y phase diagram is presented in Figure 45 according to the results of the Baikov Institute of Metallurgy [91]. It coincides actually with results of

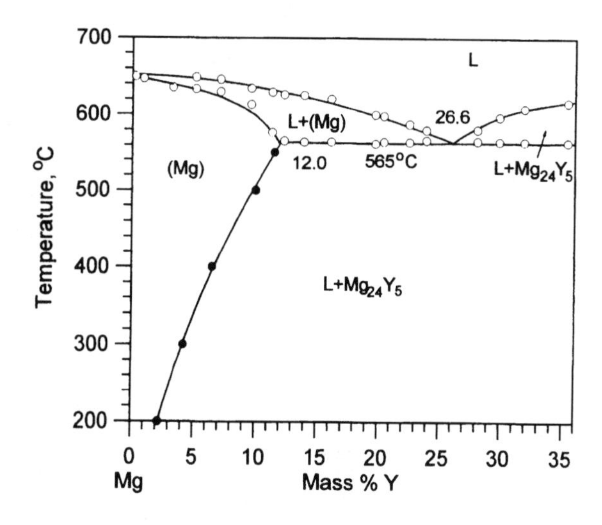

Figure 45. Mg-rich side of the Mg–Y phase diagram.

other investigations [169], except only one early work [75], where significantly less solubility of Y in solid Mg was obtained at significantly higher eutectic temperature (1.6 mass % Y (0.44 at. % Y) at 620°C). The data [75] contradict three other consistent works [74, 76, 91], and, therefore, can be considered to be mistaken. The liquidus lines, the solidus line between (Mg) and L+(Mg) fields and the eutectic temperature were determined using thermal analysis (open circles). For the eutectic temperature and the eutectic point composition the values 565°C and 26.6 mass % Y (9.01 at. % Y) were obtained, respectively [91]. In other works for the eutectic temperature and the eutectic point composition the values 567°C, 26 mass % Y (8.76 at. % Y) [74] and 565.5°C, 26 mass % (8.76 at. % Y) [76] were determined. The solubility of Y in solid Mg was determined in [91] using the electrical resistivity method and microstructure observations. Both methods gave consistent results. For separate temperatures the solubility values obtained are as follows: at 550°C – 11.5 mass % Y (3.43 at. % Y), at 500°C – 10.0 mass % Y (2.95 at. % Y), at 400°C – 6.5 mass % Y (1.87 at. % Y), at 300°C – 4.2 mass % Y (1.18 at. % Y), at 200°C – 2.2 mass % Y (0.61 at. % Y). These values are shown in Figure 45 by dark circles. Extrapolation of the solubility values at separate temperatures [91] to the eutectic one gave the maximum solubility of Y in solid Mg to be 12.0 mass % (3.59 at. %). In two other works the maximum solubility of Y in solid Mg was determined to be 9 mass % (2.63 at. %) [74] and 12.6 mass % (3.79 at. %) [76].

Mg–Sc Phase Diagram

The assessed Mg–Sc phase diagram is shown in Figure 46 [170]. It is drawn based mainly on the investigation [81] with extrapolation of their data from 60 to 100 at. % Sc and the solid solubility of Sc in Mg assumed according to the data of Baikov Institute of Metallurgy [95]. The first investigation of the Mg–Sc phase diagram was carried out within 0–40 mass % Sc (0–26.5 at. % Sc) by [80]. The Mg–Sc phase diagram differs significantly from all other Mg–RE phase diagrams. The main features of it are the peritectic character of the

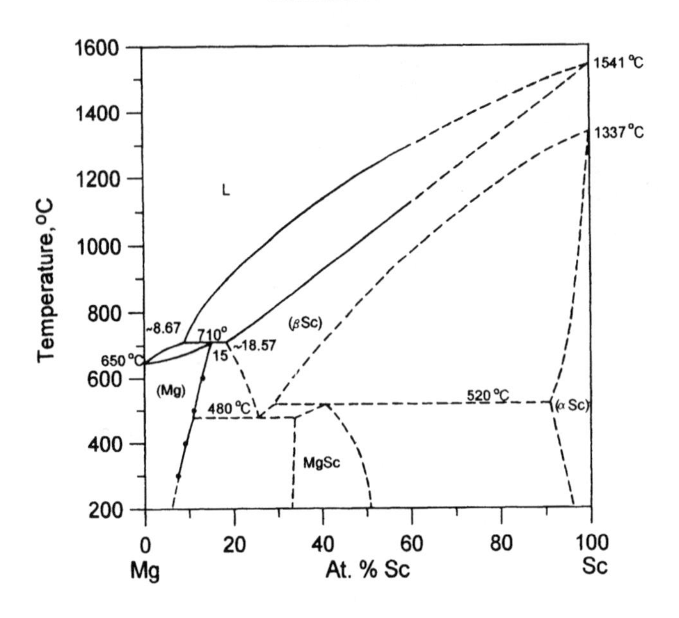

Figure 46. Mg–Sc phase diagram.

Mg-rich side and the existence of only one binary compound. Two phases are in equilibrium with Mg solid solution. One of them, β, is supposed to be the solid solution of the high temperature modification of Sc with simple body-centred cubic lattice. It exists only at high temperatures. At lower temperatures in equilibrium with Mg solid solution is the only binary compound MgSc. The MgSc homogeneity range is believed to be quite wide with extension on both sides of the stoichiometric composition. However, its boundaries are not determined reliably [170]. The MgSc crystal structure is of the body-centred cubic CsCl type (Pearson symbol is cP2) with the lattice parameter a = 0.3594 nm determined in the two-phase alloy with 31.7 at. % Sc [81] and a = 0.3597 nm reported for the stoichiometric composition [126].

The solubility of Sc in solid Mg was determined by the microscopy method [95]. The electrical resistivity method was used in [95], too, but it turned out to be less reliable because of high electrical resistivity of Mg solid solution resulting from high solubility of Sc in it. The solubility of Sc in solid Mg at separate temperatures were determined as follows: at 600°C – 22.0 mass % (13.2 at. %), at 500°C – 18.8 mass % (11.1 at. %), at 400°C – 15.7 mass % (9.1 at. %), and at 300°C – 12.8 mass % (7.4 at. %). These values (dark circles in Figure 46) are extrapolated to the maximum solubility of 24.6 mass % Sc (15 at. % Sc) at 710°C assumed by [170] according to [81]. In [95] the maximum solubility of Sc in solid Mg was determined by extrapolation to be 25.9 mass % (15.9 at.%) at the peritectic temperature 706°C assumed according to the first investigation of the Mg–Sc system [80]. The maximum solubility of Sc in solid Mg was estimated in [80] to be slightly more, 29–30 mass % (18.1–18.8 at. %) at 706°C. Composition of the liquid phase participating in the peritectic equilibrium was assumed by [170] to be 14.9 mass % Sc (8.67 at. % Sc) according to [81] as compared with 20–21 mass % Sc (11.9–12.6 at. % Sc) established by [80]. The former data are preferred by [170] because of higher purity of starting materials and quite accurate thermal analysis used by [81].

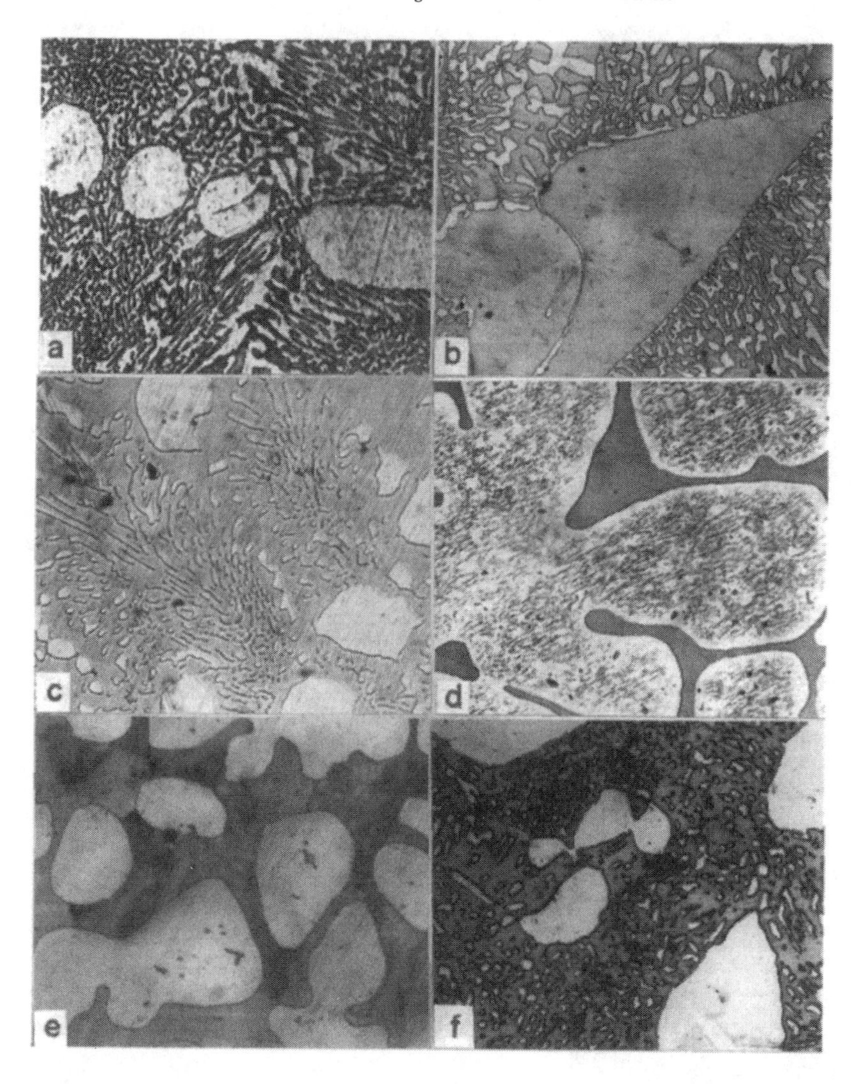

Figure 47. Microstructure of the Mg–RE alloys after slow solidification. a – Mg–15 mass % La, × 200, b – Mg–25 mass % Ce, × 120, c – Mg–20 mass % Pr, × 200, d – Mg–10 mass % Nd, × 120, e – Mg–20 mass % Nd, × 120, f – Mg–30 mass % Nd, annealed at 530°C, 5h, × 340.

Metallography

Microscopy investigation of the Mg–RE alloys solidified during thermal analysis revealed some specific peculiarities of their structures. Some of them are illustrated by the micrographs presented in Figures 47 and 48. The micrographs are taken from the alloys obtained by the author in Baikov Institute of Metallurgy[†] at approximately equal cooling rate during

[†] All other micrographs presented in the book were obtained in the investigations by the author.

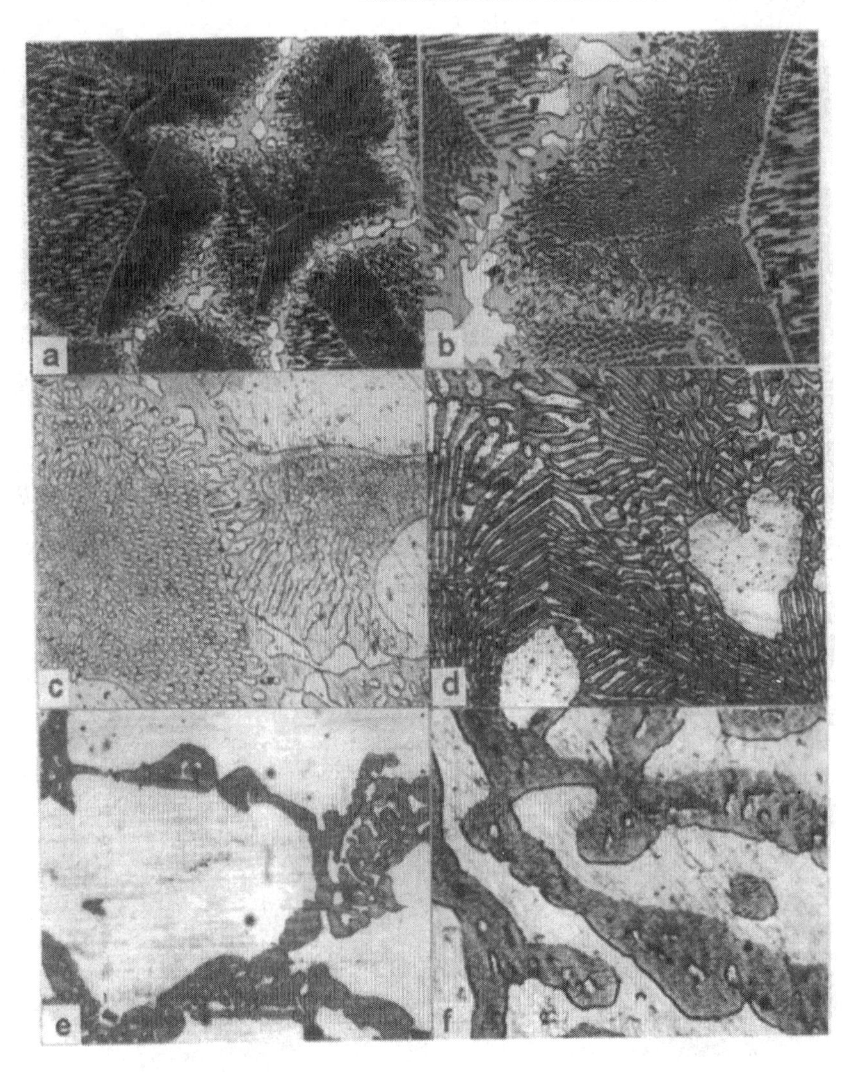

Figure 48. Microstructure of the Mg–RE alloys after slow solidification. a – Mg–37.5 mass % Nd, × 120, b – Mg–37.5 mass % Nd, × 340, c – Mg–30 mass % Sm, 420, d – Mg–40 mass % Gd, × 440, d – Mg–35 mass % Dy, × 200, f – Mg–49 mass % Tm, × 1000.

solidification. They show the presence of magnesium solid solution and the RE-rich phases observed in light microscope. Magnesium solid solution was distinguished in the microscope by white colour and weak scratches resulting from its small hardness. The RE-rich phases were in contrast a grey colour which changed from weak grey to dark grey depending on the kind and duration of etching.

Figure 47a shows the typical structure of the hypoeutectic Mg–La alloy. The eutectic is of typical plate-like view and quite fine. The next micrograph (Figure 47b) shows the microstructure of the hypereutectic alloy in the Mg–Ce systems. In this case the eutectic differs significantly from that in the Mg–La alloys. The particles of the Ce-rich phase in it are coarser and more rounded as compared with particles of the La-rich phase in the

Mg–La eutectic. The eutectic in the next system Mg–Pr (Figure 47c) is distinguished by greater size of the RE-rich phase particles as compared with the Mg–Ce system and demonstrates the evident tendency to degeneration. In the alloys with the next element of the lanthanum series Nd, this tendency increases and results in the full degenerated eutectic with quite coarse crystals of the Nd-rich phase between the dendrite branches of Mg solid solution. Such a structure is shown in Figure 47d where decomposition of Mg solid solution can also be seen. So there is a certain regularity in the eutectic constitution with increasing atomic number of the four first elements of the lanthanum series. The four next micrographs demonstrate other peculiarities of Mg–Nd alloys. At higher Nd concentrations in the hypoeutectic alloys dark areas within the Nd-rich phase can be seen (Figure 47e). Sometimes their plate-like constitution can be seen at high magnifications [89, 106]. After annealing at high enough temperature below eutectic, all areas of the Nd-rich phase become of cheese-like appearance suggesting eutectoid decomposition (Figure 47f). At higher Nd concentration the typical quite fine eutectic is observed in Mg–Nd alloys. It is shown at two magnifications. At smaller magnification the overall view of the eutectic cells can be seen suggesting the leading position of the Nd-rich phase and its dendrite form (Figure 48a). At higher magnification the constitution of the eutectic cells can be seen in detail (Figure 48b). It is characterised by the fine rod-like areas of Mg solid solution between the dendrite branches of the Nd-rich phase. These peculiarities of the Mg–Nd cast alloys were used for the Mg–Nd phase diagram in the Mg-rich part described before. The continuous Nd-rich crystals after crystallisation were considered in it to be the phase $Mg_{12}Nd$ and the Nd-rich phase in eutectoid and the fine eutectic was considered to be the phase $Mg_{41}Nd_5$.

Figure 48c demonstrates the eutectic in the hypoeutectic Mg–Sm alloy. The similarity with the Mg–Nd eutectic is evident and this fact confirms similar formulas of the RE-rich compounds forming eutectics in both systems, $Mg_{41}Ln_5$. Figure 48d clearly demonstrates a plate-like eutectic in the Mg–Gd system. In the systems with the next elements of the lanthanum series Mg–Tb, Mg–Dy, Mg–Ho and Mg–Er eutectics were of approximately the same constitution and some were coarser than that in the Mg–Gd system. As an example of them, in Figure 48e eutectic in the hypoeutectic Mg–Dy alloy is shown. The last Figure 48f demonstrates inner constitution of the Tm-rich phase in the eutectic of the Mg–Tm system. Its cheese-like character confirms eutectoid decomposition of the unknown phase Mg_yTm_x in the system.

Regularities of Mg–RE Phase Diagram Constitution

Consideration of the binary phase diagrams of the Me–RE type where Me is one of metals, except rare earth ones, and RE are various rare earth metals, shows some identical significant features of them with differences in details. This regularity was noticed first in [46] where the binary phase diagrams of Cu, Ag, Au, Mg, Al, Tl, Sn or Pb with La, Ce and Pr were considered. The phase diagrams with the same metal Me and three mentioned rare earth metals were recognised to be of the same general view and to differ from one another actually by the temperatures and the critical point concentrations of the inherent invariant equilibria. Existence of identical features amongst the binary phase diagrams with various rare earth metals and another metal was recognised also in later reviews on the rare earth metal alloys [14, 24]. Meanwhile, the similarity between the systems of the Me–RE type turned out to be more complicated than those suggested earlier by [46]. So, in many cases the differences between the systems with different rare earth metals were not limited only by characteristics of the identical invariant equilibria. They could differ also by the binary compounds formed, existence and types of separate invariant equilibria. The phase diagrams of europium and ytterbium differed significantly, as a rule, from the phase diagrams of the other rare earth metals belonging to the lanthanum series. More differences were observed commonly between the systems with the elements of the cerium or yttrium subgroups and between systems with the elements of the same subgroup, but sufficiently removed within the lanthanum series.

In general, the behaviour of the Mg systems with rare earth metals is analogous to that of systems with other metals, although some peculiarities take place. The regularities of the Mg–RE binary compound formation were considered in [46, 87, 112, 120, 164, 171, 172]. In [164] they were presented and discussed in detail using the latest data on the compounds. Moreover, [164] showed the existence of the regularities in the invariant equilibria of the Mg–RE systems as a result of the RE atomic number changing. The established regularities turned out to be quite strict so that the unknown Mg–RE phase diagram might be predicted sufficiently reliably if the phase diagrams of the systems with the closest neighbours of that rare earth metal in the lanthanum series had been previously studied.

Similarity and Difference of the Mg–RE Systems in Compound Formation

The regularities for the binary compound formation in the Mg–RE systems are seen in Figure 49 where they are summarised. Apart from Mg–Yb and Mg–Sc, all Mg–RE systems are characterised by a high number of compounds. The most remarkable feature is the existence of binary compounds with the same formula and crystal structure within a number of successive elements of the lanthanum series. Actually, such a succession exists for each type of compound, although within different numbers of rare earth elements. The successions can begin and end at certain elements, can be interrupted and pass from elements of the cerium subgroup to elements of the yttrium subgroup.

Long successions are observed for compounds with formulas MgLn, Mg_2Ln of two structure types, Mg_3Ln and $Mg_{24}Ln_5$. Along with them short successions exist. They include four compounds ($Mg_{41}Ln_5$), three compounds ($Mg_{12}Ln$ with tetragonal lattice tI26, Mg_5Ln with cubic lattice cF448) and two compounds ($Mg_{12}Ln$ with orthorhombic lattice oI338, $Mg_{17}Ln_2$ with hexagonal lattice hP38). Only the binary compounds Mg_5Eu with hexagonal lattice hP42 and Mg_4Eu of the Mg–Eu system do not have analogous compounds amongst other Mg–RE systems. $Mg_{17}Eu_2$ has analogous compounds in the Mg–La and Mg–Ce systems, but it is separated from them by four other elements of the lanthanum series. Moreover, there is a succession of the intermediate β phases with the disordered body-centred cubic lattice in the RE-rich parts of Mg–RE systems with Ho, Er, Tm and Lu. These phases form continuous succession with the RE-base body-centred cubic solid solutions existing in all other systems. These are the solid solutions of the high temperature modifications of the rare earth metals, except Eu which has only one allotropic form belonging to the body-centred cubic (bcc) type. The existence of the successions consisting of similar compounds is typical of the physico-chemical interaction between the rare earth metals and many other metals.

There are certain similarities and differences between individual Mg–RE systems by number and type of the binary compounds. In this respect the systems Mg–Eu and Mg–Yb

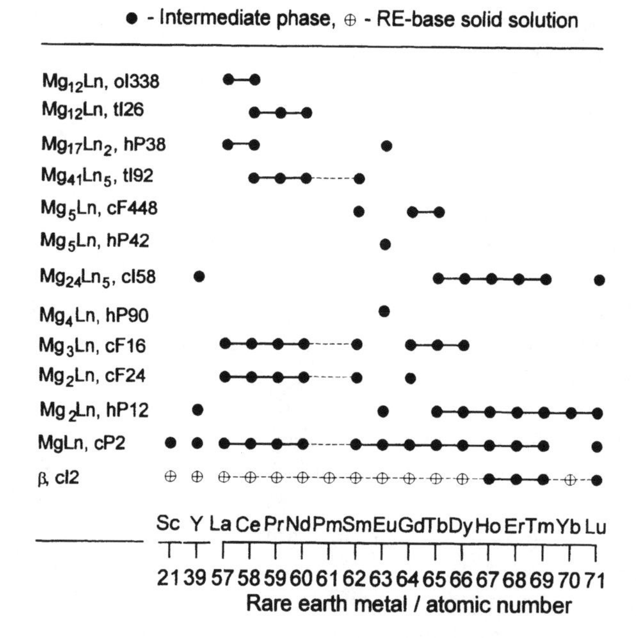

Figure 49. Summary of the intermediate phases formed in the binary Mg–RE systems.

differ more significantly from other systems with rare earth metals of the lanthanum series, although still show some similarity with them. In the former system there are two binary compounds Mg_5Eu and Mg_4Eu of the types which do not occur in magnesium systems with any other rare earth metal. Two other compounds, $Mg_{17}Eu_2$ and Mg_2Eu of the hP12 type have analogous compounds amongst other Mg–RE systems, but they are not united with them in the successions within the lanthanum series. In Mg–Yb system only one binary compound exists unlike other systems with elements of the lanthanum series. As a result, some successions of similar compounds are interrupted at Eu and Yb. Such an interruption takes place for the successions for Mg_5Ln of cF448 type, Mg_3Ln of cF16 type, Mg_2Ln of cF24 type, Mg_2Ln of hP12 type and MgLn of cP2 type. Meanwhile, both Mg–Eu and Mg–Yb have compounds belonging to similar compound successions within the lanthanum series. These compounds are MgEu and Mg_2Yb.

It is reasonable to consider the similarity and difference in compound formation between systems with rare earth metals of the lanthanum series without Eu and Yb, taking into consideration the specific position of these elements. The greatest difference between systems of both subgroups is revealed in the Mg-richest compounds. In systems with elements belonging to the cerium subgroup these compounds form short successions and tend to change with increasing atomic number successively from the $Mg_{17}Ln_2$ type to the $Mg_{12}Ln$ and $Mg_{41}Ln_5$ ones. In systems with elements of the yttrium subgroup the Mg-richest compounds are of the same $Mg_{24}Ln_5$ type except one system with the first element of the subgroup Gd. On the other hand, in systems with all elements of the cerium subgroup, there are similar compounds of Mg_3Ln, Mg_2Ln and MgLn type. In systems with elements of the yttrium subgroup there are similar compounds of Mg_3Ln type, Mg_2Ln type and MgLn type. However, in this case there is no complete similarity between the systems with elements of the yttrium subgroup. Compounds of Mg_3Ln type occur only in systems with three first elements of the subgroup, Gd, Tb and Dy. Besides, in the Mg–Gd system the Mg_2Ln compound is of cF24 type, unlike the Mg_2Ln compounds of hP12 type in all other systems with elements of the yttrium subgroup. The similarity between the Mg–RE systems with elements of different subgroups is revealed in formation of the same type compounds Mg_3Ln and MgLn. Moreover, there are compounds of the same Mg_5Ln type in the systems Mg–Sm (cerium subgroup) and Mg–Gd, Mg–Tb (yttrium subgroup). The Mg_2Ln compound in the Mg–Gd system is of the same cF24 type as the Mg_2Ln compounds in systems with elements of the cerium subgroup. This review of compound formation in Mg–RE systems with elements of the lanthanum series without Eu and Yb shows similar and different features of them. There are similar features amongst all systems, but the systems with elements of the same subgroup and disposed closer to each other in the lanthanum series show more similarity in compound formation.

The revealed similarity between the systems with elements of the cerium subgroup enable the existence of the compounds $Mg_{41}Ln_5$, Mg_3Ln, Mg_2Ln of the cF24 type and MgLn in the Mg–Pm system to be forecast quite reliably. This supposition was used earlier in the description of the proposed Mg–Pm phase diagram.

In terms of compound formation, the Mg–Y system is quite similar to other systems with elements of the yttrium subgroup, although there are some peculiarities of this system. Three compounds existing in the Mg–Y system, $Mg_{24}Y_5$, Mg_2Y of hP12 type and MgY are typical of other Mg–RE systems for the yttrium subgroup. As in systems with elements after Dy in the Mg–Y system the compound of Mg_3Ln type is absent. On the other hand, the Mg–Y system differ from the Mg–Tm and Mg–Lu systems by the absence of additional unknown compounds. So, the Mg–Y system is, at most, similar to the Mg–Ho and Mg–Er systems, where only the compounds $Mg_{24}Ln_5$, Mg_2Ln of hP12 type and MgLn are formed. However, the system Mg–Y differs from the systems Mg–Ho and Mg–Er by the absence of the intermediate body-centred cubic phase β with disordered arrangement of Mg and Y atoms. The Mg–Sc system differs significantly from all other Mg–RE systems by formation of only one binary compound. This compound is, however, of MgLn type which is observed in all Mg–RE systems, except Mg–Yb.

Formation of the compounds in the Mg–RE systems of a certain type and, as a consequence, beginning, interruption and ending of similar compound successions at certain elements of the lanthanum series are explained by the different atomic radius ratios of Mg and various rare earth metals. The important factor is also the possible electron state of rare earth atoms in their compounds [115, 128, 171]. The general view of the phase diagram depends to a great extent on the intermediate phases formed and their stability. Greater similarity in the intermediate phase formation amongst the Mg–RE systems with rare earth metals of the same subgroup promotes a certain similarity of the corresponding phase diagrams demonstrated by [87].

Lattice Parameters of the Mg–RE Compounds

The lattice parameters of the isomorphic compounds in the Mg–RE systems change regularly with increasing atomic number of the rare earth metals within the lanthanum series. The lattice parameters correlate with the atomic radii of the rare earth metals. This regularity is revealed more clearly if results of the parameters measurements of the same authors are compared. Difference between the lattice parameters determined in every work

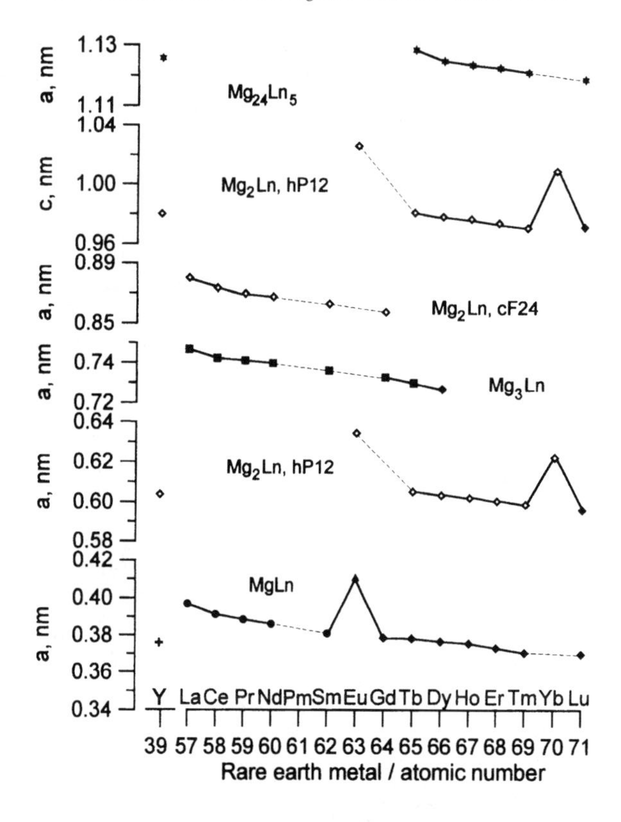

Figure 50. Lattice parameters of the compounds versus the rare earth atomic number. ● - [113], ◆ - [128], ▲ - [121], ◇ - [162], ■ - [133], + - [120], ✱ - [167].

can be significant and embarrasses establishment of the general tendency. One of the sources for scattering between lattice parameters determined for the same compounds by different authors may be the existence of some homogeneity ranges of the compounds and change of the lattice parameters within them. Figure 50 displays the lattice parameters of the Mg–RE compounds forming long successions within the lanthanum series. The presented values were taken mainly from the works where they were determined for most compounds of the same type simultaneously, and, therefore, enable us to demonstrate the regularity sufficiently reliably.

The main features of how lattice parameters change with increasing RE atomic number are as follows. Within the lanthanum series there is successive decrease of the lattice parameters with two exceptions for Eu and Yb whose compounds show anomalously high parameter values. Within the cerium subgroup such a character of the lattice parameter change is seen for Mg_2Ln of cF24 type and MgLn compounds. Within the yttrium subgroup the mentioned character of the lattice parameter change is seen for $Mg_{24}Ln5$, Mg_2Ln of hP12 type and MgLn compounds. The change of the lattice parameters correlates with the metallic radii of rare earth metals of the lanthanum series, except Ce. The more metallic radius the more lattice parameter for the same type of compound. Metallic radius of Ce is slightly less than that of Pr, but the lattice parameters of the compounds with Ce are greater than the lattice parameters of the compounds with Pr. This fact shows, as it was mentioned above, that in the Mg–RE compounds the atomic radius of cerium turns out to be larger than that of praseodymium in accordance with the "lanthanoid shortening" rule, despite the smaller metallic atomic radius of Ce, as compared with the metallic atomic radius of Pr.

The lattice parameter values presented in Figure 50 and their regular change with increasing rare earth atomic number enable us to forecast the lattice parameters of the supposed compounds Mg_3Ln-, Mg_2Ln- and MgLn-type in the Mg–Pm system.

The metallic radius of Y is near that of Gd (Table 3). Nevertheless, the lattice parameters of the Y compounds with Mg are always smaller than those of Gd. They are in the limits of the lattice parameters of the compounds formed by the rare earth metals of the middle part of the yttrium subgroup with Mg. So, the lattice parameter of $Mg_{24}Y_5$ a = 1.1257 nm is between those of $Mg_{24}Tb_5$ (a = 1.1283 nm) and $Mg_{24}Dy_5$ (a = 1.1246 nm) [167]. The lattice parameters of the hexagonal Mg_2Y a = 0.60384 nm, c = 0.98005 nm are between those of Mg_2Tb (a = 0.60512 nm, c = 0.98074 nm) and Mg_2Dy (0.60330 nm, c = 0.97786 nm) [162]. The lattice parameter of MgY a = 0.376 nm is between those of MgDy (a = 0.3767 nm) and MgHo (a = 0.3758 nm) [120, 128]. Such a behaviour of the Y in its compounds with magnesium may be explained by some change of the atomic size as compared with that in the pure metal. This change happens differently for the rare earth metals of the lanthanum series and yttrium and, as a result, the Y atoms in the compounds have a smaller radius than the Gd atoms. The only compound of scandium MgSc has a large homogeneity range, so that its lattice parameter should change within sufficiently wide limits which have not been determined exactly. However, the existing data for two separate compositions show certainly that the lattice parameter of cubic MgSc is less than the lattice parameters of the same type of compound formed by all other rare earth metals in accordance with the smallest metallic radius of scandium.

Regularities of Rare Earth Metal Solubility in Solid Magnesium

Solubility of the rare earth metals in solid Mg changes in quite wide limits. In this respect Mg differs from most other widely used metals, for example Al, where solubility of almost all rare earth metals in the solid state is insignificant without tangible difference between one another. The regular change of the rare earth metal solubility in solid magnesium was established in Baikov Institute of Metallurgy based on experiments where great attention was paid to reliability and precision of the values obtained. Such an approach was caused by the existence, at that time, of the results of the solubility determination of some rare earth metals in solid magnesium which were characterised by sufficiently high scattering. So, it was impossible to reveal any regularity. Especially it was so for the four first elements of the lanthanum series, La, Ce, Pr and Nd. The solubility of these rare earth metals in solid Mg is quite small and, therefore, is more difficult to be determined precisely by the commonly used microscopy and X-ray methods. The electrical resistivity method enabled us to determine the low solubility of the rare earth metals in solid magnesium more exactly and to reveal quite reliably its regular change with increasing the rare earth atomic number [146, 173].

Solubility in solid magnesium versus rare earth atomic number within lanthanum series is shown in Figure 51. The presented values were obtained in Baikov Institute of Metallurgy, and those of them for the rare earth metals of the cerium subgroup were determined mainly by the electrical resistivity method. The lower part of Figure 51 compares the values of the rare earth metal solubility in solid magnesium at three

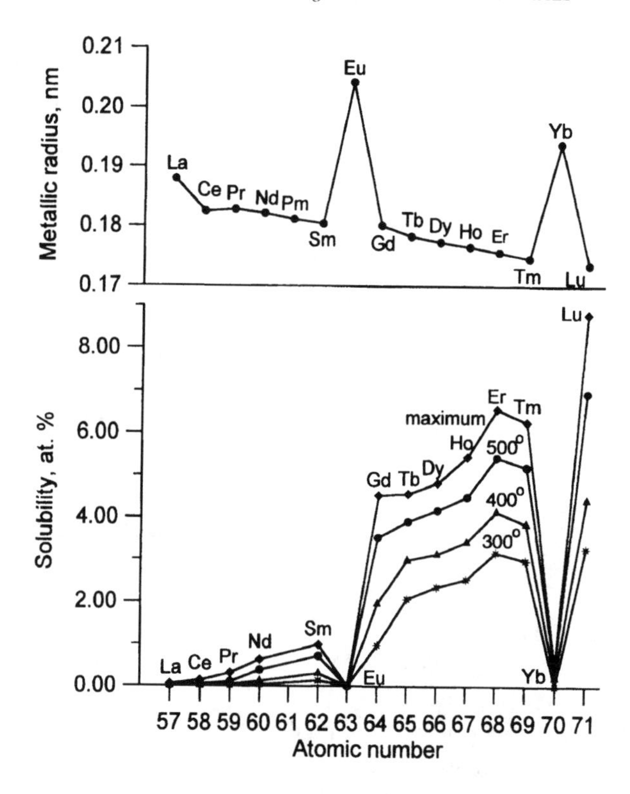

Figure 51. The RE metallic radius (upper) and the RE solubility in solid Mg (lower) versus the RE atomic number.

temperatures, 300, 400, 500°C, and the maximum solubility of them (at temperatures of the respective invariant equilibria for Mg-rich sides of the phase diagrams). The values are arranged in order of the rare earth atomic number increase, and, as one can see, their change with increasing atomic number is of the same character for every temperature and for maximum. The upper part of Figure 51 presents in the same order the change of the rare earth metallic radius derived from Table 3 for comparison with the solubility change. The presented solubility of RE in solid Mg and their regular change with increasing atomic number enable us to forecast sufficiently reliably the unknown maximum solubility and solubility at separate temperatures of Pm (atomic number 61).

For the elements of the cerium subgroup the successive growth of the solubility with increasing atomic number is observed from La up to Sm, and then for the last element of the subgroup Eu, drastic decrease of the solubility takes place. The successive decrease of the rare earth metal solubility in solid Mg within elements of the cerium subgroup from La to Sm and the least solubility of Eu became evident when the resistivity values obtained for every system after quenching from the same temperature are superposed. As an example, in Figure 52 the superposition of the resistivity values for the alloys quenched from 500°C is shown. The resistivity values for all systems within Mg solid solution coincide sufficiently well with each other and differ evidently for two-phase area. Successive increase of the resistivity values in two-phase area from Eu to La, Ce, Pr, Nd and Sm confirms successive increase of the solubility of these rare earth metals in solid Mg in the same order.

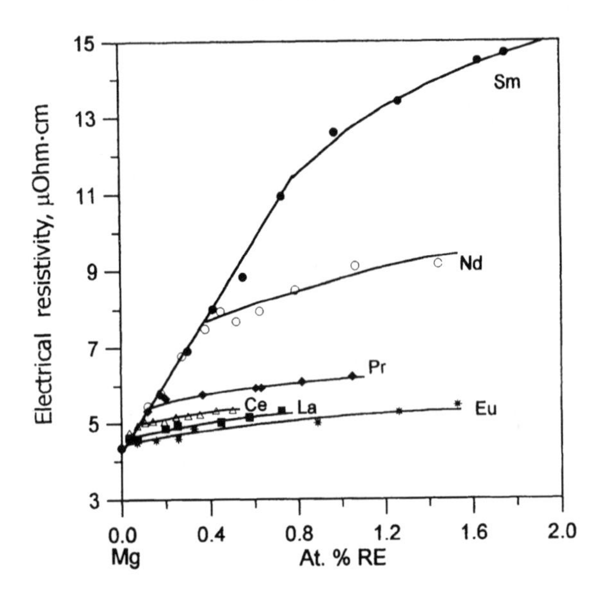

Figure 52. Electrical resistivity of Mg alloys with RE of the Ce subgroup quenched from 500°C.

With increasing atomic number after Eu the solubility of the rare earth metals in solid magnesium increases again so that the next element of the lanthanum series Gd has significantly greater solubility than Sm staying before Eu. This fact is also confirmed by the resistivity measurements. For one of the quenching temperatures, 500°C, their results are shown in Figure 53 where resistivity/concentration curves for the Mg–Sm and Mg–Gd alloys are superposed. The greater resistivity after the kinks on the curve and, therefore, the significantly greater solubility of Gd than that of Sm is evident. Figure 53 contains also the results of the resistivity measurements of the binary alloys with yttrium [68Svi] which does not belong to the lanthanum row. The resistivity of the alloys with Y shows certainly that its solubility is intermediate between those of Sm and Gd.

Beginning with Gd the solubility of the rare earth metals in solid magnesium continues to increase successively with increasing atomic number. Only the last but one element Yb ignores the general tendency. Solubility of Yb in solid Mg turns out to be significantly lower than those of the adjoining elements in the lanthanum series, Tm and Lu. Some deviation from the solubility increase with increasing the rare earth atomic number is presented also by Tm, whose solubility is determined to be somewhat lower than that of Er, although Er stays immediately before Tm in the lanthanum series. Unlike the Yb solubility, the deviation from the solubility increase between Er and Tm is small enough and, therefore, can be explained by the experimental error in the solubility determination.

The successive increase of the solubility in solid Mg for the rare earth metals of the yttrium subgroup, except Yb, was confirmed by the special investigation [173]. In this case the resistivity method was not considered to be the most exact because of the high solubility of the rare earth metals in solid Mg and, as a result, quite high electrical resistivity of Mg solid solution. Therefore, the X-ray method based on measurements of the interplane distances in Mg lattice was used. For the experiments the binary alloys of Mg with different rare earth metals of the yttrium subgroup were taken, by one alloy of each system. As was

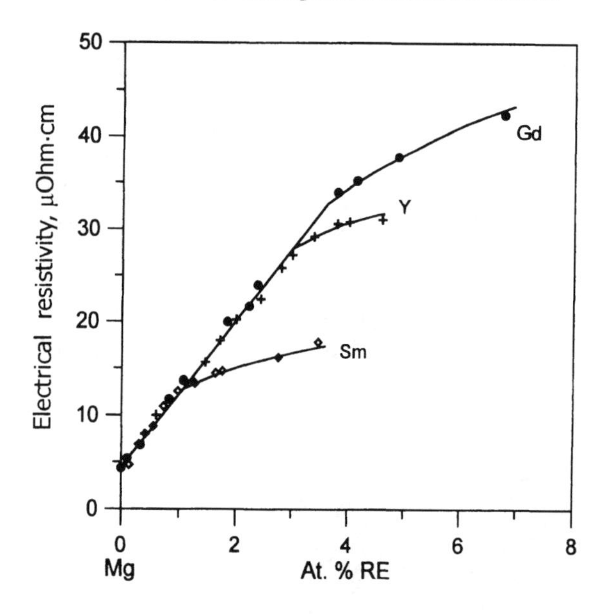

Figure 53. Electrical resistivity of Mg alloys with Sm, Gd, Y quenched from 500°C.

established in the previous microscopy investigations the alloys had a two-phase structure at the temperature 500°C opted for comparison. They contained 4.67 at. % Gd, 4.92 at.% Tb, 4.88 at. % Dy, 4.78 at. % Ho, 6.11 at. % Er, 6.34 at. % Tm or 7.28 at. % Lu. In addition to them, the alloy Mg–3.80 at. % Y with two-phase structure at 500°C and pure Mg were also taken. The alloys and pure magnesium in the form of fine filings were annealed simultaneously at 500°C for 5 h followed by quenching in water of room temperature. In this state the solid solutions concentrations in the alloys corresponded to the solubility of each rare earth metal in solid magnesium at the same temperature 500°C. In the applied then X-ray investigation of the treated alloys and pure Mg one type of interplane distances in Mg solid solution lattice could be measured quite precisely. This took place for the lattice planes of the (212) type and was performed by using the Fe radiation which gave the respective reflections at high angle of about 82° and a cylinder camera with about 86 mm in diameter.

The measured values of the (212) interplane distances [173] are presented in Figure 54. They are arranged in the order of increase of the rare earth atomic number and shows successive growth of the interplane distance d_{212} from pure Mg to the alloy with Y and then to the alloys with the rare earth metals of the lanthanum series in order of the atomic number increase from Gd up to Tm. For the alloy with Tm and the next after one rare earth metal Lu the interplane distance values are equal in the error limits. The change of the lattice parameters as a result of the solid solution formation is determined by concentration of the solid solution formed and difference between radii of the solute and solvent atoms. The larger atomic radius of the solute atoms as compared with atomic radius of the solvent promotes increase of the lattice parameters and vice versa. The greater the difference between atomic radii of the solute and solvent atoms and the higher the solid solution concentration, the greater lattice parameters and, respectively, interplane distances of the solid solution change. In accordance with this, successive increase of the interplane

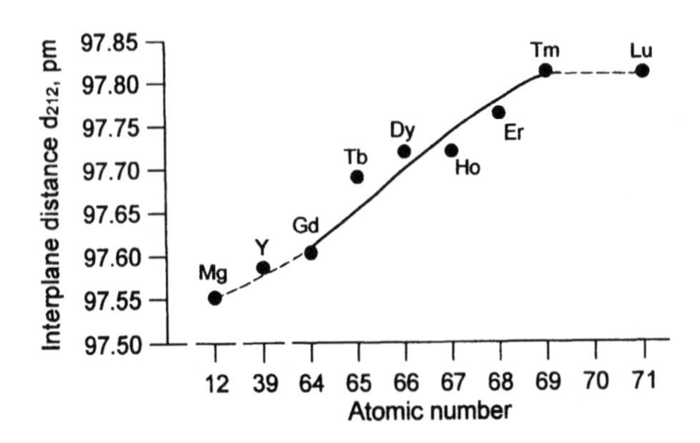

Figure 54. Interplane distance of Mg and Mg alloys with different rare earth metals quenched from 500°C.

distance d_{212} in the Mg alloys with elements from Gd to Tm confirms unambiguously the respective increase of the rare earth solubility in solid Mg with increasing atomic number, because in the same order the rare earth metallic radius decreases, but still remains greater than metallic radius of Mg. The approximately equal d_{212} values for the alloys with Tm and Lu also show the higher solubility of Lu in solid Mg compared with Tm because the metallic radius of Lu is smaller than that of Tm. Taking into account the actually equal metallic radii of Y and Gd (0.18012 nm and 0.18013 nm, respectively), the less d_{212} value for the alloy with Y than that for the alloy with Gd can be considered as confirmation of the lower solubility of Y in solid Mg as compared with solubility of Gd in it.

There is a certain connection between the solubility of the rare earth metals in solid Mg and their metallic radii. It is seen from Figure 51, where the solubility and metallic radii for the elements of the lanthanum series are shown together. The smaller metallic radius of the rare earth metal corresponds to higher solubility. In accordance with this there is the general tendency of the solubility increase with increasing atomic number of the rare earth metals within the lanthanum series with two exceptions for Eu and Yb. Meanwhile, Eu and Yb show the falling out small solubility in solid Mg as compared with their neighbours in lanthanum series which correlate with their anomalously large metallic radii. The connection of the rare earth solubility in solid magnesium with its atomic radius conforms to the reduction of the difference between the rare earth and magnesium metallic radii when the rare earth metallic radius decreases. The less the difference between the metallic radii, the more favourable the conditions are for solid solution formation. It is a well known rule in the theory of alloys [12, 174].

However, there are some exceptions to the mentioned connection between the rare earth metal solubility in solid Mg and the metallic radii. One of them is insignificant and caused by the metallic radius of Ce being less than the metallic radius of the next element of the lanthanum series, Pr. Although the metallic radius of Ce is less than that of Pr, the solubility of the former is lower than that of the latter in accordance with the lower atomic number of Ce. This disturbance may be related to the peculiarity of the metallic radius calculation from the lattice parameters of pure metals and confirms the possibility of its change when solid solutions are formed. The lower solubility of Ce than of Pr suggests the greater atomic radius of Ce in Mg solid solution, as compared with the atomic radius of Pr there. The suggestion about the greater atomic radius of Ce in Mg solid solution, as

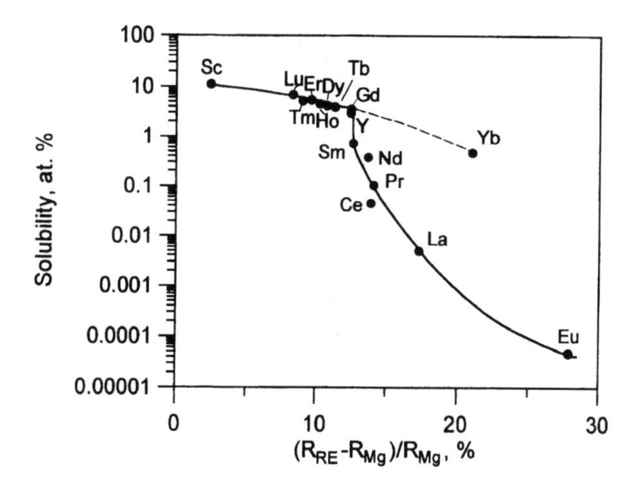

Figure 55. The RE solubility in solid Mg at 500°C versus relative difference between the RE and Mg metallic radii.

compared with that of Pr, is supported by the larger number of lattice parameters of Mg–Ce compounds than the lattice parameters of the same type Mg–Pr compounds. This fact was noted above.

The second exception to the relation between the rare earth metallic radii and the rare earth solubility in solid magnesium is more significant. It consists of the steep increase of the solubility at the transition from the elements of the cerium subgroup to the elements of the yttrium subgroup (between Sm and Gd). Meanwhile, the corresponding change of the metallic radii (from 0.18041 nm for Sm to 0.18013 nm for Gd) is very small. In this aspect, the solubility of Yb in solid Mg is quite remarkable. The metallic radius of Yb is greater than those of La, Ce, Pr and Nd, but the solubility of Yb in solid Mg is higher than the solubility of those metals at the same temperatures. The anomalous growth of the rare earth solubility in solid Mg at the transition from the cerium to yttrium subgroup becomes more evident when the solubility values are presented directly compared with the relative difference between the rare earth metal and magnesium atomic radii, $(R_{RE}-R_{Mg})/R_{Mg}$, as is shown in Figure 55 for one temperature, 500°C. In Figure 55 the solubility values of Y and Sc at the same temperature are also presented according to the data of Baikov Institute of Metallurgy [91, 95]. To help interpretation, the solubility values are displayed in the logarithmic scale.

As one can see in Figure 55 the solubility values of all rare earth metals, except Yb, lie pretty well along a single curve having two distinct branches. One of the branches corresponds to the elements of the cerium subgroup and the other one corresponds to the elements of the yttrium subgroup, including Sc. Between both branches there is a short approximately vertical piece showing the abrupt solubility change at the border of both subgroups (between Sm and Gd). The Yb solubility position corresponds well to the extrapolation of the yttrium subgroup branch, but not to the branch of the cerium subgroup. Such peculiarities of the connection between the solubility and metallic radius of the rare earth metal may be explained by the different change of the rare earth atomic size as the rare earth metals are dissolving in Mg solid solution. By this means, the atomic radii of the yttrium subgroup metals become in Mg solid solution substantially smaller than the atomic radii of the cerium subgroup metals as compared with values of metallic radii calculated from the lattice parameters of the pure metals.

The third exception to the general tendency of rare earth solubility change with changing rare earth metallic radius is the special position of yttrium. The metallic radius of Y 0.18012 nm is actually equal to the metallic radius of Gd 0.18013 nm, but the solubility of Y in solid Mg is significantly less than that of Gd. This fact is not clearly seen in the logarithmic scale in Figure 55, but it is evident from the resistivity curves and the lattice parameters presented in Figures 53 and 54. The significantly lower solubility of Y in solid Mg as compared with the solubility of Gd can be explained in the same way supposing the different change of the atomic size of Y and Gd when they dissolve in Mg solid solution. This is not unusual because Y does not belong to the lanthanum series and its atoms have a lower number of electron shells than Gd atoms.

Unlike Y, the solubility of Sc in solid Mg follows its metallic radius. The metallic radius of Sc and its solubility differ significantly from all other rare earth metals and, therefore, the possibility of Sc atoms changing somewhat in size after dissolution in Mg seems not to matter for its place in the general dependence of the solubility on the metallic radius.

Regular Change of the Invariant Equilibrium Characteristics at Mg-Rich Side

The regularity in the compound formation and solubility in solid magnesium within the rare earth metals of the lanthanum series points to the existence of certain regularities in the characteristics of the invariant equilibria in the Mg–RE systems. Along with other invariant equilibria the regularities can be revealed for the invariant equilibria closest to Mg. Similarly to other characteristics of the phase diagram, they are seen most clearly when results of the experiments conducted in the same conditions are compared. Such is the case for the results of Baikov Institute of Metallurgy where all Mg–RE phase diagrams, except that with Pm, were studied. The data of Baikov Institute of Metallurgy agree reasonably well with the majority of results of other investigations and, consequently, can be considered to be conforming to the general tendency in the characteristics change. In all Mg–RE systems for the rare earth metals of the lanthanum series, the invariant equilibria closest to Mg are of eutectic type. Their main characteristics are temperature and the eutectic point concentration. Both characteristics are shown in Figure 56 in order of increasing the rare earth atomic number. All values presented in Figure 56 were obtained in Baikov Institute of Metallurgy except the eutectic point concentration for the Mg–Yb system assumed according to [78]. For the system Mg–Gd the latest and most reliable values [108] of several data of Baikov Institute of Metallurgy were used.

The lower part of Figure 56 shows clearly the regular change of the eutectic temperature with increasing atomic number of the rare earth metals. It consists of successive decrease of the eutectic temperature within elements of the cerium subgroup from La to Sm and successive increase of the eutectic temperature within elements of the yttrium subgroup from Gd to Lu, except ytterbium. Europium and ytterbium clearly depart from the two mentioned tendencies, and such behaviour of these elements can be considered also as a regularity. In the case of the eutectic point concentration there is more scattering in the experimental values obtained, especially within elements of the yttrium subgroup, but the general tendencies have been also revealed. They consist of successive

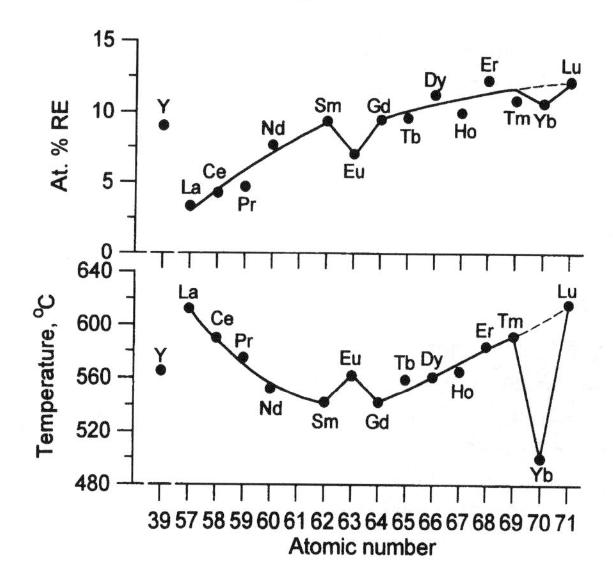

Figure 56. Eutectic point concentration (upper) and eutectic temperature (lower) in the Mg–RE systems versus the RE atomic number.

increase of the eutectic point concentration with increasing the RE atomic number within both subgroups, from La to Sm and from Gd to Lu. The eutectic point in the Mg–Eu system is an obvious exception to the two mentioned tendencies. In the Mg–Yb system, the eutectic point concentration departs from the mentioned tendencies, too, but this effect is considerably smaller and is actually in limits of the observed scattering. The eutectic temperature and eutectic point concentration in the Mg–Y system are within the respective values for the systems with the elements of the first half of the yttrium subgroup. The successive change of the invariant equilibrium characteristics of the systems with increasing atomic number of the elements within the cerium subgroup can be used to forecast sufficiently exactly the unknown characteristics of the invariant equilibrium in the Mg–Pm system.

No doubt, reasonably strict regularities in the change of the eutectic temperature and eutectic point concentration with increasing rare earth atomic number reflect the existence of the connection between the phase diagrams and the rare earth metal properties, their ability to form compounds with magnesium and the stability of these compounds. Nevertheless, analysis of this quite complicated connection and explanation of it are outside the remit of this book. It is reasonable only to note the existence of the correlation between the melting point of the yttrium subgroup metals and the eutectic point temperature in the respective Mg–RE systems. On the other hand, within the systems with rare earth metals of the cerium subgroup the successive decrease of the eutectic temperature with increasing RE atomic number correlates with successive change of the respective compound type as a result of the compound instability. The successive increase of the eutectic point concentration with increasing RE atomic number correlates with the tendency of increasing rare earth concentrations in the compounds closest to Mg within elements of the cerium subgroup and increase of the eutectic temperature within elements of the yttrium subgroup.

Unlike other rare earth metals, the closest to Mg invariant equilibrium in the system with Sc is of peritectic type, and any comparison of its characteristics with those of the eutectic equilibria in other Mg–RE systems will not be justified.

Ternary and Quaternary Phase Diagrams of Magnesium Systems Where at Least One of the Rare Earth Metals is a Component

Many of the investigations were devoted to the ternary and more complicated phase diagrams of magnesium alloys with individual rare earth metals. Most of them were undertaken in connection with the development and application of new magnesium commercial alloys with improved mechanical characteristics for application at ambient and elevated temperatures. Meanwhile, the scientific aspect in most studies was also significant. The list of the investigated ternary and quaternary phase diagrams is presented in Table 23. It includes only the systems which are known by the author as those having been investigated. At the same time it is probable that some unknown systems of Mg with rare earth metals have been investigated by others. However, the number of such investigations unknown to the author must not be great. Anyway, very many phase diagrams have been studied already, and their examinations enable the main generalisations on their constitution to be deduced reliably. Most investigations were limited by the concentration areas adjoining the Mg corners of the phase diagrams, and the main objects in them were extension of Mg solid solution region, the phases being in equilibrium with Mg solid solution, the liquidus surface and the invariant equilibria with Mg solid solution participation. Another remarkable feature of the systems chosen for the investigations is their connection with development of commercial alloys. This is the reason why the systems studied are those with the most often used rare earth elements in magnesium alloys, lanthanum, cerium, neodymium, yttrium. A few systems studied were also with samarium, gadolinium and dysprosium which were only recently recognised as quite effective in strengthening Mg.

The extensive information on the ternary and quaternary phase diagrams of Mg systems with rare earth metals would require too much space in this book to describe each of them in detail. Therefore, in this section, the author will report only the main common peculiarities of the phase diagram constitution which have been established, and illustrate them by the most typical examples. The ternary phase diagrams are divided into two main groups. One of them includes the systems, where, besides Mg, two different rare earth metals are components. The second group includes all other systems, where, besides Mg, only one rare earth metal is a component. Commonly, the third component in these systems is one of the typical alloying elements for magnesium alloys. It is reasonable to consider separately both groups of the ternary phase diagrams of Mg systems with the rare earth metals.

Ternary Mg Systems with Two Rare Earth Metals as Components

The studied ternary phase diagrams of this type enable some important conclusions to be drawn about the common features of their constitution in the Mg-rich area.

The first common feature is the existence of the compounds closest to Mg in the adjoining Mg–RE binary systems in equilibrium with Mg solid solution. No phase, besides that closest to Mg compound in the adjoining binary systems, is in equilibrium with solid

Table 23. Ternary and quaternary phase diagrams of Mg systems with rare earth metals.

System	Year	References	System	Year	References
Mg–La–Ce	1972	[93]	Mg–Y–Al	1979, 1989	[233, 234]
Mg–Nd–Pr	1972	[94]	Mg–Y–Cd	1979	[200]
Mg–La–Y	1983, 1984, 1994, 1995	[175–178]	Mg–Y–Mn	1968, 1970	[91. 232]
Mg–Ce–Y	1981	[179, 180]	Mg–Y–Zn	1978,1979, 1982	[201–203, 212]
Mg–Nd–Y	1971	[181]	Mg–Y–Si	1975, 1978	[221, 228]
Mg–Sm–Y	1983, 1984	[165, 219]	Mg–Y–Zr	1977	[204]
Mg–Gd–Y	1983	[182]			
Mg–Sm–Gd	1996, 1997	[183, 184]	Mg–Sm–Zn	1985, 1986	[205, 229]
Mg–Y–Sc	1976	[220]			
			Mg–Gd–Al	1996, 1997	[206, 207]
Mg–La–Ni	1981	[185]			
Mg–La–Al	1981, 1988	[187, 236]	Mg–Dy–Zn	1992	[208]
Mg–La–Zn	1985, 1987	[186, 196]			
			Mg–Sc–Mn	1977, 1980	[106, 231]
Mg–Ce–Mn	1949, 1957	[222, 223]	Mg–Sc–Al	1991	[237]
Mg–Ce–Al	1941, 1967, 1989	[224, 225, 234]			
Mg–Ce–Zn	1946, 1971, 1989	[188, 226, 227]	Mg–La–Y–Zn	1989, 1992	[209, 230]
			Mg–Nd–Y–Zn	1972, 1974	[210, 211]
Mg–Pr–Al	1988	[189]	Mg–Nd–Y–Zr	1978	[212]
Mg–Pr–Zn	1985	[195]	Mg–Nd–Mn–Ni	1968	[213]
			Mg–Y–Zn–Cd	1979, 1980, 1981	[214–216, 218, 240]
Mg–Nd–Al	1968, 1969, 1981, 1988	[187, 190, 191,235]	Mg–Y–Zn–Zr	1981	[217]
Mg–Nd–Mn	1962	[192]			
Mg–Nd–Ni	1966	[193, 194]			
Mg–Nd–Zn	1971, 1974, 1985	[195, 197, 198]			
Mg–Nd–Zr	1978	[199]			

Mg. In accordance with this, in the Mg corner of the phase diagram not more than one invariant equilibrium can exist. Another possible case is the absence of the invariant equilibrium, if both compounds closest to Mg in the adjoining systems form continuous solid solution.

The second common feature is an ability to form quite extended solid solutions of each of the compounds closest to Mg in the binary Mg–RE systems with another rare earth

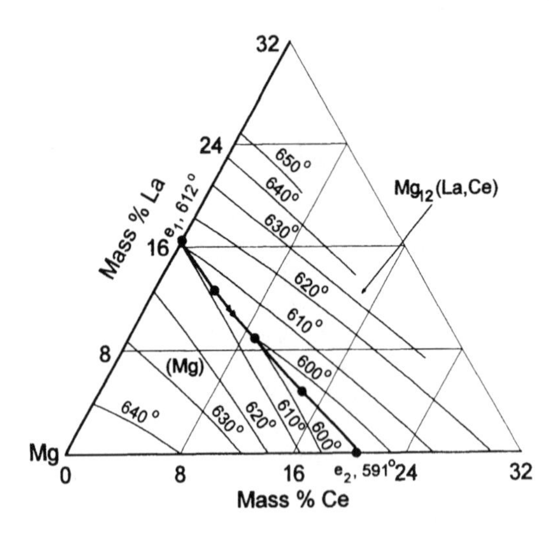

Figure 57. Partial liquidus surface of the Mg–La–Ce phase diagram.

metal. The compounds closest to Mg in both adjoining systems can form continuous solid solution if they belong to the same crystal type or the solid solutions with quite large homogeneity ranges when their crystal lattices belong to different types.

The third common feature is as follows. Each of the rare earth metals decreases solubility of another in solid Mg. As a result of this rule, the joint solubility of two rare earth metals in solid Mg is always intermediate between solubility of each of them in solid Mg. Addition of the rare earth metal with lower solubility in solid Mg to the Mg–RE alloys with the more rare earth solubility can not result in increase of the joint solubility of both rare earth metals in solid Mg and vice versa.

In each of three phase diagrams studied, Mg–La–Ce, Mg–Pr–Nd and Mg–Gd–Y, only one phase in equilibrium with Mg solid solution was revealed. Such a phase in each system was recognised to be continuous solid solution between the phases closest to Mg in each pair of the adjoining binary systems. In accordance with this, no invariant four-phase equilibrium was revealed in each system, and the three-phase eutectic equilibria in the adjoining binary systems were related to one another in the ternary system by the only monovariant three-phase eutectic equilibrium. All three phase diagrams in Mg corner have the same general view and differ from one another only by temperatures and concentrations of the surfaces, lines and points dividing the phase areas. As an example, in Figure 57 the constructed liquidus surface of the phase diagram Mg–La–Ce [93] is presented. Its main feature is the existence of the only monovariant eutectic line connecting the respective eutectic points in the binary systems Mg–La and Mg–Ce. In equilibrium with Mg solid solution in the Mg–La–Ce system is only one phase. This phase is assumed to be the continuous solid solution $Mg_{12}(La,Ce)$ between $Mg_{12}La$ and $Mg_{12}Ce$ taking into consideration the above reviewed binary systems Mg–La and Mg–Ce. The monovariant eutectic line divides the regions of the primary crystallisation of the magnesium solid solution (Mg) and the continuous solid solution $Mg_{12}(La,Ce)$.

The next Figure 58 demonstrates extension of Mg solid solution in the Mg–Pr–Nd phase diagram [94]. The isotherms of Mg solid solution region are drawn using the electri-

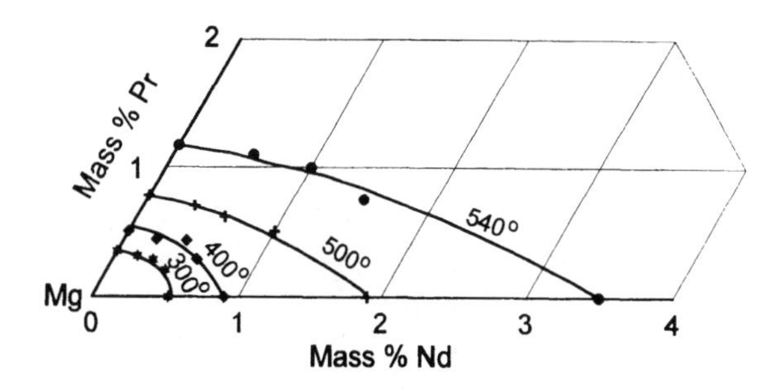

Figure 58. Isotherms of the joint solubility of Pr and Nd in solid Mg.

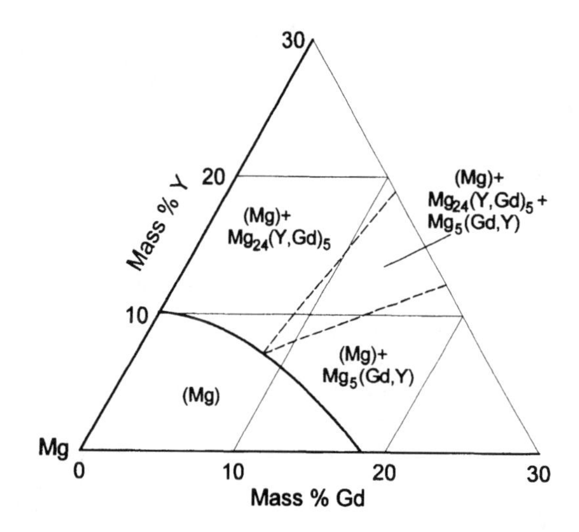

Figure 59. Partial isothermal section of the Mg–Gd–Y phase diagram at 500°C.

cal resistivity method. They show the above mentioned decrease of the solubility in solid Mg for each rare earth metal when another is added. This character of the joint solubility in solid Mg in the Mg–Pr–Nd system is of great importance for application of Nd as an alloying element in commercial Mg alloys. Technical Nd (didymium) used in industry is accompanied by Pr. Therefore, the noted decrease of the joint solubility of Nd and Pr in solid Mg with decreasing Nd/Pr ratio requires to limit reasonably the Pr contents in technical Nd. Although only the RE-rich phase was recognised in the investigation of the Mg–Pr–Nd phase diagram [94], the latest studies lead to the conclusion that the phases being in equilibrium with Mg solid solution in the systems Mg–Pr and Mg–Nd belong to different types, $Mg_{12}Pr$ and $Mg_{41}Nd_5$, respectively. Existence of continuous solid solution between these phases is not possible, although quite extended solid solutions of each rare earth metal in the compound of the other are very probable because of closeness of the electron constitution and metallic radius of the Pr and Nd atoms. This fact requires the assumption of two phases being in equilibrium with Mg solid solution in the ternary Mg–Pr–Nd system, and existence of the four-phase invariant equilibrium in the Mg corner where liquid phase L, magnesium solid solution (Mg) and the solid solutions of the binary

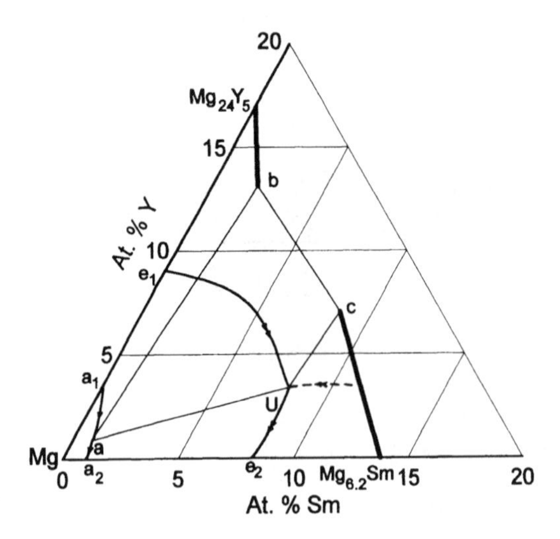

Figure 60. Mg–Sm–Y phase diagram. Projection of the four-phase equilibrium plane and the connecting double saturation lines.

compounds Mg_{12}(Pr,Nd) and Mg_{41}(Nd,Pr) are participants. Apparently, the solid solutions of the compounds could not be distinguished in the investigation by optical microscopy [94] because of the close RE contents in them.

Figure 59 demonstrates the isothermal section of the Mg–Gd–Y phase diagram [182]. As in the system Mg–Pr–Nd, it shows a decrease of each rare earth solubility in solid Mg when another rare earth metal is added. In this system two phases being in equilibrium with Mg solid solution were not distinguished, either, but they should be assumed taking into consideration the different crystal structures of both compounds closest to Mg in each of the binary systems. Most likely, the binary compounds form extended solid solutions with formulas Mg_5(Gd,Y) and $Mg_{24}(Y,Gd)_5$ for the same reasons as in the Mg–Pr–Nd system. The solid solutions of the compounds can have close contents of the rare earth metals and, therefore, are difficult to distinguish in the light microscope. Consequently, in the isothermal section of the phase diagram shown in Figure 59, two different two-phase fields and one three-phase field are assumed. Besides, an invariant four-phase equilibrium between L, (Mg), Mg_5(Gd,Y) and Mg_{24}(Y,Gd) should be assumed in the Mg corner, although in the experiments [182] it was not revealed.

In other investigated ternary systems with two rare earth metals as components the existence of two phases in equilibrium with Mg solid solution was revealed reliably by light microscopy. The binary compounds differed significantly in the rare earth contents and were of different darkness after etching the polished surface of the samples. In some of the systems solid solution formation on the basis of the binary compounds was also proved and determined by the methods of local spectral analysis. Typical representatives of this kind of system are the Mg–Sm–Y and Mg–Sm–Gd ones. The characteristic fragments of their phase diagrams are presented in Figures 60 and 61.

Figure 60 displays the projection of the four-phase equilibrium plane Uabc and the connecting double saturation lines of the Mg–Sm–Y phase diagram redrawn in at. % from [165]. The only invariant equilibrium in this part of system is of transition type, $L+Mg_{24}(Y,Sm)_5 \rightleftharpoons (Mg)+Mg_{6.2}(Sm,Y)$, with compositions of the phases corresponding

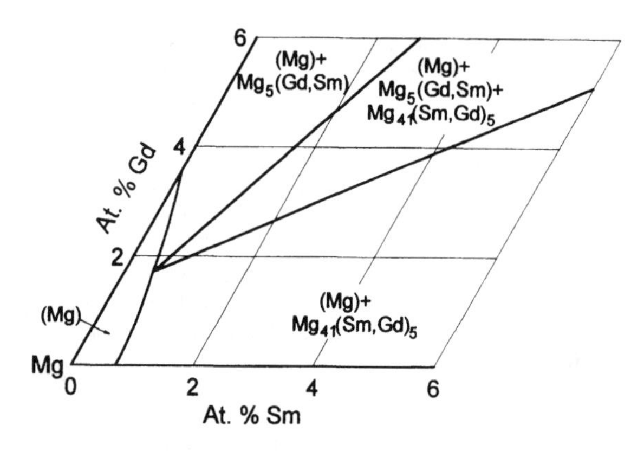

Figure 61. Partial isothermal section of the Mg–Sm–Gd phase diagram at 500°C.

to the points U, b, a and c, respectively. Compositions of the solid solutions of the compounds participating in the equilibrium, b and c, were determined based on the results of local spectral analysis. As one can see, the solubilities of Sm in $Mg_{24}Y_5$ and Y in $Mg_{6.2}Sm$ are limited, but sufficiently high.

Figure 61 shows the partial isothermal section of the Mg–Sm–Gd phase diagram [184]. Its appearance is typical of the relevant ternary systems. The remarkable features of the section are two two-phase fields and one three-phase field adjoining the Mg solid solution field and the form of the Mg solid solution field. The existence of only two two-phase and one three-phase fields shown in the section conforms to only two solid phases being in equilibrium with Mg solid solution. Both phases are the solid solutions on the basis of the compounds closest to Mg in the respective Mg–RE systems, Gd in $Mg_{41}Sm_5$ and Sm in Mg_5Gd. Formation of compound solid solutions is reflected in directions of the straight lines dividing the three-phase and each of the two-phase fields. The straight lines, which must end at the solid solutions of the compounds, deflect significantly from the concentration axes of the section. The form of the Mg solid solution field shows decrease of the solubility of each rare earth metal in it when another is present, otherwise the Mg solid solution field would have been of a parallelogram form. The joint solubility of Sm and Gd in solid Mg was determined by light microscopy and electrical resistivity measurements, and both methods confirmed unambiguously the successive decrease of RE solubility with increasing Sm/Gd ratio [184].

In the above considered systems, the solid solution formation of the compounds and character of the joint solubility of two rare earth metals in solid Mg were specially studied and established most reliably. In some other systems they were not so clear. So, in the ternary system Mg–La–Y the solubility of La in the closest to Mg compound of the Mg–Y system and that of Y in the closest to Mg compound in the Mg–La system were established to be insignificant [177, 178]. This fact could be explained by the large difference between atomic radii of La and Y. The joint solubility of the rare earth metals in the systems Mg–La–Y and Mg–Ce–Y [178, 180] was not established to follow strictly the above mentioned rule, but this fact could be explained by insufficient exactness of the methods used. The only exception to the common character of the joint solubility in the investigations was observed in the system Mg–Y–Sc. In this system the Mg solid solution field was established to be close in form to a parallelogram (Figure 62) showing approximately

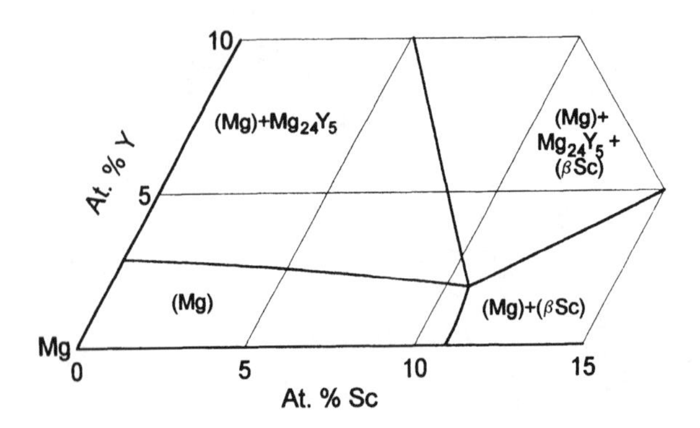

Figure 62. Isothermal section of the Mg–Y–Sc phase diagram at 500°C.

independent solubility of Y and Sc in solid Mg [220]. In many positions the Sc behaviour is quite different from that of all other rare earth metals, and the joint solubility of Sc with Y in the Mg–Y–Sc system seems to be one of the examples of such a specific behaviour of Sc.

Ternary Mg Systems with One Rare Earth Metal as a Component

As far as there are certain similar features in the constitution of the binary phase diagrams of different rare earth metals with any other metal, it is reasonable to expect also similar features in the constitution of Mg–RE–X phase diagrams, where RE are different rare earth metals and X is a metal other than a rare earth one. In general, such a conclusion can indeed be made comparing the ternary phase diagrams of the Mg–RE–X type studied, although certain differences do exist between the phase diagrams Mg–RE–X, where X is the same metal being other than RE. Therefore, it is reasonable to carry out consideration of the ternary phase diagrams of Mg with one rare earth metal as a component in the order of the opted third components X. Mainly they are Zr, Mn, Al, Zn which are widely used in commercial magnesium alloys as alloying elements. Three other third components chosen for studies are Ni, Cd and Si. They are also used for alloying of Mg alloys, but much more seldom.

In both phase diagrams studied, Mg–Nd–Zr and Mg–Y–Zr, only two solid phases, each of the Mg binary systems were revealed in Mg corner. They were the closest to Mg compound of the Mg–RE system and (αZr) of Mg–Zr one. In Figure 63 the isothermal section of the Mg–Y–Zr phase diagram at 500°C [204] is reproduced. It demonstrates the minor effect of the small Zr contents on the solubility of Y in solid Mg and significant decrease of the Y solubility in solid Mg at the higher Zr contents. Solubility of Zr in solid Mg decreases with increase of Y content up to the point of double saturation. In the investigated Mg–Nd–Zr system, actually the same character of the joint solubility in solid Mg was observed [199]. Invariant equilibria in the Mg corner of the Mg–Nd–Zr and Mg–Y–Zr phase systems were not studied in [199, 204]. However, it can be supposed that one four-phase invariant equilibrium in each of the ternary systems exists with liquid phase L, (Mg), (αZr) and the closest to Mg compound in the Mg–RE system as participants taking into consideration the phase rule and two solid phases being in equilibrium with

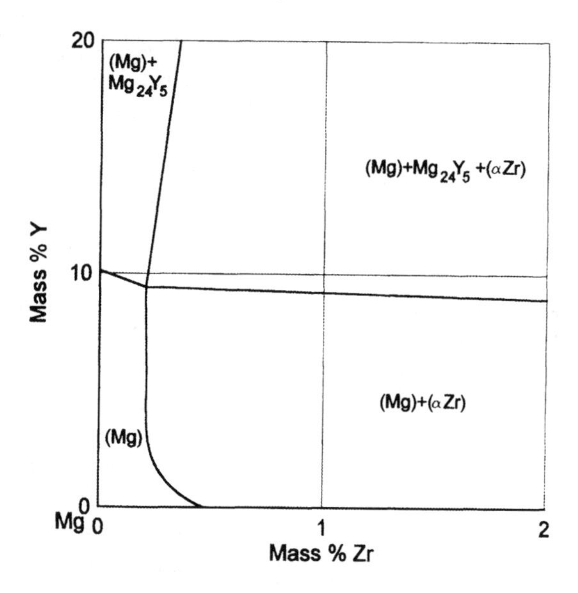

Figure 63. Partial isothermal section of the Mg–Y–Zr phase diagram at 500°C.

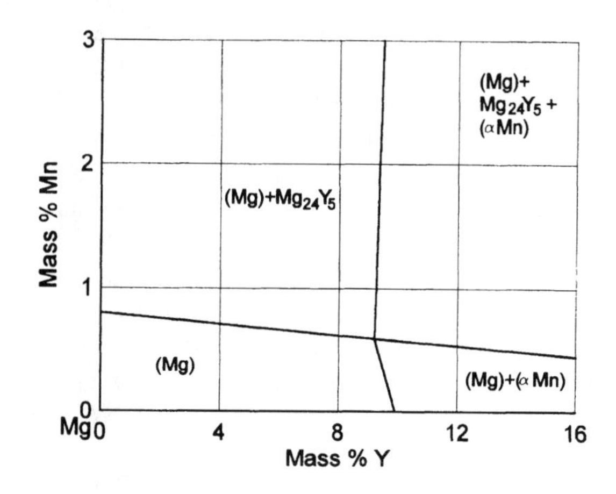

Figure 64. Partial isothermal section of the Mg–Y–Mn phase diagram at 500°C.

solid Mg.

As in the Mg–RE–Zr systems, investigations of the ternary Mg–RE–Mn phase diagrams showed the existence of only two solid phases being in equilibrium with solid Mg. Both phases belonged to the relevant binary systems, Mg–RE and Mg–Mn. They were the closest to Mg compound of Mg–RE system and (αMn) of the Mg–Mn system. Within the limits of experimental error, Mn and RE did not affect the solubility of each other in solid Mg. It is demonstrated in Figure 64 by isothermal section of the Mg–Y–Mn phase diagram [232] presented as an example. In accordance with two phases being in equilibrium with solid Mg, there was only the invariant four-phase equilibrium in Mg corner of each Mg–RE–Mn system. Its type could be various. In the Mg–Nd–Mn and

Mg–Sc–Mn systems the invariant four-phase equilibria were recognised to be of transition type [106, 192, 231], but in the Mg–Ce–Mn and Mg–Y–Mn systems the invariant four-phase equilibria were recognised to be eutectic ones [91, 222, 223].

A great number of investigations were devoted to the Mg–RE–Al phase diagrams. La, Ce, Pr, Nd, Gd, Y and Sc were used as RE. The phase diagrams were studied experimentally in full or quite large concentration ranges. The results of the investigations showed a few similar features of the phase diagrams with different rare earth metals. In all systems studied, except Mg–Sc–Al, the ternary compounds were established in approximately the same concentration area. Crystal structure of the compounds was established to be of the same $MgZn_2$ type. In the systems with La, Ce, Pr, Nd and Y, extended solid solutions of the compounds MgLn and Mg_2Ln were revealed with substitution of Mg in them by Al. In systems with rare earth metals of the cerium subgroup (La, Ce, Pr, Nd) three solid phases were established to be in equilibrium with Mg solid solution along with the compounds of the adjoining binary systems Mg–RE and Mg–Al. The phases belonged to the opposite Mg binary systems RE–Al. They were of the $LnAl_2$, $LnAl_3$ and $Ln3Al_{11}$ ($LnAl_4$) type. In the studied Mg–RE–Al ternary diagrams with Y and Gd belonging to the yttrium subgroup and Sc only one compound of the $LnAl_2$ type was established in equilibrium with Mg solid solution apart from the phases belonging to the adjoining binary systems Mg–RE and Mg–Al. All phases of the $LnAl_2$ type were congruently melting ones with quite high melting points (in limits 1405–1525°C [140]). This fact pointed to the existence of the pseudobinary sections of the Mg–$LnAl_2$ type in the ternary systems. In the investigation of the ternary system Mg–Sc–Al, the pseudobinary section Mg–$ScAl_2$ was confirmed and constructed [237]. In investigation of the Mg–Gd–Al phase diagram the pseudobinary section Mg–$GdAl_2$ was constructed partially [206]. Existence of congruent compounds of the $LnAl_2$ type with quite high melting temperatures in equilibrium with Mg solid solution suggests the possibility of significant decrease of the rare earth metal solubility in solid Mg with increasing Al contents. The joint solubility of a rare earth metal and Al in solid Mg was studied specially in two systems, Mg–Nd–Al [235] and Mg–Y–Al [233]. In the Mg–Y–Al systems the significant decrease of the Y solubility in solid Mg with increasing Al contents was observed indeed. It can be seen clearly in Figure 65, where the partial isothermal section of the Mg–Y–Al phase diagram at 400°C for Mg-rich alloys is shown. The section is redrawn from [233] with designation of the phase YAl_2 in equilibrium with Mg solid solution according to [234]. In the investigation of the Mg–Nd–Al system [235] small Al additions decreased insignificantly the solubility of Nd in solid Mg, but at higher Al contents the solubility of Nd increased to a great extent and became more than without Al. After the Nd solubility maximum was reached, it decreased gradually. This result is, however, doubtful because it contradicts the abrupt decrease of the hardening effect during solid solution decomposition in Mg–Nd alloys when Al is added [238, 239]. The change of the Nd solubility in solid Mg when Al is added needs to be checked by another investigation.

In accordance with three solid phases being in equilibrium with solid Mg two invariant four-phase equilibria were established in the Mg corner of the Mg–Y–Al system [233]. Both of them are of transition type and take place at 631°C for $L+YAl_2 \rightleftharpoons (Mg)+Mg_{24}Y_5$ and at 438°C for $L+YAl_2 \rightleftharpoons (Mg)+Mg_{17}Al_{12}$. Taking into consideration the same number of solid phases being in equilibrium with Mg solid solution, as in Mg–Y–Al system, in Mg corners of the ternary systems with Gd and Sc two four-phase invariant equilibria may be also expected. In accordance with five solid phases being in equilibrium with solid Mg in

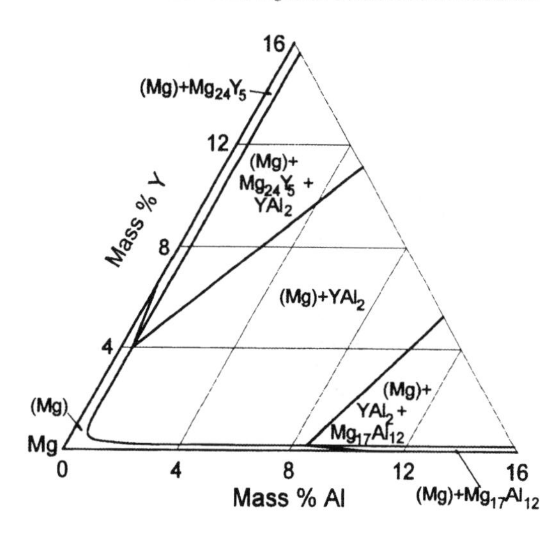

Figure 65. Partial isothermal section of the Mg–Y–Al phase diagram at 400°C.

the Mg–Nd–Al system four invariant four-phase equilibria were established in it [191]. The same number of the invariant four-phase equilibria may be expected in Mg corners of the ternary systems with other rare earth metals of the cerium subgroup, Mg–La–Al, Mg–Ce–Al and Mg–Pr–Al.

The ternary phase diagrams with Zn were studied for La, Ce, Pr, Nd, Sm, Y and Dy. The main similar feature of this kind of phase diagram is a great number of solid phases in equilibrium with Mg solid solution. Apart from the phases belonging to the binary Mg–RE and Mg–Zn systems, the solid phases which were analysed turned out to be ternary compounds. Mg–RE compounds can dissolve some Zn. In accordance with many solid phases being in equilibrium with Mg solid solution many three-phase and four-phase invariant equilibria were established in each system. The number of the four-phase invariant equilibria equals the number of the solid phases including those belonging to the binary Mg systems minus one conforming to the theory. The majority of the three-phase invariant equilibria are eutectic ones and they take place in the pseudobinary or, at least, partially pseudobinary systems. The four-phase invariant equilibria could be of the transition or eutectic type. As far as in the binary Mg–Zn system the eutectic equilibrium closest to Mg is at quite low temperature 340°C [139], there is a tendency for lowering temperatures of the invariant four-phase equilibria with increasing Zn / RE ratio amongst the alloys. The joint solid solubility of a rare earth metal and Zn was studied in detail in the systems Mg–Nd–Zn, Mg–Sm–Zn (the rare earth metals belonging to the cerium subgroup) and Mg–Y–Zn, Mg–Dy–Zn (the rare earth metals belonging to the yttrium subgroup). Results of the investigations of these four systems show a decrease of the rare earth metal solubility in Mg solid solution with addition of Zn. Nevertheless, there is some difference in the joint solubility of Zn and rare earth metals in solid Mg if the rare earth metals belong to different subgroups. The difference is revealed mainly at low temperatures where Mg solid solution is in equilibrium with solid phases, but not with liquid one (at about 300°C and lower). In systems with rare earth metals of the cerium subgroup, Zn addition decreases the rare earth metal solubility gradually. Unlike this, in systems with rare

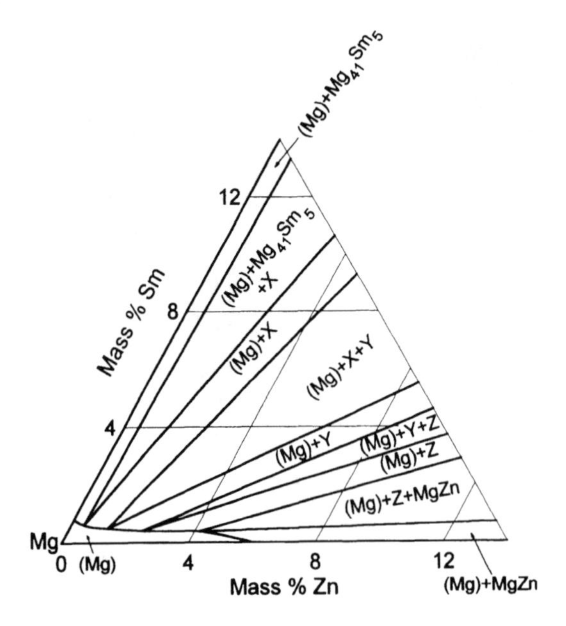

Figure 66. Partial isothermal section of the Mg–Sm–Zn phase diagram at 300°C.

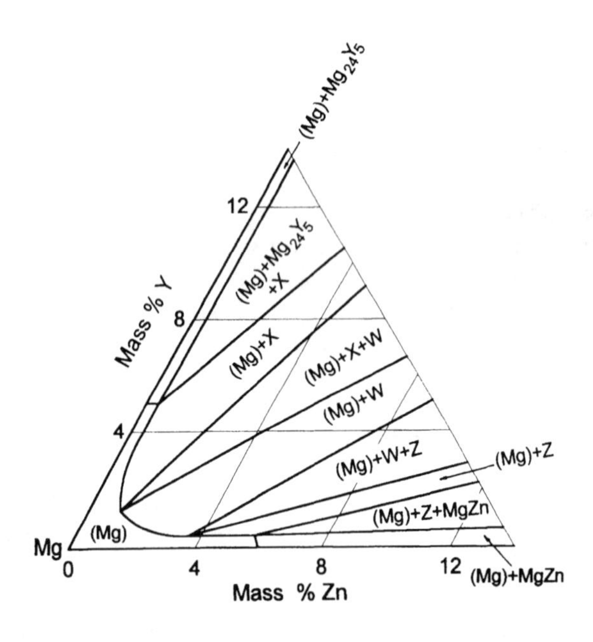

Figure 67. Partial isothermal section of the Mg–Y–Zn phase diagram at 300°C.

earth metals of the yttrium subgroup, small Zn addition practically does not change the rare earth solubility in solid Mg. Further increase of the Zn content results in abrupt decrease of the rare earth metal solubility in solid Mg and at most Zn contents up to the binary Mg–Zn side the rare earth metal solubility in solid Mg continues to decrease gradually remaining

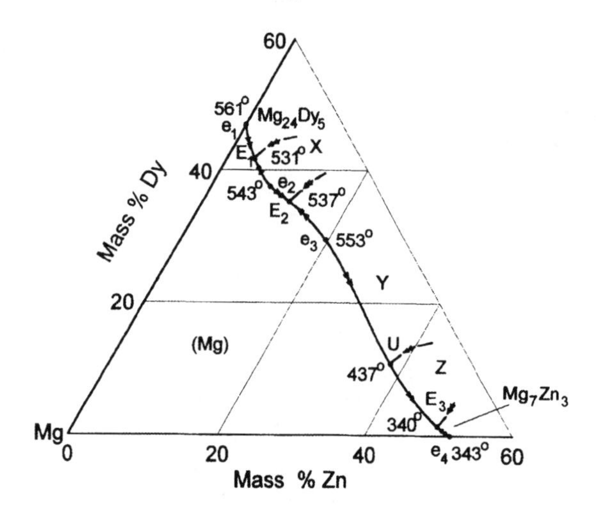

Figure 68. Projection of the monovariant lines in the liquidus surface of the Mg–Dy–Zn phase diagram.

at quite low level. The above mentioned peculiarities of the joint solid solubility of Zn and rare earth metals in Mg are illustrated in Figures 66 and 67 where the partial isothermal sections of the Mg–Sm–Zn [229] and Mg–Y–Zn [201] phase diagrams are presented as examples. The sections demonstrate also the existence of many solid phases in equilibrium with Mg solid solution and many phase fields adjoining the Mg solid solution field in Mg corner of the Mg–RE–Zn phase diagrams. The new solid phases established in the investigations are marked in the sections by X, Y, Z and W. The different forms of Mg solid solution fields in the phase diagrams of the systems with rare earth metals of the cerium or yttrium subgroup seem to be connected to a great extent with significantly different solubility of the elements each subgroup in solid Mg. Figure 68 demonstrates, as an example, the monovariant lines in the liquidus surface of the Mg–Dy–Zn phase diagram, showing existence of many invariant equilibria in Mg corner of the system [208]. The critical points of the three-phase invariant equilibria are marked by e_1, e_2, e_3, e_4, including those in the binary systems. The critical points of the four-phase invariant equilibria are marked by E_1, E_2, E_3 (eutectics) and U (an equilibrium of the transition type). In Figure 68 the fields of the primary crystallisation of Mg solid solution (Mg), binary phases $Mg_{24}Dy_5$, Mg_7Zn_3, and the unknown earlier phases X, Y, Z are shown, as well.

The only studied phase diagram of those with Cd was the Mg–Y–Cd one [200]. Investigation of the Mg–Y–Cd phase diagram was carried out in the concentration area of the Mg-rich alloys containing up to 28 mass % Y and 20 mass % Cd. No phase, in addition to $Mg_{24}Y_5$, was revealed in equilibrium with Mg solid solution. In accordance with formation of the Mg–Cd continuous solid solutions Cd was present in solid Mg. Besides, a certain solubility of Cd in $Mg_{24}Y_5$ was established by local spectral analysis. Cd decreased the liquidus temperature and temperature of the eutectic reaction $L \rightleftharpoons (Mg) + Mg_{24}Y_5$ of the Mg–Y binary alloys. The gradual decrease of Y solubility in solid Mg with increasing Cd content was established.

Two ternary phase diagrams with Ni were studied. They are Mg–La–Ni and Mg–Nd–Ni. There is some difference between the phase diagrams in the area of Mg corner. The Mg–La–Ni phase diagram was investigated up to the compounds closest to Mg in the adjoining binary systems which were assumed to be $Mg_{17}La_2$ and Mg_2Ni [185]. No additional phase was revealed in equilibrium with Mg solid solution in the system. The three-phase invariant eutectic equilibrium between $Mg_{17}La_2$ and Mg_2Ni at 535°C and the four-phase invariant eutectic equilibrium $L \rightleftharpoons (Mg)+Mg_{17}La_2+Mg_2Ni$ at 495°C were established. The joint solid solubility of La and Ni in Mg was not studied. The Mg–Nd–Ni phase diagram was studied at the concentrations 0–40 mass % Nd and 0–55 mass % Ni [193, 194]. Unlike Mg–La–Ni, in the Mg–Nd–Ni system an unknown phase was observed in equilibrium with Mg solid solution in addition to the Mg-richest compounds of the binary systems Mg–Nd and Mg–Ni. In accordance with this two four-phase invariant equilibria in the Mg corner of the Mg–Nd–Ni were established. One of them was of transition type (at 468°C) and the second equilibrium was of eutectic type (at 455°C). Solubility of Nd in solid Mg decreased when Ni was added to the Mg–Nd alloys.

Investigation of the ternary system Mg–Y–Si [221, 228] showed the existence of a solid phase X in equilibrium with Mg solid solution in addition to the phases $Mg_{24}Y_5$ and Mg_2Si of the binary systems Mg–Y and Mg–Si. The polythermal section Mg–X was suggested to be the pseudobinary system of eutectic type with eutectic temperature 552°C. Moreover, two invariant four-phase equilibria were established, $L+X \rightleftharpoons (Mg)+Mg_{24}Y_5$ at 636°C and $L \rightleftharpoons (Mg)+Mg_2Si+X$ at 637°C. Solubility of Y in solid Mg decreased with increasing Si content in the Mg–Y–Si alloys.

Quaternary Mg Systems with at Least One Rare Earth Metal as a Component

All known quaternary phase diagrams of Mg with rare earth metals were studied within the limits of the concentration area near Mg corner. Their number is rather small. Nevertheless, some generalisations from them are possible. One of the remarkable features of the quaternary phase diagrams studied is the absence of any quaternary phase in equilibrium with Mg solid solution. Only phases of the binary and ternary systems adjoining Mg in the quaternary systems are observed in equilibrium with solid Mg. In general, the main peculiarities of the interaction between components in the ternary systems appear in the corresponding quaternary systems, as well.

As in the ternary Mg–RE–Zr systems, in the quaternary system Mg–Nd–Y–Zr studied there is no phase formed by Zr, except the αZr-base solid solution [212]. In this quaternary system Zr does not actually affect temperatures of the critical points and the relations between the Mg–Nd and Mg–Y phases. In the other quaternary system with zirconium, Mg–Y–Zn–Zr, it forms the same αZr-base solid solution and the phase Zr_3Zn_2 of the binary system Zr–Zn. The phases formed by Zr exist in equilibrium with Mg solid solution along with all phases in Mg corner of the ternary system Mg–Y–Zn [217]. Almost similar behaviour is observed for Mn in the quaternary system Mg–Nd–Mn–Ni system [213]. In this system Mn forms two phases which are in equilibrium with Mg solid solution. One of them, the αMn-base solid solution, occurred in the binary Mg–Mn system and the other phase, the ternary compound A, occurred in the ternary Mg–Mn–Ni system. These phases exist in the structure of the alloys together with all solid phases occurring in the Mg corners of the ternary Mg–Nd–Mn and Mg–Nd–Ni systems without destroying their formation.

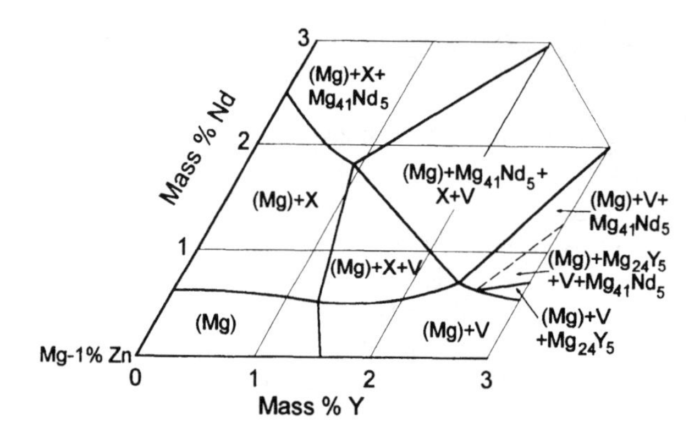

Figure 69. Partial isothermal section of the Mg–Nd–Y–Zn phase diagram for 300°C and 1 mass % Zn.

Investigation of the Mg–Nd–Y–Zn phase diagram [210, 211] revealed that in equilibrium with Mg solid solution only the solid phases occurred in the Mg corners of the adjoining ternary phase diagrams Mg–Nd–Y, Mg–Nd–Zn, Mg–Y–Zn. The total number of solid phases observed in the concentration area studied was nine including Mg solid solution, three binary phases of the systems Mg–Nd, Mg–Y, Mg–Zn, three ternary phases of the system Mg–Nd–Zn marked by X, Y, Z and two ternary phases of the system Mg–Y–Zn marked by V, W. The phase V of the Mg–Y–Zn system actually coincides with the phase X shown above in Figure 67. Coexistence of the solid phases was studied for two temperatures, 500 and 300°C at various concentrations of the components and represented by the sections of the isothermal concentration tetrahedrons characterising the quaternary phase diagram. The sections crossed one-, two-, three- and four-phase volumes inside the isothermal tetrahedrons corresponding to the phase coexistence. One of the represented sections of the Mg–Nd–Y–Zn isothermal tetrahedron for 300°C is shown in Figure 69. It is accepted from [210] with correction of the formula Mg_9Nd assumed there for the solid phase closest to Mg in the binary Mg–Nd system for the formula $Mg_{41}Nd_5$ according to the latest data. The section depicted in Figure 69 shows quite complicated phase relations in the Mg-rich alloys of the Mg–Nd–Y–Zn system. Investigation of the quaternary Mg–La–Y–Zn phase diagram [209, 230] revealed that in equilibrium with Mg solid solution were also only the solid phases which occurred in the adjoining ternary phase diagrams Mg–La–Y, Mg–La–Zn and Mg–Y–Zn, and quite complicated relations between the phases.

Investigation of the Mg–Y–Zn–Cd phase diagram was carried out for the concentration ranges 0–15 mass % Y, 0–25 mass % Zn, and 0–10 mass % Cd [214–216, 218, 240]. The results obtained led to the conclusion that, in the concentration area studied, Cd does not change the solid phases which could be in equilibrium with Mg solid solution in the ternary Mg–Y–Zn system. Meanwhile, Cd results in some shift of the phase field boundaries to fewer Y contents and some fall of the phase transformation temperatures as compared with those in the Mg–Y–Zn phase diagram. In this respect, the Cd behaviour in the quaternary Mg–Y–Zn–Cd system is similar to that in the ternary Mg–Y–Cd system considered above.

Structure of Mg–RE Alloys After Rapid Solidification

The above considered phase diagrams describe the phase structure of the Mg–RE alloys in equilibrium conditions. Such a state is reached either after slow enough solidification followed by slow enough cooling or after long enough annealing of the alloys at a chosen temperature. Meanwhile, from the applied and scientific viewpoint it is important to know what phase structure of the alloys may be anticipated if they are obtained in nonequilibrium conditions. Such a case takes place when the solidification rate of the alloys increases significantly and does not guarantee equilibrium conditions to be kept. Increase of the solidification rate is a quite effective tool for structure change, and often is used aiming to improve mechanical and some other properties of the alloys. Special attention has been paid in the recent years to the structure of the alloys obtained in conditions of rapid solidification. Parameters of "rapid solidification" are not strictly defined yet, but commonly this term is used for the processes where special efforts are made to increase substantially the solidification rate, at least up to about 10^5–10^6 K/s. One of the well-known processes of this kind is solidification of the melts on a rapidly rotating disk to produce thin strips (spinning). The possibility to reach an especially high solidification rate in this case is connected to a great extent with the thin cross-section of the strips. There are many other processes where the same high solidification rate is reached to produce various pieces with quite thin cross-sections. The thin strips and other pieces obtained during rapid solidification demonstrate quite high strength properties, but they are not suitable to make most things used in practice. They may be, however, compacted and worked to obtain bulk materials, retaining high enough strength after treatment. Investigations show that all this is justified for thin strips and bulk materials prepared from them if magnesium alloys are used [241–247].

The first visible effect of rapid solidification is substantial refinement of the alloy structure. Meanwhile the phases and their relations in the structure may also be changed after rapid solidification. They can be revealed by comparison between the structure of the alloys obtained after rapid solidification and that corresponding to the respective equilibrium phase diagram. Such investigations were carried out on three systems of Mg binary alloys with the rare earth metals, Mg–La [248, 249], Mg–Nd [250] and Mg–Y [251–252].

The alloys of the Mg–La system with 0.1–24 mass % La were studied after rapid solidification performed by quenching melts on the surface of high speed rotating Cu disk [248]. Thickness of the strips obtained was about 90 μm. The solidification rate was estimated to be about 10^5 K/s. The structure of the alloys consisted of dendrites of Mg solid solution and crystals of the Mg–La compound. Dendrite parameter decreased with increasing solidification rate and amounted after rapid solidification to about 2 μm as compared with 40 μm after solidification in common conditions at a rate of about 10^1 K/s. Solubility of La in solid Mg was checked by X-ray measurements of the lattice parameters. A special precise technique was used for registration of the X-ray reflexes aiming to provide more reliable data. Nevertheless, no visible extension of the La solubility in solid Mg was established in the experiments after rapid solidification as compared with solidification in normal conditions at 10^1 K/s. Investigation of the rapidly solidified alloys using transmission electron microscopy and electron diffraction [249] showed that the Mg–La compound visible in the structure had a hexagonal crystal lattice. Its parameters were $a = 1.178$ nm and $c = 1.020$ nm. These values were close to those of the equilibrium phase $Mg_{17}La_2$, but a few additional reflexes in the electron patterns indicated that the lattice of the Mg–La compound in the rapidly solidified alloys differed from that of $Mg_{17}La_2$. The Mg–La compound in the rapidly solidified alloys did not coincide, either, with the other equilibrium phase $Mg_{12}La$ established in the Mg-rich Mg–La alloys because the lattice of the latter was of the orthorhombic type.

The rapidly solidified alloys of the Mg–Nd system were prepared also by quenching melts on the high speed rotating disk in He atmosphere [250]. The estimated solidification rate was about 10^6 K/s. The alloys contained up to 38 mass % Nd. Solubility of Nd in Mg solid solution during rapid solidification was checked by X-ray determination of the Mg solid solution lattice parameters. The obtained results of the experiments demonstrated the increase of the lattice parameters with increasing content of Nd in the alloys up to 15 mass %. The Mg–Nd compound was observed in the structure of alloys with lower Nd concentrations than 15 mass %. Nevertheless, the compound quantities were small and, therefore, the authors [250] assumed the solubility of Nd in solid Mg during rapid solidification to be about 15 mass %. Although this value is quite approximate, it is, however, significantly higher than the maximum solubility of Nd in solid Mg in equilibrium conditions (3.6 mass % according to the phase diagram described above). Therefore, the conclusion can be made that, unlike Mg–La alloys, in the Mg–Nd alloys rapid solidification results in extension of the rare earth metal solubility in solid Mg.

The Mg–Y alloys obtained after rapid solidification were studied for the concentration range 0–31 mass % Y [251]. Rapid solidification was performed by quenching droplets of the melted alloys on the rotating disk in He atmosphere. Thickness of flakes obtained was about 20 μm. The solidification rate was estimated to be about 10^6 K/s. Beginning with 13.74 mass % the structure of the alloys consisted of Mg solid solution and the Mg–Y compound. The compound was identified as the equilibrium phase $Mg_{24}Y_5$ using the X-ray diffraction method and taking electron diffraction patterns during observation of the structure in electron microscope. Solubility of Y in solid Mg was checked by X-ray measurements of the Mg lattice parameters and by resistivity measurements. Both methods showed extension of the solid solubility in Mg during rapid solidification. The estimated solubility of Y in solid Mg after rapid solidification was 17 mass % as compared with maximum solubility of 12 mass % Y in the above described Mg–Y phase diagram for equilibrium conditions. Extension of the Y solubility in solid Mg during rapid solidification was confirmed in [252], where it was determined to be 39 mass % Y.

Chapter 2

Decomposition of Supersaturated Solid Solutions in Mg–RE Alloys

Formation of the Mg-base supersaturated solid solution and its decomposition are of great importance for behaviour of Mg–RE alloys during treatment and their properties. Ability to form supersaturated solid solution is connected with existence of the RE solubility in solid Mg and its decrease with lowering temperature. In accordance with this, the effect of the solid solution decomposition and the resulting change of the alloy properties depend on the solubility of the individual rare earth metals in solid Mg. Along with this aspect, decomposition of the supersaturated solid solution in Mg–RE alloys is interesting as a process in solids from the scientific viewpoint. Decomposition of the supersaturated solid solution in magnesium alloys with different rare earth metals is characterised by both similar and different features.

Kinetics of Solid Solution Decomposition in Mg–RE Alloys

Kinetics of solid solution decomposition in metal alloys is commonly studied using hardness and electrical resistivity measurements. These are the most convenient methods. Both properties are quite sensitive to the main changes in the structure of the alloys during decomposition and can be simply determined experimentally. Samples are not damaged when the hardness and electrical resistivity measurements are carried out. Therefore, the whole series of tests can be performed using the same sample. By this method the inevitable difference between separate samples in the start condition may be ignored, and, as a consequence, the peculiarities of solid solution decomposition can be established more clearly and reliably. At the same RE concentration electrical resistivity is higher when the alloying element is dissolved in Mg solid solution as compared with its presence in the form of the second phase. In accordance with this, decrease of electrical resistivity during solid solution decomposition testifies directly to its depletion by the alloying element atoms and, vice versa, increase of electrical resistivity testifies to the enrichment of Mg solid solution by the alloying element. In several Al alloys, for example in Al–Cu, Al–Cu–Mg, Al–Mg_2Si systems, some increase of electrical resistivity is observed at earlier stages of solid solution decomposition connected with formation of GP zones. So far this effect, however, in Mg alloys has not been established.

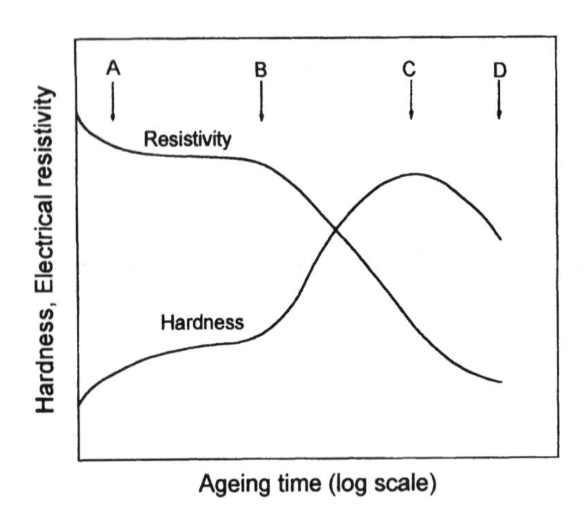

Figure 70. Generalized curves of the hardness and electrical resistivity change during decomposition of Mg supersaturated solid solution.

The hardness measurements can be considered as an indirect method of the decomposition registration. Nevertheless, it is quite indicative and, moreover, enables one to establish an ability of the alloy for strengthening (or hardening) when decomposition of the supersaturated solid solution takes place. This is the important characteristic of the process for its application in practice. In the book, hardness is given in HB or HV (kg/mm^2).

Figure 70 displays the electrical resistivity and hardness curves for solid solution decomposition in the most generalised form. As an example, decomposition is considered to run during isothermal ageing. Electrical resistivity decreases with increasing ageing time from beginning of the initial stage (A) up to full decomposition (D). Hardness increases at first, reaches maximum (C) and then decreases with increasing ageing time. The condition before the hardness maximum is often named as "underaged", and that after the hardness maximum is often named as "overaged". As a rule, the hardness maximum corresponds to the greatest part, but not the whole decrease of the electrical resistivity. Both electrical resistivity and hardness curves are not smooth, and this suggests several stages of solid solution decomposition. The curves in Figure 70 show only two stages (AB and BC) of solid solution decomposition. Transition between both stages corresponds to kinks (B) on the resistivity and hardness curves. Depending on ageing temperature, rare earth metal, supersaturation rate and other factors, every stage of solid solution decomposition can be expressed differently.

Decomposition of supersaturated solid solutions in Mg–RE alloys is commonly observed after ageing at elevated temperatures (artificial ageing). In [253] an ability of solid solution decomposition was specially checked for ageing at room temperature (natural ageing) using the Mg–1.1 mass % Nd alloy. The alloy samples were solution treated by quenching from 535°C in water of room temperature and then naturally aged. No tangible change of electrical resistivity and hardness of the samples was observed for ageing from 20 min till 101 days. This result led to the conclusion about actual absence of solid solution decomposition in the alloy during natural ageing, at least up to about 3.5 months. The author does not know another publication on the special check of the ageing at a room temperature of the Mg–RE alloys, but he can inform additionally that in his

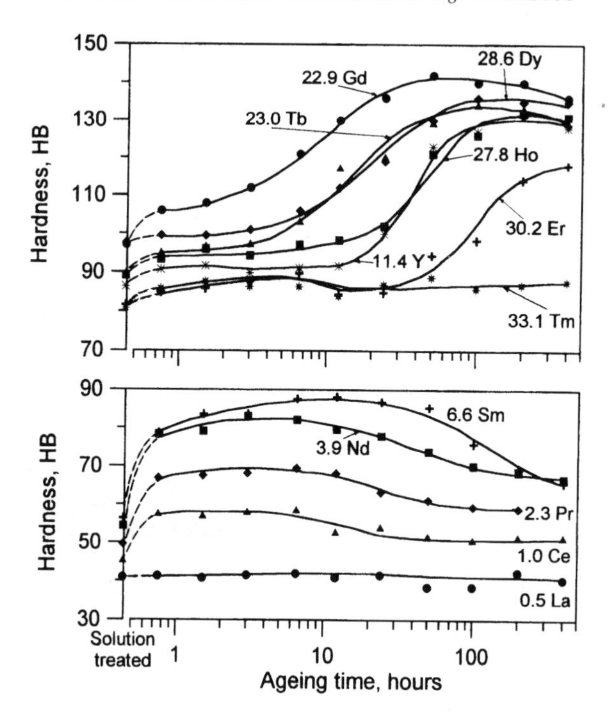

Figure 71. Hardness of the Mg–RE alloys versus ageing time at 200°C. RE content is shown in mass %.

experiments a few samples of the Mg–RE alloys were naturally aged after solution treatment for more than a year retaining actually the same electrical resistivity and hardness as after several minutes immediately after solution treatment.

Decomposition of supersaturated solid solution in Mg–RE alloys at elevated temperatures was studied in many works [106, 253–287] and the main regularities of this process were established. Meanwhile, a few features of solid solution decomposition in the Mg–RE alloys are unknown yet and seem to be discovered in future.

Comparison of the Decomposition Kinetics in the Binary Mg Systems with Various Rare Earth Metals

There are similarities and differences between the binary magnesium alloys with various rare earth metals in the kinetics of solid solution decomposition. The most impressive features of them were established during isothermal ageing [106, 259]. The results of respective experiments are presented in Figure 71 (hardness) and Figure 72 (electrical resistivity). The alloys used in the experiments were in the form of extruded rods. They contained various rare earth metals of concentrations close to the maximum solubility in solid Mg in every system. Samples prepared from the rods were solution treated and then aged at the same temperature 200°C. Solution treatment was performed by heating of the samples at temperatures slightly below eutectic ones in every system followed by quenching in water of room temperature. Samples of all the alloys were aged

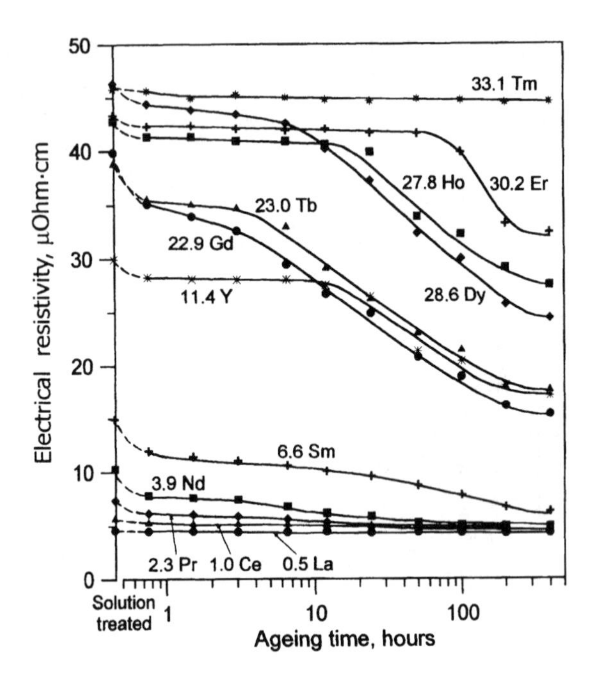

Figure 72. Resistivity of the Mg–RE alloys versus ageing time at 200°C. RE content is shown in mass %.

simultaneously aiming to improve the reliability of comparison between them as much as possible. The ageing temperature of 200°C was chosen because that is the often used ageing temperature for Mg alloys.

The results of the experiments presented in Figures 71 and 72 indicate the greater difference in ageing behaviour between the alloys with rare earth metals of the cerium and yttrium subgroups. On the other hand, there is more similarity in the ageing behaviour between alloys with rare earth metals belonging to the same subgroup. All the alloys, except those with La and Tm, indicate perceptible hardening effect during ageing. Within the alloys with elements of the cerium subgroup (from Ce to Sm) the hardness increase is observed beginning with the lowest ageing time of 0.75 h and continues gradually up to the maximum. After maximum there is a hardness fall with increasing ageing time. Only one stage is seen clearly in the hardness curves for solid solution decomposition in the alloys with elements of the cerium subgroup. The hardness maximum and the hardening effect during ageing increase successively with increasing atomic number of the rare earth elements. This fact is explained by the respective increase of the rare earth metal solubility in solid Mg. Actual absence of hardening effect in the Mg–La alloys results evidently from the insignificant solubility of La in solid Mg.

Unlike the alloys with elements of the cerium subgroup, in the alloys with elements of the yttrium subgroup (from Gd to Tm and Y) two stages are demonstrated evidently in the hardness curves of isothermal ageing. At the first stage, hardness increases slowly with increasing ageing time, and then remains approximately at a constant level. At the second stage hardness begins to ascend steeply, reaches a maximum and descends after maximum with increasing ageing time. The first stage is seen clearly in all the alloys studied, but the second one is seen fully only in alloys with Gd, Tb, Dy, Ho and Y. In the Mg–Er alloy the steep increase of the hardness of the second stage is seen only as an evident result of insufficient ageing time used in the experiments. In the alloy with Tm, being the last of the elements used of the lanthanum series, the second stage in the hardness change is not seen at all and this fact is also considered to be the result of insufficient ageing time in the experiments. In alloys with elements of the yttrium subgroup the maximum hardness reached is substantially more as compared with the hardness maximum in the alloys with elements of the cerium subgroup. Meanwhile, there is no visible increase of the hardness maximum within alloys with elements of the yttrium subgroup with increasing atomic number of the rare earth metals and their solubility in solid Mg, as is observed in alloys with elements of the cerium subgroup. On the contrary, there is a weak tendency for maximum decrease with increasing atomic number from Gd to Tb, Dy and Ho.

The electrical resistivity change with increasing ageing time corresponds, in general, to the hardness change in accordance with the scheme considered in Figure 70. No increase of electrical resistivity at the early stage of solid solution decomposition is observed. It either decreases, indicating solid solution depletion by rare earth metal, or remains constant in the limits of experimental error, indicating absence of depletion. In alloys with elements of the cerium subgroup, there is a decrease of the electrical resistivity beginning with the least ageing time of 0.75 h which continues up to the most used duration. This effect is not clearly expressed, however, in Figure 72 for alloys with lower resistivity because of the too small scale used. The absolute values of the electrical resistivity and effects of the resistivity change after solid solution decomposition increase successively with increasing atomic number of the rare earth metals resulting from growth of the solubility in solid Mg in the same order. In accordance with this, the change of the resistivity in Mg–La alloy during ageing is insignificant because of quite small solubility of La in solid Mg. Unlike the hardness curves, in the electrical resistivity curves of the Mg–Sm and Mg–Nd alloys, weak kinks are observed suggesting the existence of two stages in solid solution decomposition. The discrepancy between both methods may be explained by the higher sensitivity of electrical resistivity. Existence of several stages in solid solution decomposition was revealed in Mg–2.9 mass % Nd alloy by electrical resistivity measurements also in [257].

In alloys with rare earth metals of the yttrium subgroup, electrical resistivity decreases in two clearly seen stages. The first stage is characterised by insignificant resistivity change and ends at the kink on the resistivity curve. The second stage is characterised by significant decrease of electrical resistivity after the first stage is ended. Both stages are observed for all the alloys studied with elements of the yttrium subgroup, except that with Tm. For the Mg–Tm alloy the second stage is not seen, but it may be supposed at more ageing exposure than the most of those used in the experiments. In accordance with higher solubility in solid Mg the absolute values of electrical resistivity and their changes during solid solution decomposition are significantly larger for alloys with elements of the yttrium subgroup as compared with those of the alloys containing elements of the cerium subgroup.

One of the important results of the experiments considered is the regular change of the decomposition rate with increasing atomic number of the rare earth metals within the lanthanum series. This regularity is seen better for alloys with elements of the yttrium subgroup and can be estimated by comparison of the kink and maximum positions on the hardness and resistivity curves. They are shifted successively to longer ageing times from Gd to Tb, Dy, Ho and Er (Figures 71, 72). In this aspect, absence of the second stage on the hardness and resistivity curves for the Mg–Tm alloy corresponds to the atomic number of

Tm being the largest amongst the rare earth metals used. The partial second stage for the alloy with Er corresponds also to the mentioned regularity as far as Er occupies place before Tm in the lanthanum series. It is possible to note the successive increase of the ageing time corresponding to the hardness maximum for the alloys with Ce, Pr, Nd and Sm (Figure 71). The shift of the maximum positions on the hardness curves of the alloys with elements of the cerium subgroup confirms decrease of the decomposition rate with increasing atomic number of the rare earth metals, although this effect is weaker as compared with that of alloys with the yttrium subgroup elements. The significantly lower decomposition rate in the alloys with elements of the yttrium subgroup as compared with that of elements of the cerium subgroup of the lanthanum series also corresponds to the general rule.

Behaviour of the Mg–Y alloy is similar to that of alloys with other elements of the yttrium subgroup. The hardness and resistivity curves for the Mg–Y alloy indicate its decomposition rate to be the closest to that of the Mg–Ho alloy. In general, there is a correlation between the decomposition rates in the alloys and melting points of the rare earth metals. The higher melting point of the rare earth metal, the slower solid solution decomposition.

The same peculiarities of solid solution decomposition in Mg alloys with different rare earth metals were noted for the stepped ageing with successive ascend of the ageing temperature [106, 261]. In this series of experiments, the expected decomposition of the supersaturated solid solution in the alloy with Tm was observed additionally. Its kinetics and hardening effect were close to those of the alloy with Er in accordance with the general dependence of the decomposition rate on the atomic number of the rare earth metal.

The alloys with four rare earth metals, except Pm, are not presented in the experiments described. They are scandium, europium, ytterbium and lutetium. There is no information about the solid solution decomposition in the binary Mg–Sc alloys, at least the author does not know of any publication on this issue. According to [106], the ageing behaviour of the Mg–Eu alloys was studied, but no change of electrical resistivity and hardness during it was revealed. This fact was expected and it confirmed the actual absence of Eu solubility in solid Mg. Decomposition of the supersaturated solid solution in Mg–Yb alloy was investigated specially in [279]. The main characteristics of its kinetics are shown in Figure 73 where isothermal curves of hardness and electrical resistivity for an ageing temperature of 200°C are presented. The alloy studied shows a small, but tangible hardening effect at that temperature. Meanwhile, the resistivity decrease confirms depletion of Mg solid solution quite evidently. On both hardness and electrical resistivity curves two stages of solid solution decomposition may be seen. The decomposition kinetics in the Mg–Yb alloys are closer to the decomposition kinetics in the alloys with the rare earth metals of the cerium subgroup rather than that of the alloys with the rest of the rare earth metals of the yttrium subgroup.

Kinetics of solid solution decomposition in the Mg alloy with 39 mass % Lu is described in [106]. It was studied separately from the alloys with other rare earth metals and the ageing temperatures used were more than 200°C applied for the comparison experiments of Figures 71 and 72. Nevertheless, the results obtained enable conclusions to be drawn about correspondence of the Mg–Lu alloys to the described regularities in the Mg–RE solid solution decomposition and the last position of Lu in the lanthanum series. The Mg–39 mass % Lu alloy studied showed significant strengthening effect during ageing at 250, 300 and 350°C with clearly visible two stages of the process at 250 and 300°C, which is characteristic of alloys with the rare earth metals of the yttrium subgroup,

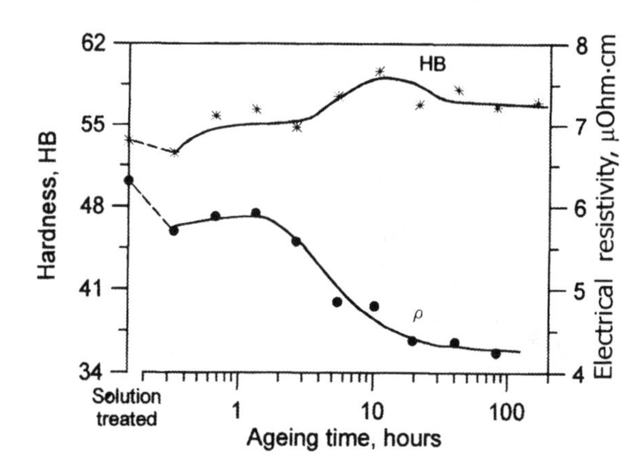

Figure 73. Hardness (HB) and electrical resistivity (ρ) of Mg–2.9 mass % Yb versus ageing time at 200°C.

except Yb. Beginning at the second stage, a steep increase of hardness (kink on the hardness curve) took place for Mg–39 mass % Lu at 250°C at ageing time of more than ageing time of the second stage beginning for Mg–30.2 Er at 200°C. So, the solid solution in Mg–39 mass % Lu decomposes significantly more slowly than in Mg–30.2 mass % Er in accordance with the last position of Lu in the lanthanum series as compared with Er staying two positions before in it, except Yb.

Effect of Temperature on Kinetics of the Solid Solution Decomposition

Similar features and their successive change with increasing atomic number of the rare earth metals are revealed in the dependence of solid solution decomposition on ageing temperature. The main action of the ageing temperature is acceleration of solid solution decomposition with increasing temperature of ageing and delay of the solid solution decomposition when it lowers. This action is specific in different Mg–RE systems. In alloys with rare earth metals of the cerium subgroup, only the acceleration of solid solution decomposition was revealed. Its magnitude may be seen in Figure 74, where dependence of the hardness and electrical resistivity on ageing time are presented for the alloys Mg–2.3 mass % Pr [272], Mg–3.4 mass % Nd [287] and Mg–5.65 mass % Sm [265]. All the alloys were in the form of extruded rods. They were solution treated by quenching in cold water from temperatures slightly below eutectic ones and then aged isothermally at different temperatures. As one can see in Figure 74a, where results for the alloy Mg–2.3 mass % Pr are reproduced there is shift of the hardness curves in the direction to the less ageing time with increasing ageing temperatures from 150 to 250°C. In the limits of the ageing exposures used (0.5–200 h) the hardness maximum is revealed for ageing at 175, 200 and 225°C. At 150°C it is not reached, although it may be quite near. At 250°C the hardness maximum seems to be overpassed after the least used ageing time of 0.5 h, yet it exists. Magnitude of the hardness maximum is practically the same for the ageing temperatures when it is reached. Similar character of isothermal hardness curves was observed also for ageing of the alloy Mg–1.1 mass % Nd [253].

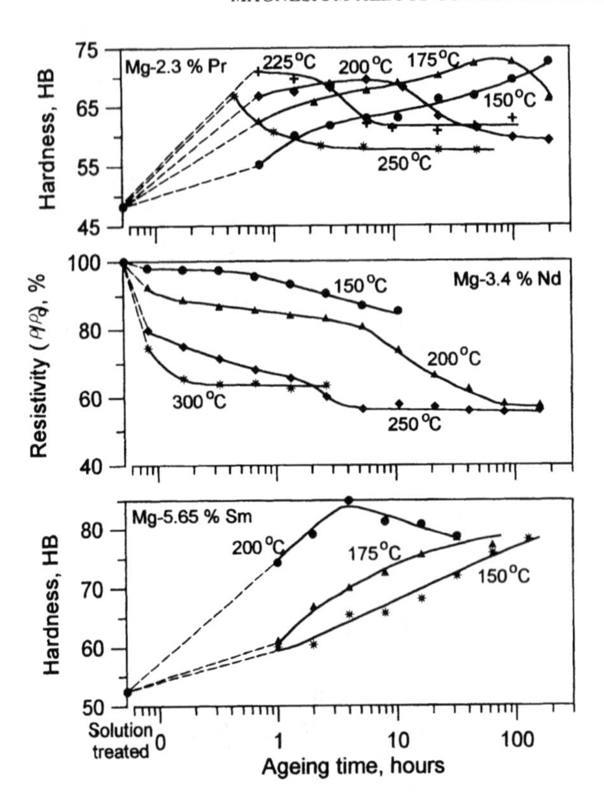

Figure 74. Effect of temperature on the ageing kinetics of the Mg–Pr, Mg–Nd and Mg–Sm system alloys.

Figure 74b demonstrates the effect of ageing temperature on the electrical resistivity change of the more enriched Mg–Nd alloy Mg–3.4 mass % Nd. Following [287] electrical resistivity is presented as ρ/ρ_0, where ρ is electrical resistivity of the sample after ageing and ρ_0 is electrical resistivity of the same sample after quenching. Decomposition is manifested best in the temperature range 150–300°C. The electrical resistivity curves show evidently two stages of solid solution decomposition. With increasing temperature the curves shift to a shorter ageing time and at highest temperature 300°C the first stage can not be revealed. In general, the character of the electrical resistivity change during ageing of the alloy Mg–3.4 mass % Nd corresponds to the mentioned hardness change during ageing of the alloys Mg–2.3 mass % Pr and Mg–1.1 mass % Nd. For the Mg–Sm alloys effect of ageing temperature was studied only by hardness measurements in the narrow range of 150–200°C (Figure 74c). The hardness curves for ageing of the alloy Mg–5.65 mass % Sm shift to a shorter ageing time with increasing the ageing temperature following the same tendency and approximately with the same rate as for alloys of the Mg–Pr and Mg–Nd systems.

Figure 75 shows the representative results of isothermal ageing studies at different temperatures for Mg alloys with two rare earth metals of the yttrium subgroup. The alloys were in the form of extruded rods as the above considered alloys with Pr, Nd and Sm. The rods were cut into pieces which were solution treated by quenching in cold water from

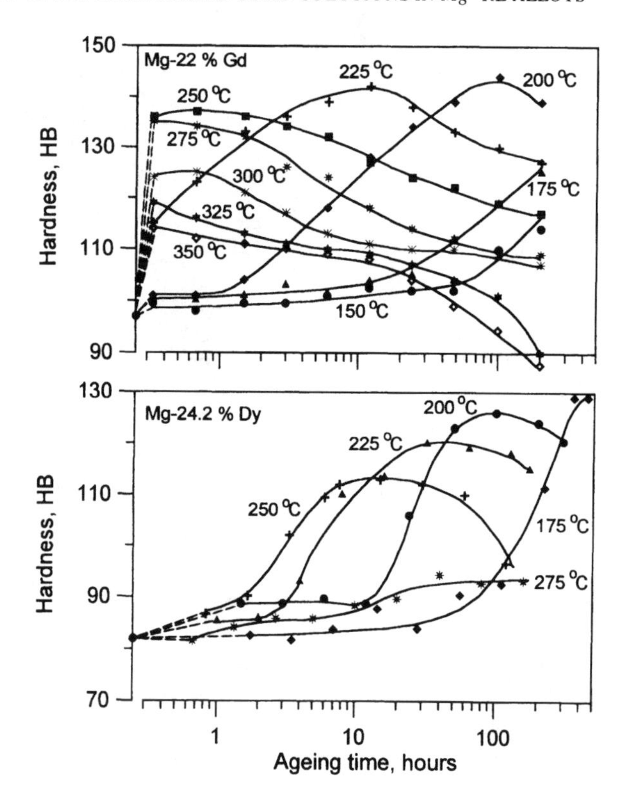

Figure 75. Effect of temperature on the ageing kinetics of the Mg–Gd and Mg–Dy system alloys.

temperatures slightly below temperatures of the respective eutectics and then aged. The hardness curves for the alloy with the first element of the subgroup, Mg–22 mass % Gd, are shown in Figure 75 (upper part) [269, 280]. They demonstrate successive acceleration of the solid solution decomposition with increasing ageing temperature by shift to shorter ageing times. The size of the hardness maximum does not change actually with increasing temperature as far as it is seen from comparison of curves for 200 and 225°C. At higher temperatures the size of the hardness maximum could not be determined because it is reached after times being shorter than those used in the experiments. However, the curves show certainly existence of the noticeable hardening effect during ageing up to the highest of the temperatures used, 350°C. In general, character of the ageing kinetics change with increasing temperature for the Mg–Gd alloy turns out to be similar to that for Mg alloys with rare earth metals of the cerium subgroup.

Another case occurs for alloy with the third element of the yttrium subgroup of the lanthanum series, Mg–24.2 mass % Dy. Its hardness curves for ageing at different temperatures are shown in Figure 75 (lower part) [266]. For the Mg–Dy alloy there is also the shift of the hardening curves with increasing ageing temperature pointing to acceleration of the solid solution decomposition, but, unlike Mg–22 mass % Gd, size of the hardness maximum decreases successively with increasing temperature, and at 275°C the hardening effect is actually absent. Investigation of the ageing kinetics in the alloy with the

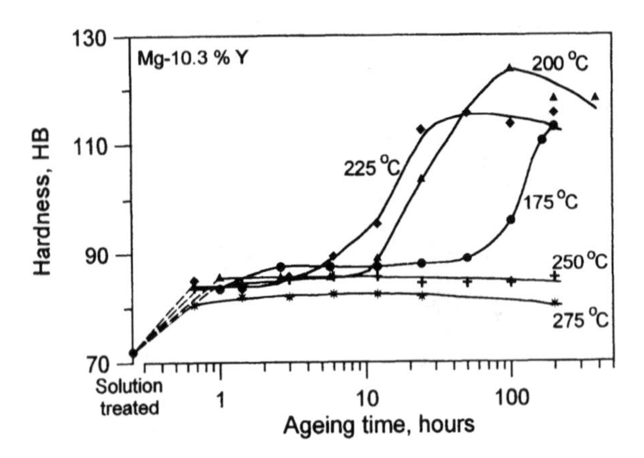

Figure 76. Effect of temperature on the hardness change during ageing of the Mg–Y system alloy.

second element of the yttrium subgroup, Mg–23 mass % Tb [268], showed that its hardening effect changed with increasing temperature intermediately between those for the alloys Mg–22 mass % Gd and Mg–24.2 mass % Dy. Unlike Mg–22 mass % Gd, in the alloy Mg–23 mass % Tb the successive decrease of the hardness maximum was observed certainly with increasing ageing temperature, but it was less than that in the alloy Mg–24.2 mass % Dy. These experiments enable us to conclude that change of the hardening effect with increasing ageing temperature depends regularly on the atomic number of the rare earth metals of the yttrium subgroup.

To a certain extent, the regularity in dependence of the hardening effect change with increasing ageing temperature on increase of the RE atomic number within elements of the yttrium subgroup is confirmed by behaviour of the Mg–Y alloys. As was noted above in comparison of binary Mg alloys with different rare earth metals, the rate of the solid solution decomposition in Mg–Y alloys was the nearest to that of the Mg–Ho alloys. Therefore, dependence of the Mg–Y alloy ageing kinetics on temperature is expected to be similar at most to that of the Mg–Ho alloys. As far as Ho is the next element in the lanthanum series after Dy, the noticed regularity in character of the hardness maximum change with increasing ageing temperature for the Gd, Tb and Dy should successively continue in the Mg–Y alloys. It is truth indeed. Figure 76 shows the hardness curves obtained during ageing of the alloy Mg–10.3 mass % Y at different temperatures [267]. The alloy was delivered and treated in the same conditions as the alloys considered before. The results of the experiments show a certain decrease of the hardening maximum with increasing ageing temperature from 200 to 225°C and its actual disappearance at 250°C. In the Mg–10.3 mass % Y alloy the hardening maximum decreases with increasing ageing temperature faster than in the Mg–24.2 mass % Dy alloy where disappearance of the hardening effect is observed during ageing only at higher temperature 275°C. This fact follows the general tendency in dependence of the hardening effect on the ageing temperature with increasing the rare earth atomic number, if Y is assumed to occupy the place of Ho.

The Mg–Y alloy was used also to analyse the effect of temperature on the ageing kinetics by the electrical resistivity method. The results obtained were rather unusual. The

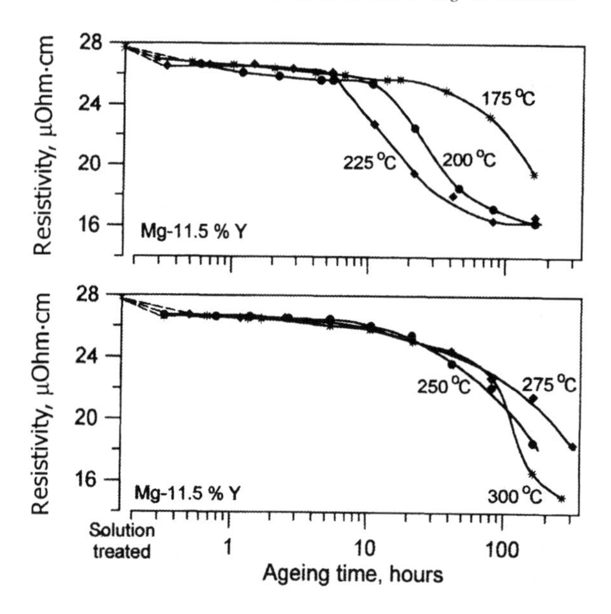

Figure 77. Effect of temperature on the resistivity change during ageing of the Mg–Y system alloy.

main part of them is presented in Figure 77 [274, 288]. Composition of the alloy Mg–11.5 mass % Y was close to that used in the above described hardness experiments. Its conditions of delivery and treatment were actually also the same. For clearness the resistivity curves are shown in two plots. The upper plot contains the resistivity curves for the ageing temperatures 175, 200 and 225°C corresponding to the existence of the hardening effect, and the lower plot contains the resistivity curves for the ageing temperatures 250, 275 and 300°C when the hardening effect is absent. The curves are divided into two groups inasmuch as they are superposed on each other and become hardly distinguished in the single plot. The upper plot of Figure 77 shows the common acceleration of the solid solution decomposition with increasing ageing temperature from 175 to 225°C accompanied by the respective shift of the resistivity curves to shorter times. Unusual behaviour of the alloy Mg–11.5 mass % Y is revealed at the next ageing temperature 250°C (lower plot), when certain delay of the solid solution decomposition is observed with shift of the respective resistivity curve to the longer times as compared with the curve for 225°C. The resistivity curve for the still higher ageing temperature of 275°C shows continuation of the solid solution decomposition delay and turns out to be shifted to longer ageing times as compared with that for 250°C. Only at the next higher temperature of 300°C some acceleration of the solid solution decomposition begins, but even then the steep depletion of Mg solid solution takes place at longer ageing times than that for 200°C. Further rise of the ageing temperature above 300°C results in acceleration of the solid solution decomposition in the ordinary way [274]. Although, during ageing, with delayed solid solution decomposition, this process takes place, yet, at longer times it is not accompanied with alloy hardening. This conclusion could be drawn from comparison of the hardness and resistivity curves in Figure 76 and Figure 77 (lower part) for ageing temperatures 250 and 275°C. The results of the electrical resistivity experiments conducted with the alloy

Mg–11.5 mass % Y show the quite complicated nature of the hardening effect during solid solution decomposition in Mg alloys with rare earth metals and its dependence on the ageing temperature.

It is natural to suppose that the regular acceleration of the hardening effect decrease with rising ageing temperature in the order of the atomic number increase established for Gd, Tb, Dy continues in the next rare earth metals of the lanthanum series. To a certain extent, experiments with Mg–Y confirm even this supposition. Nevertheless, there is not enough assurance that it is so. Moreover, some experiments on the Mg–Er and Mg–Lu alloys show the existence of a significant hardening effect in them after ageing at 250 and 300°C [106, 273] contradicting the established tendency for Gd, Tb, Dy. It is possible that at one of the elements of the yttrium subgroup after Dy the established temperature dependence of the hardening effect during ageing on the rare earth atomic number changes into an opposite one. A number of additional experiments are required to answer this question.

Influence of the Grain Size and Supersaturation

Influence of the grain size on kinetics of the solid solution decomposition in the Mg–RE alloys was checked experimentally using the alloy Mg–16 mass % Gd. The different grain size in the alloy was obtained after quenching its samples after annealing at different temperatures in the range 485–545°C. All annealing temperatures enabled us to obtain the structure of the Mg homogeneous solid solution with the same Gd concentration. The grain size was in the limits from 17 to 56 μm. Both hardness and electrical resistivity measurements showed no change of decomposition kinetics with changing the grain size in the limits studied [106].

Investigation of binary Mg–RE alloys with different contents of the alloying element revealed the regular change of the decomposition kinetics with the rare earth metal concentration in the supersaturated Mg solid solution. They were demonstrated on the Mg–Y [289] and Mg–Gd alloys [106]. The results obtained by the hardness measurements for the Mg–Y alloys are presented in Figure 78. The respective experiments were conducted on Mg–Y rods of about 11 mm in diameter produced by hot extrusion of the ingots with cross section reduction of 87%. Extrusion temperature was about 400°C. After extrusion the rods were cooled in air, and this cooling rate was high enough to prevent Mg solid solution decomposition. This fact was proved by the respective response of the hardness and electrical resistivity during the subsequent ageing of the alloys and corresponded to the thermal resistance of the Mg–Y supersaturated solutions against decomposition which will be described in the third part of this chapter. Ageing of the alloys was conducted isothermally at 200°C.

As one can see in Figure 78, the alloys with low Y contents from 1.0 to 3.2 mass % Y demonstrate only insignificant increase of hardness at the first short ageing time which can be ascribed to the possible first stage of the solid solution decomposition. The steep hardness increase during ageing showing the second intense stage of the solid solution decomposition in the alloys with 1.0–3.2 mass % Y is not observed at all. This stage of decomposition is observed only beginning with the alloy containing 4.9 mass % Y and manifested by the characteristic kink on the hardness curve. Position of the kinks on the hardness curves is shifted successively to the shorter ageing times with increasing Y content in the alloys from 4.9 to 12.6 mass % showing evidently acceleration of the solid

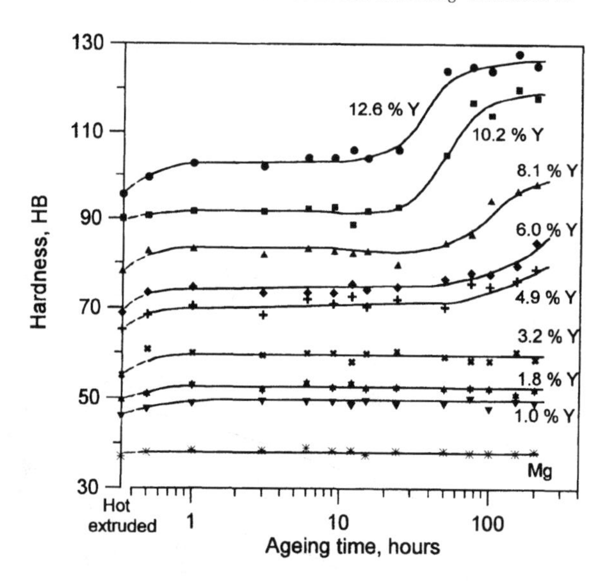

Figure 78. Hardness of the hot extruded Mg and Mg–Y alloys versus ageing time at 200°C.

solution decomposition. Actually, the solid solution decomposition during ageing is observed if there is a certain supersaturation of Mg solid solution, and decomposition is noticeably accelerated with increasing supersaturation. The hardening effect also increases with increasing supersaturation of Mg solid solution. A similar conclusion could be drawn from results of the electrical resistivity measurements during ageing of the Mg–Y alloys [289] and the hardness and electrical resistivity measurements of the Mg–Gd alloys [106]. Investigation of the Mg–Gd alloys [106] was conducted using samples after special solution treatment which consisted of annealing at temperature near the Mg–Gd eutectic temperature followed by quenching in cold water.

The considered dependence of the decomposition kinetics on the rare earth metal content in Mg solid solution and its supersaturation do not contradict the above noticed regularity in the decomposition deceleration with increasing atomic number of the rare earth metals. The regular deceleration of the Mg solid solution decomposition with increasing the rare earth atomic number was established on alloys with successively more concentration of the rare earth elements in solid solutions in accordance with change of their solubility in solid Mg. The successive higher concentration of the Mg solid solutions with increasing the rare earth atomic number in these experiments could only accelerate their decomposition according to the former rule, and, therefore, weaken only the general effect of the decomposition deceleration which was revealed.

Phase Transformations During Decomposition of Mg–RE Solid Solutions

Investigations of the phase transformations during decomposition of the supersaturated solid solutions in the Mg–RE alloys revealed their complicated nature. The accompanying

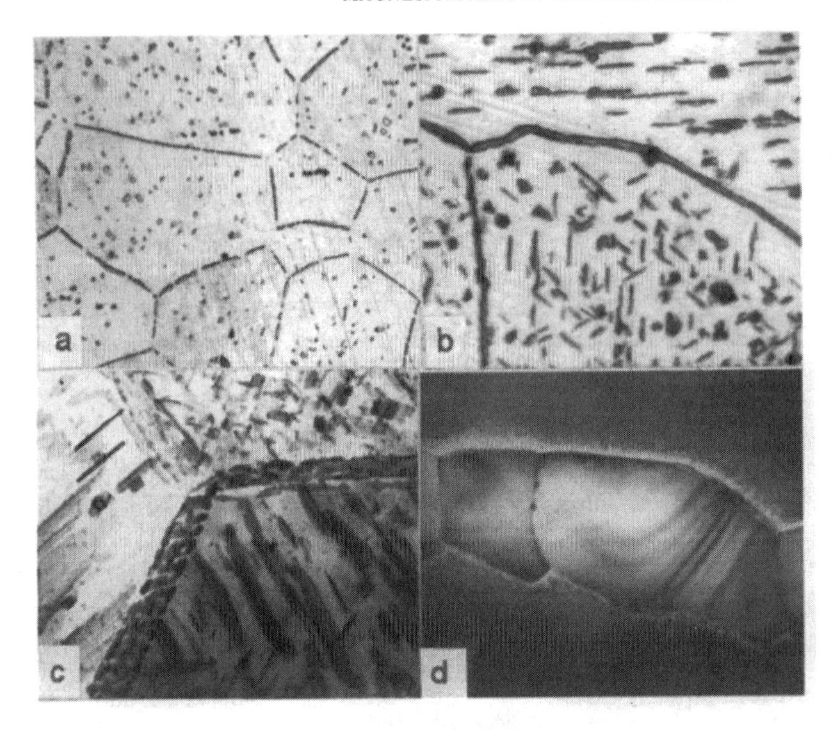

Figure 79. Light micrographs (a, b) and transmission electron micrographs (c, d) of Mg–RE alloys after hot extrusion, solution treatment and ageing. a – Mg–3.4 mass % Nd, aged 450°C, 24 hours, × 250, b – Mg–16 mass % Gd, aged 350°C, 6 hours, × 1000, c – Mg–5.65 mass % Sm, aged 250°C, 6 hours, × 14000, d – Mg–24.2 mass % Dy, aged 400°C, 5 hours, × 13000.

change in the structure during solid solution decomposition was studied by common light microscopy in all the binary Mg–RE systems, except the systems Mg–Sc and Mg–Eu. The latter system was hardly suitable for such a study because of the practical absence of Eu solubility in solid Mg. For all the systems studied, except Mg–Ho, Mg–Tm and Mg–Lu, transmission electron microscopy was applied additionally. Both similar and different features of phase transformation during decomposition were observed when the different Mg–RE systems were compared.

The most common and evident change in the structure during decomposition is the preference for precipitation of the RE-rich phases along the grain boundaries of Mg solid solution which is accompanied by formation of the precipitate free zones (PFZ). This phenomenon was observed in alloys with all rare earth metals and is typical of decomposition of supersaturated solid solutions of metallic systems. At earlier or middle stages of solid solution decomposition, precipitates along the grain boundaries are often observed in light microscope as continuous layers. At the latest decomposition stages taking place at the highest ageing temperatures, the precipitates along the grain boundaries coarsen significantly and crystals of the RE-rich phases of them turn out to be usually separated. Figures 79a and 79b demonstrate typical micrographs with the precipitates along the grain boundaries of Mg solid solution obtained using the light microscope. The first micrograph with low magnification (Figure 79a) demonstrates the general view of the structure after decomposition. It was taken from the alloy aged at quite high temperature resulting in significant coarsening of the precipitates. Their quantity is rather small because of the small solubility of Nd in solid Mg and coarsening. The second micrographs with higher magnification demonstrate the structure after decomposition with more quantity of precipitates in the Mg–Gd alloy and the precipitate free zones (PFZ) along the grain boundary. Transmission electron microscopy (Figures 79c and 79d) revealed a complicated constitution of the precipitates along the grain boundaries. Even in the continuous layers they consist of individual crystals in contact with one another and are arranged in a certain order. The arrangement of the precipitated crystals in the grain boundaries changes with increasing ageing temperatures, whilst the crystals become larger in size.

Differences between the different Mg–RE systems were observed mainly for phase transformations within the grains of Mg solid solution. Investigations showed that in this case there are more different features in the phase transformations when alloys with the rare earth metals of different subgroups, cerium or yttrium ones, are compared. On the other hand, alloys with rare earth metals belonging to the same subgroup, except Yb, demonstrate mostly similar features in the solid solution decomposition. The system Mg–Yb was established to differ significantly from systems with other rare earth metals of the yttrium subgroup in phase transformations during solid solution decomposition, but also differs significantly in this aspect from systems with rare earth metals of the cerium subgroup. Taking into consideration the similar features during solid solution decomposition, the phase transformations within grains of Mg solid solution are considered together for the Mg–RE systems with rare earth metals of the cerium subgroup, for systems with the rare earth metals of the yttrium subgroup, except Yb, and separately for the system Mg–Yb.

Systems with Rare Earth Metals of the Cerium Subgroup

Phase transformations during solid solution decomposition were studied using transmission electron microscopy in the systems Mg–La [262], Mg–Ce [258, 262, 284], Mg–Pr [272], Mg–Nd [257, 263, 284] and Mg–Sm [265], just as in all of the systems with the rare earth metals of the cerium subgroup, where the process can proceed. In [263] the ternary Mg–Nd–Zr alloy was studied, but Zr seemed not to take part in and not to hinder Mg–Nd solid solution decomposition. Although there are some discrepancies between results of the separate works and some details need to be checked or studied additionally, the main features of the decomposition process have been well established, and there are reasons to assert that it is similar in all systems in general. The process includes successive formation of several products of precipitation. They are the GP zones, two metastable phases and the stable equilibrium phase. Its sequence may be presented as follows:

$$\text{Mg supersaturated solution} \rightarrow \text{GP zone} \rightarrow \beta'' \rightarrow \beta' \rightarrow \beta,$$

where β'', β' are the metastable phases and β is the stable one. The different stages of decomposition are superposed on one another. So, after a certain ageing two successive products of precipitation can coexist in the alloy structure.

The GP zones are the earliest products of precipitation. They are formed after ageing corresponding to hardening of the alloys to less than the maximum. The GP zones are revealed in electron diffraction patterns by straight diffuse streaks. In the differently

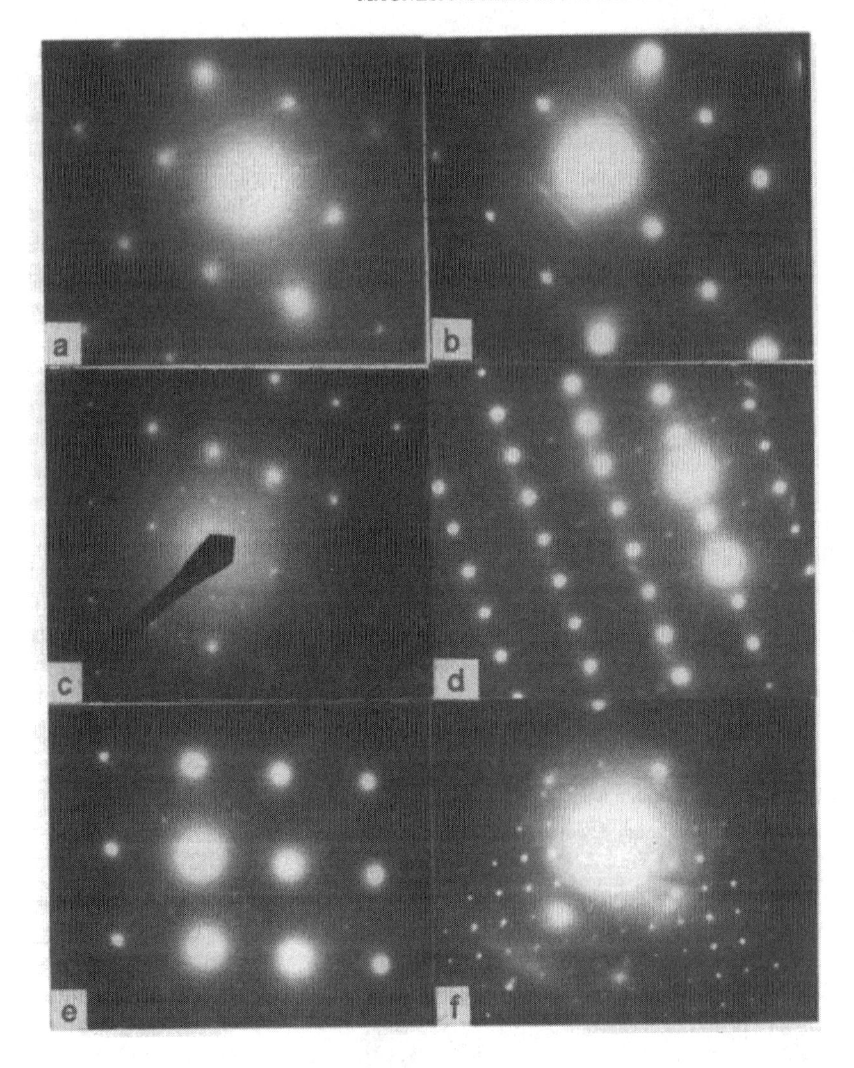

Figure 80. Electron patterns from the Mg–RE alloys. a, b – Mg–2.3 mass % Pr, ageing at 200°C for 6 h, c – Mg–3.8 mass % Sm, ageing at 225° C for 6 h, d – Mg–5.65 mass % Sm, ageing at 200°C for 16 h, e – Mg–2.3 mass % Pr, ageing at 200°C for 24 h, f – Mg–2.3 mass % Pr, ageing at 350°C for 6 h.

oriented grains, the streaks always coincide with traces of basal planes of the reciprocal lattice from the Mg solid solution which is also hexagonal. The streaks are directed to the reflexes from the planes of the (uv0) type if these reflexes are present in the electron diffraction pattern. The typical electron diffraction patterns with streaks from the GP zones for this decomposition stage are presented in Figures 80a and 80b[†] for two different orientations of the Mg solid solution grains. In Figures 81a and 81b the corresponding

---

[†] All electron diffraction patterns as with all micrographs presented in the book were obtained in the investigations of the author.

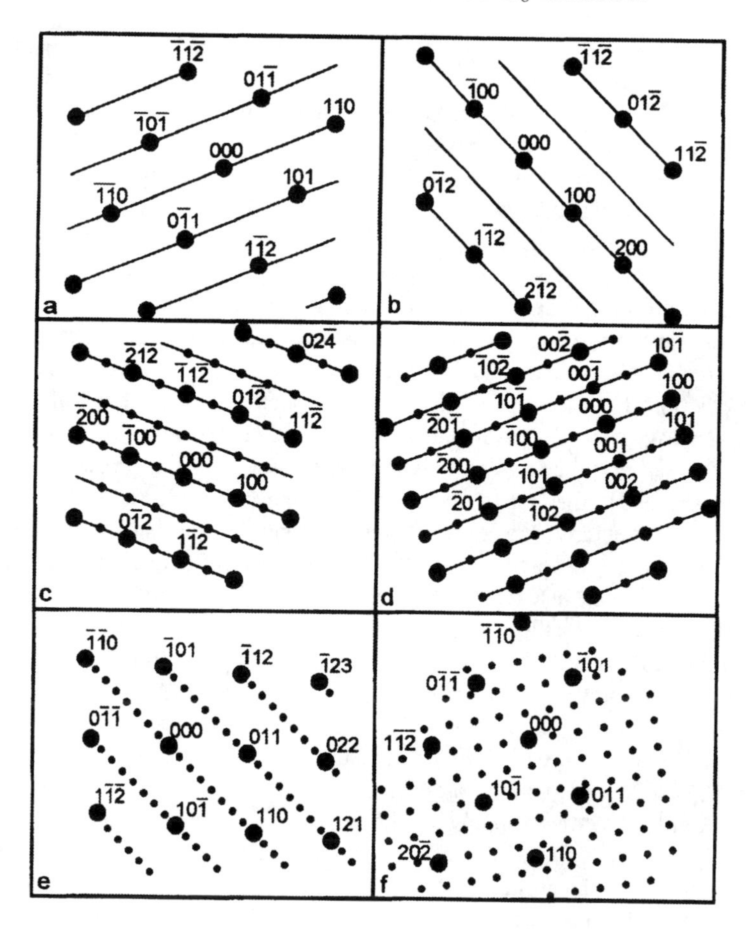

Figure 81. Schemes of the respective electron patterns presented in Figure 80.

schemes of the patterns are shown. This position of streaks in the electron diffraction patterns suggests that they are the sections of the diffraction planes from the GP zones coinciding with the basal planes of the reciprocal lattice from Mg solid solution. Consequently, the GP zones are of the needle form with the long axes directed along the hexagonal axis [001] of the Mg solid solution lattice. Such a conclusion was drawn in investigations of the Mg–Pr, Mg–Nd and Mg–Sm alloys [257, 263, 265, 272]. Estimation of the GP dimensions gave about 20 nm long with about 0.7 nm in diameter for the largest ones in the Mg–Nd alloy after ageing at 180°C for 8 h. Their density in structure was estimated to be more than 10^{22} m^{-3} then [257].

With progress of precipitation superstructure reflexes arise in the electron diffraction patterns. They are disposed in the streaks and coexist with them. The typical electron diffraction patterns for two different orientations of Mg solid solution grains are presented in Figures 80c and 80d with their respective schemes in Figures 81c and 81d. Position of the superstructure reflexes corresponds only to the middles of the reciprocal lattice radius-vectors of the [100]* type and may be, therefore, considered as a result of ordering in the Mg solid solution lattice of the Mg_3Cd type (DO_{19} type). The ordering phase is considered to be a metastable one marked by β''. The β'' phase is coherently connected with the matrix. Its crystal lattice is hexagonal close-packed with parameters $a_{\beta''} \cong 2a_{Mg}$, $c_{\beta''} = c_{Mg}$. The particles of the β'' phase remain extended along the hexagonal axis [001] of the Mg solid solution lattice. Their form was analysed in the Mg–Nd alloy [257] and recognised to be plate-like one with {110} habit planes. Such a structure corresponds approximately to the hardness maximum. According to [265], the β'' particles are of 10–100 nm in size. The described features of the solid solution decomposition connected with formation of the β'' phase were established in the investigations of the Mg–Pr, Mg–Nd and Mg–Sm alloys [257, 265, 272]. The results of these works are consistent on the β'' phase and can be considered to be the most reliable. In the investigation of the Mg–Ce alloy [250] the precipitates extended along the [001] axis were also observed at an earlier stage of decomposition. They were plate-like particles coherently connected with the matrix as in the Mg–Pr, Mg–Nd and Mg–Sm alloys [257, 265, 272]. However, the hexagonal crystal lattice of the earlier precipitates in the Mg–Ce alloy was recognised in [262] to be differently oriented in the matrix to that of the β'' phase established in Mg–Pr, Mg–Nd and Mg–Sm alloys in [257, 265, 272]. Besides, the plate-like particles in the Mg–Ce alloys were established to be disposed on the {100} prism planes of matrix as compared with the {110} prism planes established in the Mg–Nd alloy [257]. The data obtained in the investigation of the Mg–Ce alloy [262] are, however, less reliable because of the significantly less quantity of precipitates which could be formed and analysed then as a result of the lower solubility of Ce in solid Mg as compared with that of Pr, Nd and Sm. In the latest investigation [284], where conventional and high resolution transmission electron microscopy (CTEM and HRTEM) were used, the β'' phase in the Mg–Ce alloy was recognised to be "a stepped quasi-DO_{19} structure based on the DO_{19} superlattice" at the earlier stage of formation. Unlike the conventional one, the established stepped quasi-DO_{19} lattice of the β'' phase also contains periodic atomic steps. As decomposition continues, the stepped quasi-DO_{19} particles transform into the conventional phase of the DO_{19} type[†]. According to [284], the β'' precipitates with the stepped DO_{19} type lattice are thin platelets which form on the {100} prism planes of matrix. The β'' precipitates with the conventional DO_{19} type lattice are approximately of plate-like form, but they are thicker than those with the stepped DO_{19} type lattice and tend to be arranged along the {110} type of the matrix planes. These matrix planes are also prism ones with threefold symmetry. For the Mg–Nd alloy the same results were obtained in [284]. In general, the conclusions [284] are close to those made for the Mg–Ce alloy in [262] for the earliest stage of the β'' formation and agree quite well with those made for the Mg–Nd alloy in [257] for the later stage formation when the particles with conventional DO_{19} lattice form.

The next metastable phase β' begins to form just as after the hardness maximum. Softening after maximum is accompanied with further formation and growth of the β' precipitates. The β' phase is only partially coherent with the Mg solid solution matrix. This conclusion was drawn from some displacement of the β' electron diffraction patterns from that from matrix [257] and is confirmed by the more clearly outlined boundaries of the β' particles observed by electron microscope. On the other hand, in other instances the strict relation between the reflexes from β' and matrix was observed, suggesting a coherency

[†] Unlike the previous publications [257, 265, 272] and the designation assumed in this book, in [284] the stepped DO_{19} type phase seems to be marked by β'' and the conventional DO_{19} type phase seems to be marked by β'.

between both phases along certain atomic planes. Besides, at some inclinations of the grains, the deformation contrast was seen near the β' particles [272]. One of the main features of the β' phase are more complicated electron diffraction patterns from it as compared with that from the β'' phase. The superstructure reflexes from β' phase appear more frequently between the reflexes from Mg solid solution than the reflexes from the β'' phase. A typical example of the electron diffraction patterns with reflexes from the β' phase is presented in Figure 80e with its scheme in Figure 81e. The pattern is taken from the alloy Mg–2.3 mass % Pr [272]. The superstructure reflexes in it are disposed along the traces of the basal planes of the reciprocal lattice of Mg solid solution with the distance between them being less by eight than the radius-vector [110]*. The β' particles were established to be plate-shaped with approximately rectangular form of the large facet. They lie on the prism planes {100}of Mg solid solution with threefold symmetry. The typical micrographs with the β' precipitates are shown in Figures 82a and 82b. The β' particles are disposed approximately uniformly within grains. Sometimes there are bands of them of the same size and orientation with constant distance between them. This instance is presented in Figure 82b. The crystal structure of the β' phase turned out to be different in different systems. In the Mg–Pr alloy it was established to be an orthorhombic one with parameters $a_{\beta'} \cong 4d_{(100)Mg}$, $b_{\beta'} \cong 8d_{(110)Mg}$, $c_{\beta'} \cong c_{Mg}$. According to this, $a = 1.111$ nm, $b = 1.284$ nm, $c = 0.521$ nm. The orthorhombic lattice of the β' phase in the Mg–Pr alloy is oriented in the Mg solid solution matrix by three kinds with threefold symmetry around the hexagonal axis [272]. In the Mg–Nd alloy the crystal lattice of the β' phase was identified to be a hexagonal one with parameters $a = 0.52$ nm, $c = 1.30$ nm, $c/a = 2.52$. The orientation relationship between the β' phase and the Mg matrix in the Mg–Nd alloys is [257]:

$$(1\bar{2}0)_{\beta} \parallel (100)_{Mg},\ (\bar{1}04)_{\beta} \parallel (001)_{Mg}.$$

The β' phase remains in the structure up to the stage when the alloys could be considered to be completely softened after hardening during ageing.

The equilibrium phase β appears in the structure of the alloys beginning at ageing temperatures of about 300–350°C with an exposure of several hours. The β crystals are bigger in size and thicker than the β' crystals. There is no visible connection between β crystals and matrix and they are assumed to be incoherent. At first the β crystals coexist in the structure along with the β' crystals and seem to grow at their expense. This conclusion can be drawn from the existence of the β' precipitate free zones around the β crystals as is shown in the electron micrograph in Figure 82c. With elevating ageing temperature the β' precipitates disappear completely and only the β crystals may be seen in the structure. They become bigger and polyhedral in form. An example of an electron micrograph is shown in Figure 82d. The equilibrium phases give electron patterns with more frequent reflexes as compared with the electron patterns from the metastable phase in accordance with their more complicated crystal structure. One of them is shown in Figure 80f with its scheme in Figure 81f. The reflexes from the β phase correspond to the zone axis coinciding with the tetragonal axis of the equilibrium phase $Mg_{12}Pr$ lattice. No orientation relation between the β phase and matrix was established in the Mg–Pr, Mg–Nd and Mg–Sm alloys [257, 265, 272]. However, in the Mg–Ce alloy the orientation relationship between the equilibrium β phase and matrix was revealed. This relationship is obtained in the form [284]:

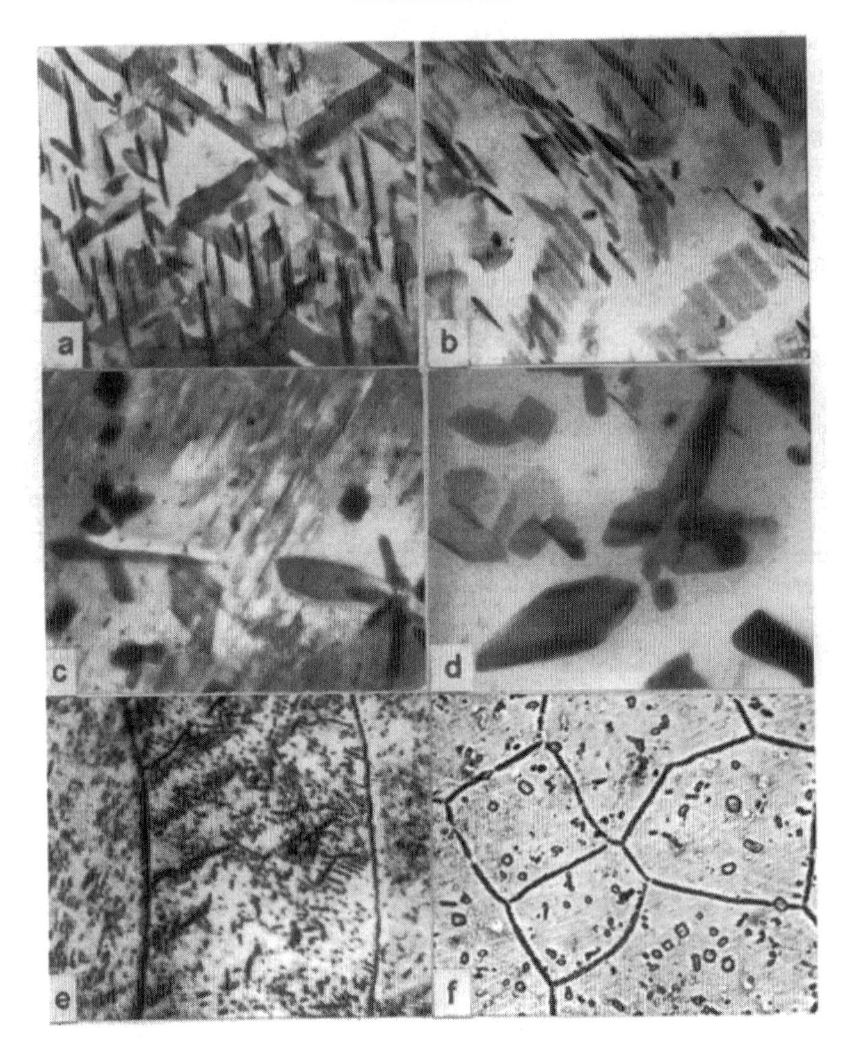

Figure 82. Electron (a–d) and light (e, f) micrographs of the Mg–RE alloys. a – Mg–5.65 mass % Sm, ageing at 275°C for 5 h, × 6500, b – Mg–3.8 mass % Sm, ageing at 250°C for 6 h, × 12500, c – Mg–2.3 mass % Pr, ageing at 300°C for 6 h, × 7800, d – Mg–3.8 mass % Sm, ageing at 325°C for 6 h, × 12600, e – Mg–3.8 mass % Sm, ageing at 250°C for 6 h, 1000, f – Mg–5.65 mass % Sm, ageing at 400°C for 6 h, × 240.

$$[100]_\beta \parallel [110]_{Mg}, [001]_\beta \parallel [1\bar{1}0]_{Mg}, [010]_\beta \parallel [001]_{Mg}.$$

Light microscope investigations reveal the RE precipitates within grains in the structure of the alloys only at the stages of overageing (after hardness maximum). Two typical micrographs are presented in Figures 82e and 82f. The first of them shows the structure with plate-like precipitates of the β' phase. The β' particles are not distributed uniformly within the grains and tend to gather into band-like collections. The second of them, Figure 82f, shows the structure after ageing at high temperature resulting in formation of the equilibrium phase and significant coarsening of its particles. Figure 82f demonstrates the approximately equiaxial form of the β crystals.

Systems with Rare Earth Metals of the Yttrium Subgroup, except Ytterbium

The phase transformations during the solid solution decomposition were studied using light and transmission electron microscopy for most of these systems, Mg–Y [254, 255, 267], Mg–Gd [269, 271, 280], Mg–Tb [268], Mg–Dy [266] and Mg–Er [273]. The studies revealed that there were the same transformations in all the systems at all stages of decomposition. For the rest of the systems, Mg–Ho, Mg–Tm and Mg–Lu, the structure of the alloys after the solid solution decomposition was not studied by transmission electron microscopy, and only the light microscopy method was applied. This method could reveal only the latest stages of solid solution decomposition. Meanwhile, the limited data obtained in the Mg–Ho, Mg–Tm and Mg–Lu by light microscopy confirmed the similarity of phase transformations in all systems.

The phase transformations within the Mg solid solution grains for alloys with rare earth metals of the yttrium subgroup, except ytterbium, differ significantly from those for the alloys with the rare earth metals of the cerium subgroup, although there are also some similar features between them. Unlike the alloys with the rare earth metals of the cerium subgroup, no GP zones form during solid solution decomposition in alloys with rare earth metals of the yttrium subgroup. Even immediately after solution treatment, the Mg solid solution in them turned out to be inhomogeneous with signs of ordering. By analogy with the alloys with the cerium subgroup elements, the ordering phase may also be marked by β'' in this case. Such a state is replaced by formation of the metastable phase β', when decomposition progresses. The morphology and crystal structure of the β' precipitates are similar for all Mg–RE systems with RE belonging to the yttrium subgroup, except ytterbium. Next, the β' phase is replaced by the equilibrium phase β with the same morphology for all these systems. In general, the sequence of solid solution decomposition in binary Mg alloys with rare earth metals of the yttrium subgroup, except ytterbium, may be depicted by the scheme:

$$\text{Mg solid solution with } \beta'' \rightarrow \beta' \rightarrow \beta.$$

As in alloys with rare earth metals of the cerium subgroup, the next stage of the precipitation is generally superposed on the previous one, so that two successive kinds of precipitates can exist in the structure simultaneously.

The ordering in Mg solid solution immediately after solution treatment is revealed in the electron diffraction patterns by existing superstructure reflexes. These superstructure reflexes and, correspondingly, the ordering are retained after ageing for a stage when changes of hardness and electrical resistivity are insignificant (first stage). They become, however, better outlined after such ageing suggesting some rearrangement in the structure. The superstructure reflexes are diffuse ones. They are disposed on the basal planes of the reciprocal lattice from Mg solid solutions in the places corresponding to the middles of the [100]* type radius-vectors. The superstructure reflexes are extended on the basal planes of the reciprocal lattice and are of the round disk form with the flat round surface being parallel to the basal planes of the reciprocal lattice. The typical electron diffraction patterns

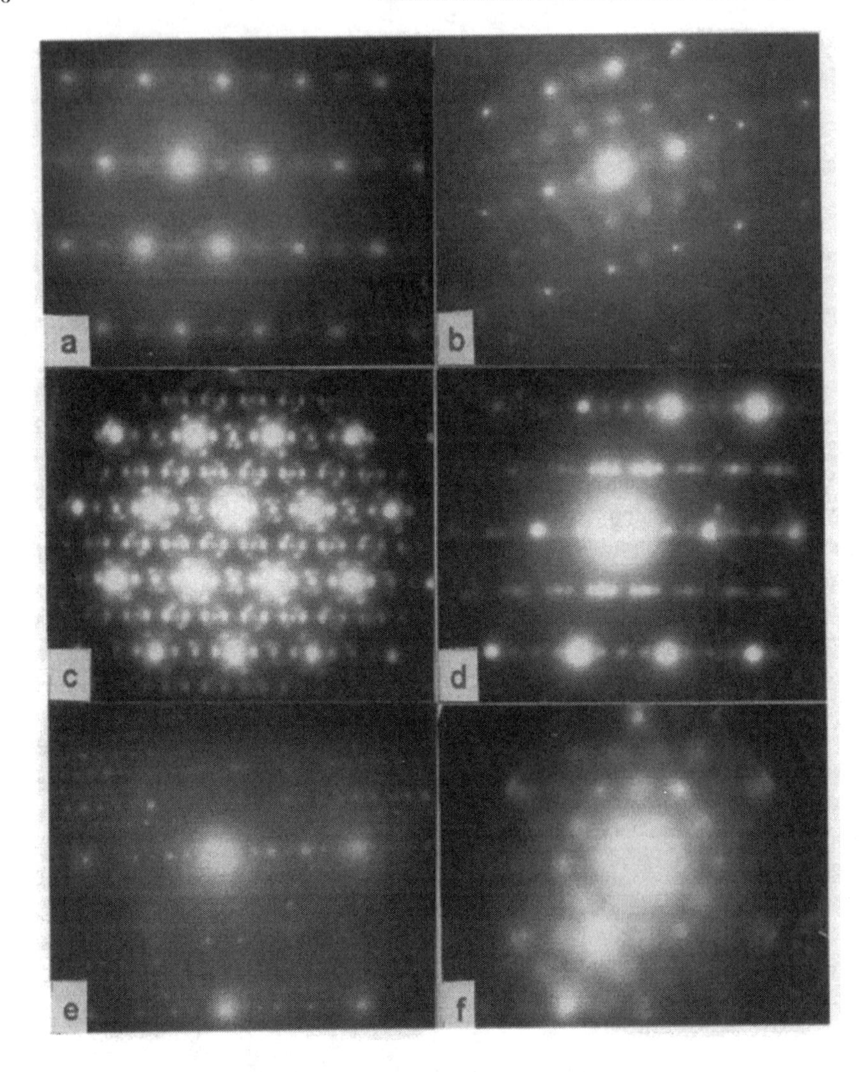

Figure 83. Electron patterns from the Mg–RE alloys. a – Mg–22 mass % Gd, ageing at 200°C for 1.5 h, b – Mg–29.3 mass % Er, ageing at 200°C, 5h, c – Mg–22 mass % Gd, ageing at 200°C for 100 h, d – Mg–24.2 mass % Dy, ageing at 200°C for 50 h, e – Mg–22 mass % Gd, ageing at 275°C for 48 h, f – Mg–23 mass % Tb, ageing at 350°C for 6 h.

with such superstructure reflexes are shown in Figures 83a and 83b for two different orientations of the Mg matrix lattice. In Figures 84a and 84b the respective schemes of electron diffraction patterns are presented. The superstructure reflexes in them are visible as streaks along the traces of the basal planes of the reciprocal lattice from the Mg matrix lattice. The streaks are the sections of the round disks in the reciprocal lattice from the Mg matrix lattice. The arrangement of the superstructure reflexes corresponds to the ordering of the Mg_3Cd type or DO_{19} type, as in the alloys of Mg with rare earth metals of the cerium subgroup. Meanwhile, the extended form of the superstructure reflexes in alloys with the rare earth metals of the yttrium subgroup (without ytterbium) suggests short ordering in

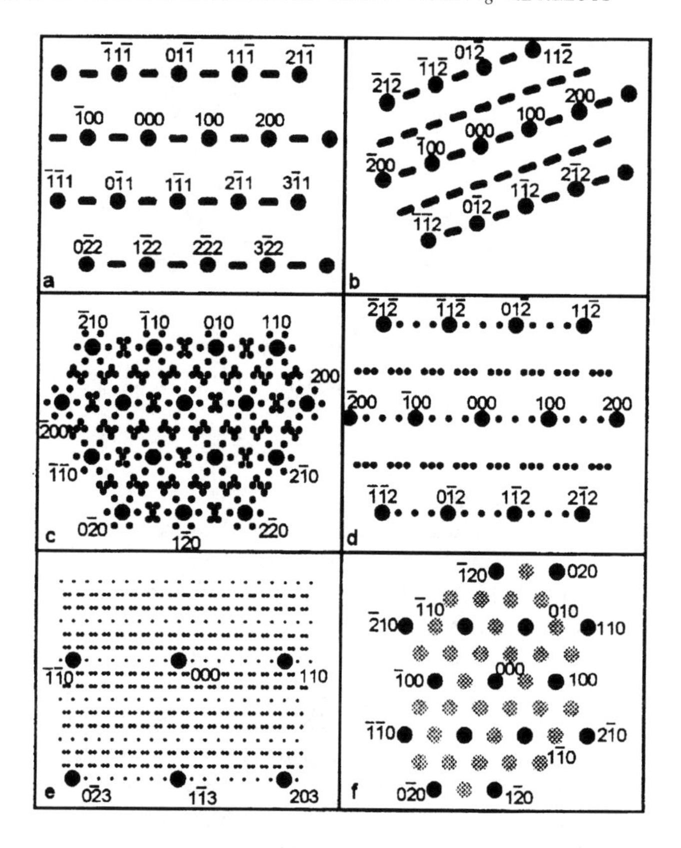

Figure 84. Schemes of the respective electron patterns presented in Figure 83.

them, as compared with long ordering in the alloys with rare earth metals of the cerium subgroup where the respective superstructure reflexes are visible as small sharp points. Dimensions of the domains with the ordering structure (β'' phase) were not determined in the alloys with the yttrium subgroup elements, but they are believed to be quite small. The lattice parameters of the ordering phase β'' are assumed to be $a_{\beta''} \cong 2a_{Mg}$, $c_{\beta''} = c_{Mg}$. The phase β'' domains have to be closely coherent with the matrix.

With progress of solid solution decomposition during ageing, metastable phase β' arises in the structure. Its first appearance corresponds to the beginning of the steep increase of hardness. The β' phase then exists in the structure of the alloys up to the hardness maximum and is retained in the structure also in the overaged state up to ageing temperatures of about 300°C when the hardening effect from the solid solution decomposition ceases completely. Appearance of the β' phase in the structure is accompanied in the electron diffraction patterns with transformation of every β'' phase superstructure reflex into five sharp superstructure reflexes forming a point figure resembling the letter X. Along with them two sharp additional reflexes appear in the positions of 1/4 and 3/4 distances of the [100]* type radius-vectors. Meanwhile, all superstructure reflexes are arranged within the basal planes of the reciprocal lattice from the Mg solid solution matrix. Electron patterns from the differently oriented grains enabled the construction of the reciprocal lattice of Mg solid solution containing β' precipitates. It is of hexagonal

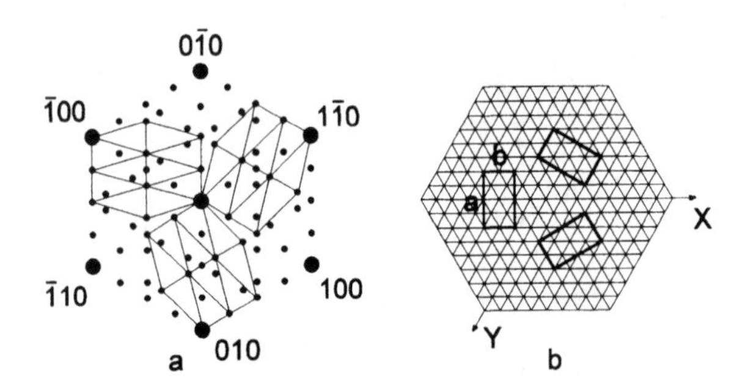

Figure 85. Basal planes of the reciprocal (a) and crystal (b) lattices of Mg solid solution with the β′ precipitates (scheme).

symmetry and consists of identical basal planes of the reciprocal lattice at the Mg solid solution places. One of the basal planes of the reciprocal lattice is shown in Figure 83c with its scheme in Figure 84c. It is of the zero order. Figure 83d shows one of the characteristic electron patterns from the Mg solid solution lattice with β′ precipitates which cross the basal planes of the reciprocal lattice. Its scheme is presented in Figure 84d. The main features of this electron diffraction pattern are 1) three superstructure reflexes at equal distances between the matrix reflexes along the [100]* type direction, 2) existence of the superstructure reflexes only along the traces of basal reciprocal lattice planes of Mg solid solution, and 3) the superstructure reflexes in the traces of the basal planes of the reciprocal lattice without the matrix reflexes are distributed into groups of every three of them. Every group of three superstructure reflexes in the latter electron pattern conforms to the section of the "X" cross of them in the respective basal plane of the reciprocal lattice.

Actually the described type of basal plane of the reciprocal lattices and the reciprocal lattice of the Mg solid solution with β′ precipitates were obtained in the earliest investigation of the Mg–Y alloy [254]. Treating them the authors [254] concluded that the β′ phase had a monoclinic lattice with the *b* axis directed along the *c* axis of the matrix. However, later treatment of the same reciprocal lattice revealed the higher symmetry of the β′ lattice [266, 267] and this treatment is elucidated by Figure 85 where the schemes of the basal planes for the common and reciprocal lattices of the Mg solid solution with the β′ precipitates are depicted. According to Figure 85a, all reflexes in the basal plane of the reciprocal lattice can be grouped into three identical systems arranged with threefold symmetry around the hexagonal axis. As far as the whole reciprocal lattice consists only of the same basal planes it means that all reflexes in it are grouped into three identical space systems consisting of the respective layers shown in the basal plane of the reciprocal lattice. The type of each system of the reflexes suggests it to conform to the orthorhombic lattice with parameters $a_{\beta'} \cong 4d_{(100)\mathrm{Mg}}$, $b_{\beta'} \cong 2a_{\mathrm{Mg}}$, $c_{\beta'} \cong c_{\mathrm{Mg}}$. The orthorhombic lattices of the β′ precipitates are arranged in a matrix with threefold symmetry around the hexagonal axis as it is shown schematically in Figure 85b. Parameters *a* and *b* of the β′ cell lie in the basal plane of the matrix lattice. Parameter *c* of the β′ cell coincides with parameter *c* of the matrix lattice. The character of the electron patterns suggests close coherency between the β′ phase and the matrix.

With the beginning of the β′ phase formation, the structure of the alloys observed by electron microscope acquires "ripple" appearance without clear interphase boundaries

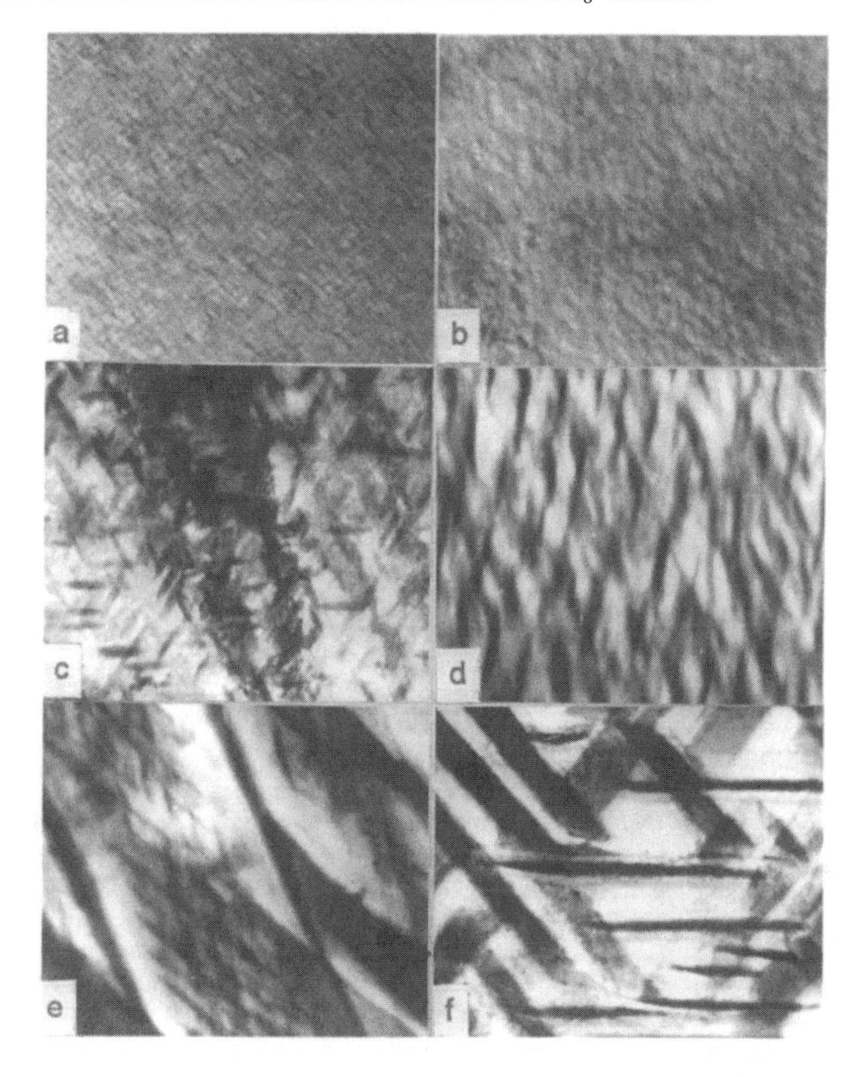

Figure 86. Electron micrographs of the Mg–RE alloys. a – Mg–23 mass % Tb, ageing at 200°C for 50 h, × 97000, b – Mg–24.2 mass % Dy, ageing at 200°C for 200 h, × 42500, c – Mg–24.2 mass % Dy, ageing at 250°C for 32 h, × 31000, d – Mg–22 mass % Gd, ageing at 250°C for 50 h, × 30000, e – Mg–22 mass % Gd, ageing at 300°C for 3 h, × 18000, f – Mg–22 mass % Gd, ageing at 325°C for 3 h, × 12000.

(Figures 86a and 86b), and this fact confirms the close coherency between the β' phase and the matrix. With progress of the β' phase precipitation, the "ripple" character of the structure becomes coarser, and the separate β' particles become visible (Figures 86c and 86d). They are of plate-like form and are arranged with the largest side on the {110} planes Mg solid solution in accordance with threefold symmetry and three possible orientations of the β' lattice in matrix. The distances between the same oriented particles are approximately equal. For suitable orientation of the Mg solid solution grain, three directions of the β' platelet arrangement in the matrix may be seen (Figure 86c). Meanwhile, for some other orientations only two of three directions of the β′ platelets are visible (Figure 86d). The third direction turns out to be unfavourable for observation in these cases.

The β phase forms in the structure of the alloys at the highest ageing temperatures of the Mg solid solution decomposition. The β particles are significantly bigger in size as compared with the β′ particles and differ from them by the different orientation in the matrix. At the beginning of its precipitation β particles can coexist in the structure together with the β′ particles. One such structure is shown in Figure 86e. As decomposition progresses the β′ particles disappear in the structure and only the β precipitates remain in it. An electron micrograph of the structure containing only β precipitates is shown in Figure 86f. As Figures 86e and 86f demonstrate, the β particles are rather coarse platelets. They are thicker in the middle than on the edges. No deformation contrast is visible around the β particles in the matrix suggesting actual absence of coherency between them. Analysis of the β platelets orientation in the matrix leads to the conclusion that they are arranged by the larger sides along the {100} type planes of the Mg solid solution lattice with threefold symmetry around the hexagonal axis of matrix.

In electron diffraction patterns, β phase is distinguished by more frequent reflexes as compared with those from the β′ phase. This fact is explained by the more complicated lattice of the β phase. As an example, one of the electron diffraction patterns with β phase reflexes is shown in Figure 83e with its scheme in Figure 84e. For the system Mg–Gd the relationship between lattices of the β phase and matrix was specially investigated [271]. The investigation established parallelism between certain atomic planes of the β particles and matrix lattices. Also, the electron diffraction patterns suggested that β particles consist of twins. The electron diffraction pattern presented in Figure 83e conforms to this conclusion. Every second and third row of β reflexes in it includes pairs of them, so that all β reflexes group into two mirror-like symmetrical systems. Such an arrangement of reflexes results from the same two lattices being in the twin position [271]. The β precipitates were identified as the respective equilibrium phases in the systems Mg–Y [255], Mg–Tb [268], Mg–Dy [266] and Mg–Er [273]. In the investigation of the system Mg–Gd [271], interplane distances of the β phase lattice derived from electron diffraction patterns did not agree with the then known lattice of the phase being in equilibrium with Mg solid solution. But they agree, in general, with the more complicated lattice established later for this equilibrium phase in [135]. The equilibrium nature of the β precipitates is confirmed by retention of them in the structure after heating up to high temperatures.

The β precipitates formed at high ageing temperatures turn out to be separated by large areas of Mg solid solution. The electron diffraction patterns taken from these areas show that Mg solid solution is not a homogeneous one in this case and includes the same short ordering as after special solution treatment or the first stage of ageing. One such electron pattern is presented in Figure 83f with Figure 84f as its scheme. By the way, Figure 83f illustrates the round form of the superstructure reflexes from the short ordering on the basal plane of the reciprocal lattice from Mg solid solution. The short ordering in Mg solid solution observed in the structures with only β precipitates suggests existence of this inhomogeneity in Mg solid solution at high temperatures when the respective ageing is carried out. Existence of short ordering in the Mg–Gd and Mg–Er solid solutions at different temperatures was confirmed by investigation of X-ray diffuse scattering [290].

Light microscopy reveals only the larger precipitates in the structure of the alloys after solid solution decomposition. The visible precipitates are commonly the β phase, although at the highest magnifications the β′ particles can also be distinguished. At first

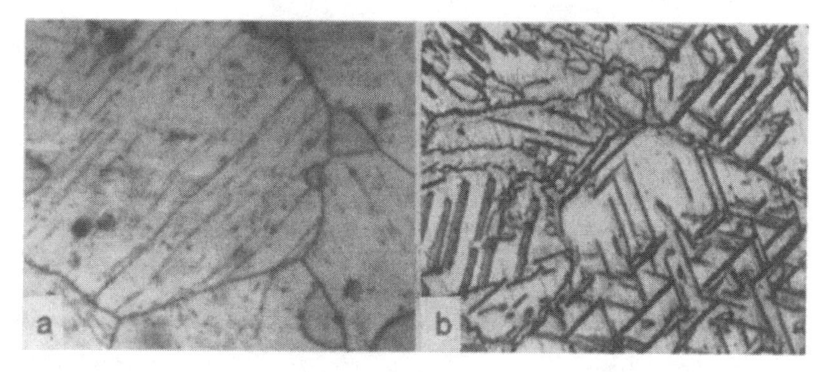

Figure 87. Light micrographs of the Mg–RE alloys. a – Mg–32 mass % Er, ageing at 300°C for 6 h, × 1000, b – Mg–28 mass % Ho, ageing at 350°C for 6 h, × 1000.

the precipitates may be observed as straight or wavy parallel lines within grains of Mg solid solutions (Figure 87a). As decomposition progresses, the quantity of the precipitates increases and they are evidently seen to be extended along one, two or three directions within grains depending on the grain orientation (Figure 87b). This Widmanstatten type of structure was observed by the author in alloys with elements of the lanthanum series up to such high annealing temperatures as 425°C for annealing time of 6 h. However, in the investigation of the Mg–Y alloy [255] long-time ageing at 260°C, 144 h + 315°C, for 144 h resulted in a structure with approximately equiaxial coarse precipitates.

The Magnesium–Ytterbium System

Solid solution decomposition in the Mg–Yb alloys shows certain specific features [279]. The initial stage of solid solution decomposition in this system is accompanied within grains of Mg solid solution by formation of fine particles oriented along the hexagonal axis of the lattice. The electron diffraction patterns display the superstructure reflexes resulting from these particles, but the connection between them and the matrix reflexes was observed only at several matrix orientations. Based on these observations, the authors [279] supposed only the existence of partial coherency between the precipitated particles and the matrix. No lines showing formation of the GP zones in the structure of the decomposed solid solution were observed in the electron diffraction patterns.

Progress in solid solution decomposition results in increase of the Yb-rich precipitates, which appear in a plate-like form with three different orientations within the Mg solid solution grains. The typical electron micrograph for this stage of the solid solution decomposition is shown in Figure 88a. Analysis of the plate-like particle orientations reveals that they lie on three prismatic (100) planes of the matrix. A part of the plate-like precipitates shows the inner fine constitution. It is possible to see them consisting of separate small particles arranged as flat spirals. With increasing ageing temperature and ageing time, the number of the precipitates arranged into flat spirals becomes larger, the separate particles in them increase in size and the areas occupied by them grow. The electron micrograph in Figure 88b shows the spirals of the precipitates at this stage of solid solution decomposition. It is not established exactly what phase forms the flat spirals in the structure, but

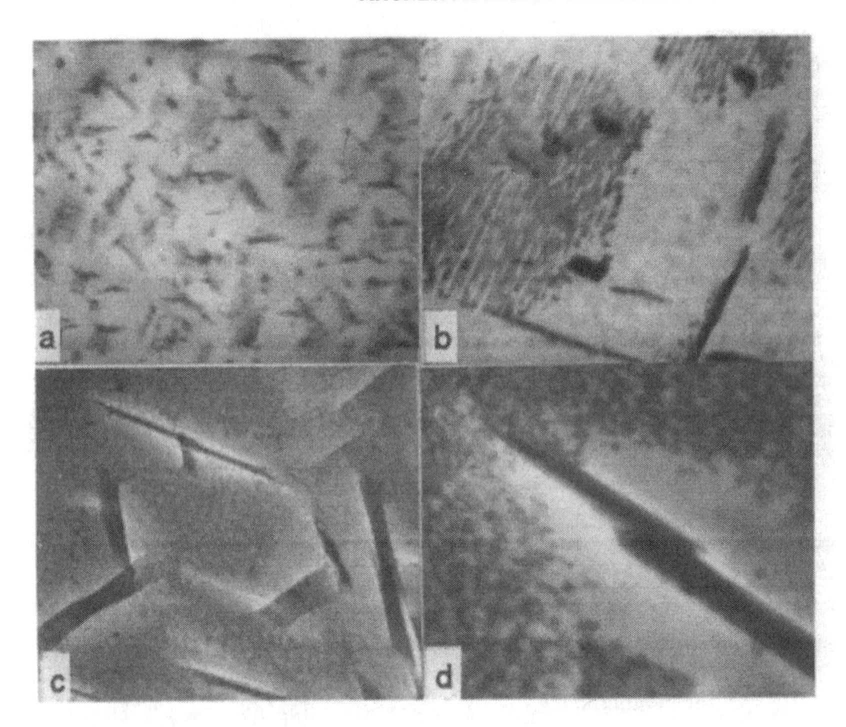

Figure 88. Electron micrographs of the Mg–RE alloys. a – Mg–2.9 mass % Yb, ageing at 225°C for 3 h, × 10000, b – Mg–2.9 mass % Yb, ageing at 250°C for 4 h, × 32500, c – Mg–5.2 mass % Sm–5.2 mass % Y, ageing at 220°C for 200 h, × 14500, d – Mg–5.2 mass % Sm–5.2 mass % Y, ageing at 220°C for 200 h, × 36000.

the order of magnitude of the distances between superstructure reflexes in the electron diffraction patterns suggested it to be the equilibrium phase Mg_2Yb.

The actual absence of the GP zones formation and weak coherency between precipitates and the matrix at the earlier stages seem to be responsible for a small hardening effect during solid solution decomposition in the Mg–Yb alloys.

Solid Solution Decomposition in Ternary Mg Alloys Containing Rare Earth Metals

From the practical viewpoint, solid solution decomposition in Mg alloys with rare earth metals of ternary and more complicated systems is of great importance in that commercial alloys contain, as a rule, several alloying elements. Such investigations were carried out, however, only in some ternary systems [264, 267, 275, 278, 280, 282, 285, 291].

In the ternary systems with two different rare earth metals as components, solid solution decomposition reflects, to a certain extent, the peculiarities of the respective ternary phase diagrams. If both rare earth metals belong to the same subgroup (except Yb), kinetics of the solid solution decomposition is similar in character to those of the separate

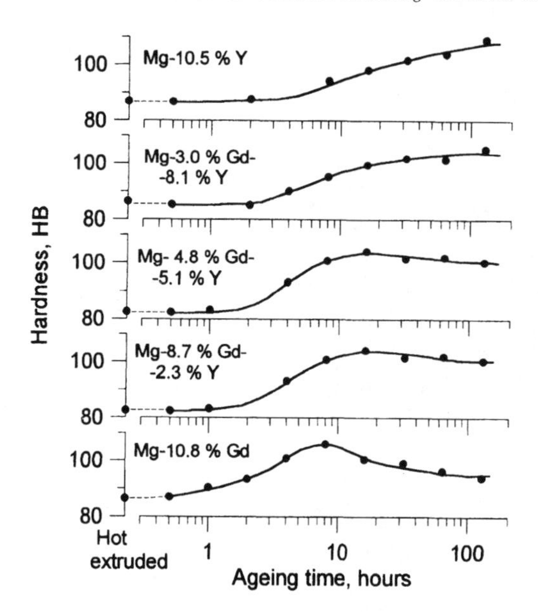

Figure 89. Hardness of the Mg–Gd–Y alloys versus ageing time at 225°C. (Contents in mass %.)

rare earth metals. Also the decomposition rate turns out to be intermediate between the decomposition rates of the separate rare earth metals. For example, Figure 89 shows the hardness curves for solid solution decomposition in the alloys of the Mg–Gd–Y system [280]. All alloys studied were aged immediately after hot extrusion followed by cooling at a common rate. In this condition they retained the supersaturated solid solution and, as the hardness curves show, a significant strengthening effect could be observed during ageing. The alloys contained Gd and Y in different ratios with approximately the same sum of them in mass %. For comparison the results of simultaneous tests of the binary Mg–Gd and Mg–Y alloys are also presented in Figure 89. As one can see, the character of the hardness curves and, consequently, the character of the decomposition kinetics, are the same for all the alloys studied. Nevertheless, the hardness maximum on the curves shifts successively to longer ageing time with increasing Y content in the alloys in accordance with the lower decomposition rate in Mg–Y alloys as compared with that in the alloys Mg–Gd. The similar phase transformation during solid solution decomposition in the alloys with rare earth metals of the same subgroup, at least at the earlier stages, led to the supposition of the same phase transformations in ternary alloys. The formed products of Mg solid solution decomposition in the ternary system seem to be the solid solutions between the respective products of the Mg solid solution decomposition in the binary systems.

If each of two rare earth metals belongs to different subgroups, the process of decomposition in the alloy of the Mg ternary system with two RE is expected to proceed in a more complicated manner because of the significant difference in kinetics and phase transformation during the process in each of the respective binary systems. Such an Mg solid solution decomposition was studied in the alloys Mg–Sm–Gd [282], Mg–Nd–Y [275, 285] and Mg–Sm–Y [278]. Investigations of all three systems showed gradual transition in the ternary alloys from the decomposition kinetics inherent in the alloys with

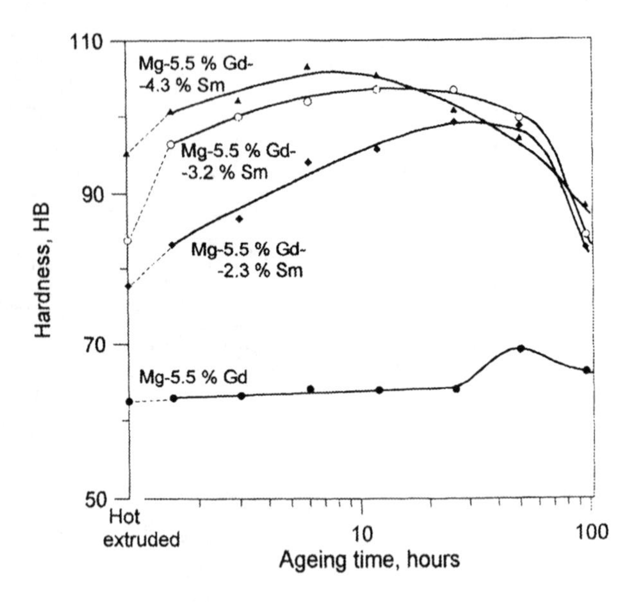

Figure 90. Hardness change of the Mg–Sm–Gd alloys during ageing at 200°C. (Contents in mass %.)

rare earth metals of one of the subgroups, to the kinetics inherent in the alloys with rare earth metals of the other subgroup. This feature of the solid solution decomposition is illustrated in Figure 90 where results of the hardness measurements of some Mg–Sm–Gd alloys are reproduced [282]. The alloys were hot extruded into round rods and in this condition had sufficient supersaturated Mg solid solution. Ageing of the hot extruded rods resulted in appearance of the hardness maximum conforming to the Mg solid solution decomposition. As one can see in the binary Mg – 5.5 mass % Gd alloy, there is a quite extended first stage of decomposition in which the hardness changes insignificantly. The hardness maximum is small as a result of the small Gd content in the alloy if it is compared with the rather large solid solubility of Gd in Mg according to the equilibrium phase diagram. Addition of Sm to the binary alloy with 5.5 mass % Gd results in the appearance of a higher hardness maximum and its shift to the shorter ageing times. The first stage of decomposition shortens and disappears. In this respect the kinetics of solid solution decomposition become similar to that of Mg–Sm alloys and Mg binary alloys with other rare earth metals of the cerium subgroup.

Figure 90 illustrates also another important feature of solid solution decomposition in the ternary system with two rare earth metals belonging to different subgroups. It consists of a substantial increase of hardness with increasing sum of the rare earth metals in the alloy. No doubt the hardening effect in the ternary Mg–Sm–Gd alloys presented in Figure 90 is caused mainly by the Mg–Sm type precipitates. Nevertheless, the existence of Gd in Mg solid solution can also act on the precipitation process. Keeping in mind the Mg–Sm–Gd phase diagram with significant solubility of Gd in the equilibrium Mg–Sm compound, it is reasonable to assume significant solubility of Gd in the Mg–Sm precipitates. In turn, the significant solubility of Gd in Mg–Sm precipitates should act on Mg solid solution decomposition to form these precipitates. On the other hand, the Gd atoms dissolved in the Mg solid solution lattice may influence the decomposition process by themselves. As alloys with rare earth metals of different subgroups have different products of solid solution decomposition, at certain ratios between these rare earth metals in the alloy the precipitates inherent in each of the Mg–RE systems are expected to form in the structure simultaneously. Such a case was observed indeed in the Mg–Sm–Y alloys [278]. In Figures 88c and 88d the electron micrographs are presented which show two types of precipitates at lower and higher magnifications in the alloy with equal contents of Sm and Y (in mass %). The coarse plate-like crystals are the Sm-rich precipitates formed first and being, therefore, bigger in size. They are arranged on three lattice planes of the matrix as in binary Mg–Sm alloys. The Y-rich precipitates are formed later and, therefore, are significantly smaller in size. They are arranged as background between the coarse crystals of the Sm-rich precipitates (Figure 88c). At high magnification the precipitate free zones are visible around the Sm-rich crystals. Their formation along the grain boundaries in the aged alloys is explained commonly by diffusion of alloying element to the precipitates there. So, absence of the Y-rich particles in the precipitate free zones around the Sm-rich crystals may be considered as indirect proof of the Y dissolution in them.

The effect of a third alloying element on solid solution decomposition in a ternary system of Mg with only one rare earth metal are estimated by taking into account the equilibrium phase diagrams. The significant features of them are the ability of the third alloying element to form compounds with the rare earth metal and the effect of the third alloying element on the solubility of the rare earth metal in solid Mg. So, in systems where Zr and Mn are the third elements, their influence on solid solution decomposition is assumed, as a rule, to be insignificant without special experimental confirmation. Such conclusions are quite justified, because both Zr and Mn do not form compounds with rare earth metals and do not change significantly the solubility of the rare earth metals in solid Mg. The addition of Al and Sn to the Mg–Nd alloys decreases abruptly the hardening effect during ageing. This fact is reasonably explained by the abrupt decrease of Nd solubility in solid Mg when Al and Zn are added [238]. Addition of Zn decreases also the hardening effect in Mg–Nd alloys, but more gradually [238].

In [264] substantial different phase transformations during solid solution decomposition in the alloy Mg – 2.8 mass % Nd – 1.3 mass % Zn were established as compared with the binary Mg–Nd alloys. No GP zones were revealed at the earliest stage of solid solution decomposition (up to ~ 200°C) in this ternary Mg–Nd–Nd alloy, although they could be expected taking into account the kinetics of the process. One cause of this fact was considered to be the spherical form of GP zones and their small size. The second stage (~250°) was accompanied by formation of the metastable phase γ'' disposed by the thin plates along the (001) planes of the matrix lattice. The phase γ'' had the hexagonal crystal structure of the $MgCd_3$ type (DO_{19} type) with lattice parameters $a_{\gamma''} = \sqrt{3}a_{Mg}$, $c_{\gamma''} = 3c_{Mg}$. Between lattices of the γ'' phase and matrix there was the orientation relation:

$$[001]_{Mg} \parallel [001]_{\gamma''}, [100]_{Mg} \parallel [110]_{\gamma''}, [2\bar{1}0]_{Mg} \parallel [100]_{\gamma''}.$$

At high temperatures of solid solution decomposition (~300°C) the phase γ formed. The phase γ consisted of rods lying on the basal planes of the matrix with the long axis extended along the matrix directions $\langle 100 \rangle$ and $\langle 110 \rangle$. The crystal structure of the γ phase was of the fcc type with the lattice parameter a = 0.72 nm. Between the lattices of the γ

phase and Mg matrix the orientation relation was also established. It was as follows:

$$[001]_{Mg} \parallel [011]_{\gamma},\ [2\bar{1}0]_{Mg} \parallel [\bar{1}\bar{1}1]_{\gamma},\ [0\bar{1}0]_{Mg} \parallel [\bar{2}11]_{\gamma}.$$

In [267] kinetics and the phase transformations of the solid solution decomposition were investigated in the alloys Mg – 11.5 mass % Y – 0.6 mass % Zn and Mg – 10.1 mass % Y – 1.2 mass % Zn. The alloys with Zn demonstrated in general the same character of the decomposition kinetics as the binary Mg–Y alloy containing about the same Y. However, the second stage of the decomposition, which resulted in a steep hardness increase, was observed in the alloy with 1.2 mass % Zn only at 175 and 200°C, as compared with 175, 200 and 225°C for the binary Mg–Y alloy, and began much later. The hardness maximum decreased with addition of Zn. The only new feature noted in [267] in the phase transformation during ageing of the Mg–Y alloys with addition of Zn was the appearance of flat stacking faults on the basal planes (001) of Mg matrix.

Resistance of Mg–Re Solid Solutions Against Decomposition During Continuous Cooling and Isothermal Ageing

For the practical use of the alloys, where solid solution decomposition can take place, resistance against decomposition during cooling from temperatures of the possible solution treatment and possible heating is of great importance. Low resistance against the solid solution decomposition requires the use of high cooling rates if the supersaturated solid solution has to be obtained. In such alloys short accidental heating may bring about undesirable decomposition of the solid solution accompanied by decrease of corrosion resistance, plasticity or less strength after subsequent ageing. The high cooling rate during solution treatment is difficult to carry out for massive parts and it can result in high internal stresses in the parts accompanied by distortion or cracking. On the contrary, the high resistance against decomposition enables us to obtain the supersaturated solid solutions at low cooling rates which do not require special cooling media and, therefore, can be easily used in industrial conditions. Besides, at low cooling rates the arising internal stresses are less.

The above mentioned different rates of solid solution decomposition in Mg alloys with different rare earth metals during isothermal ageing also suggests their different behaviour during continuous cooling from solution treatment temperatures. The response of the supersaturated solution formation to cooling rate was specially studied [106]. The study was carried out by using one binary alloy of five systems, Mg–Nd, Mg–Sm, Mg–Gd, Mg–Dy and Mg–Er. The RE contents in the alloys were chosen to be close to the maximum RE solubility in Mg solid solution in each of the systems. The alloys were hot extruded into round rods 11 mm in diameter, and then cut into small specimens. The specimens were annealed at 500°C for 2 h and then cooled down to room temperature at different rates. The different average cooling rates were obtained by quenching in water at room temperature (~$3 \cdot 10^5$ degree/min), in air by throwing about the separate specimens (~64 degree/min), in air with the specimens enveloped by asbestos (~17 degree/min), in air with the specimens placed into massive block of stainless steel (~2.4 degree/min) and in the furnace (~0.56 degree/min). The solid solution decomposition was checked by electrical resistivity measurements. The results of the experiments are shown in Table 24 where the values of the electrical resistivity ρ are presented as the relative shares (in %) of the respective values ρ_0 for every alloy after quenching in water. After quenching in water ρ/ρ_0 was assumed to be 100%.

The data in Table 24 confirm the different resistance of alloys with different rare earth metals to solid solution decomposition during cooling and how it increases in general with increasing atomic number of the rare earth metals. Increase of resistance to solid solution

Table 24. Electrical resistivity of Mg–RE alloys cooled differently from 500°C.

Contents of RE, mass %	Electrical resistivity after QW*, ρ_0, μOhm·cm	Electrical resistivity ρ in ρ/ρ_0, %, after different cooling rates:				
		~3×10^5 degree/min	~64 degree/min	~17 degree/min	~2.4 degree/min	~0.56 degree/min
4.7 Nd	8.23	100.0	90.0	82.7	68.3	66.3
5.8 Sm	11.19	100.0	96.2	93.0	77.6	63.5
22.8 Gd	32.2	100.0	105.3	98.7	79.5	61.8
30.0 Dy	41.0	100.0	99.3	97.3	90.2	81.7
30.8 Er	39.0	100.0	99.5	97.4	92.8	86.4

*QW is quenching in water.

decomposition is especially significant at the transition from elements of the cerium subgroup (Nd, Sm) to elements of the yttrium subgroup (Gd, Dy, Er). In the alloys with Nd and Sm, solid solution decomposition was already visible after cooling at ~64 degree/min, whereas in the alloys with Gd, Dy and Er at this cooling rate solid solution decomposition did not actually happen. Solid solution decomposition in alloys with the rare earth metals of the yttrium subgroup became significant only at a cooling rate of ~2.4 degree/min. The alloy with Er showed the greatest resistance to decomposition of all the alloys studied in accordance with the lowest decomposition rate during isothermal ageing.

Even the slowest furnace cooling of ~0.56 degree/min did not yield the full solid solution decomposition. Additional ageing of the alloys Mg–5.8 mass % Sm and Mg–30.0 mass % Dy at 200°C for 256 h resulted in a further decrease of electrical resistivity. The attained values ρ/ρ_0 were actually the same for the specimens cooled at all cooling rates and amounted to 45.0% and 61.7%, respectively as compared with 63.5% and 81.7% for the slowest furnace cooling. So, even after furnace cooling a significant proportion of the supersaturated Sm and Dy remained in Mg solid solution.

These features of the alloys' behaviour during cooling from high temperatures suggest the use of a high cooling rate for alloys with rare earth metals of the cerium subgroup in order to prevent undesirable solid solution decomposition. In such alloys the cooling in air can be insufficient for this. For alloys with rare earth metals of the yttrium subgroup, including Y, the cooling rates preventing solid solution decomposition turn out to be much lower and, as a rule, normal cooling in air is sufficient for full retention of supersaturated solid solution.

Additional characteristics of resistance to solid solution decomposition were revealed in the experiments involving isothermal ageing [274, 287]. The investigations were carried out using the binary alloys Mg–3.4 mass % Nd [287] and Mg–11.5 mass %Y [274]. The

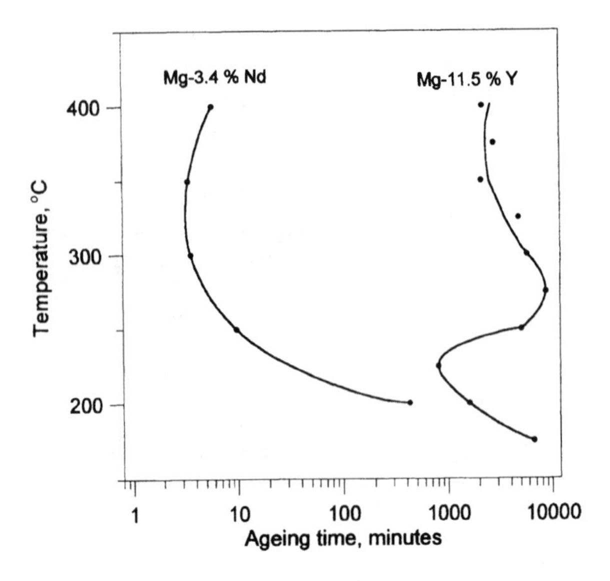

Figure 91. TTT-diagrams of the solid solution decomposition. Fraction transformed equals 0.5.

alloys were delivered as hot extruded round rods of 10.5 mm in diameter. They were solution treated by quenching in water from temperatures 510°C (Mg–Nd) and 565°C (Mg–Y). After quenching the specimens were isothermally aged at a wide range of temperatures with the decomposition of Mg solid solution in them being checked by measurements of electrical resistivity. Taking into account the results of the experiments in both investigations, TTT diagrams[†] of solid solution decomposition were developed. For the alloy Mg–3.4 mass % Nd, the fraction transformed during solid solution decomposition was characterized by the quantity $X = (\rho_0 - \rho)/(\rho_0 - \rho_f)$, where ρ, ρ_0 and ρ_f are the electrical resistivity of the specimen after a certain ageing, after solution treatment and after full decomposition of the solid solution, respectively. The value ρ_f was chosen as the least of those observed for the aged specimens. It amounted to 55.5% of the value ρ_0 [287]. For the alloy Mg–11.5 mass % Y the TTT diagram was developed with estimation of the solid solution decomposition by quantity ρ/ρ_0. However, the values ρ/ρ_0 could be recalculated into values of the fraction transformed X of Mg–11.5 mass % Y choosing for ρ_f the least of the electrical resistivity values reported for the aged condition. It amounted to 54.5% of the value ρ_0 [274].

The TTT diagrams of both alloys for solid solution decomposition are presented in Figure 91. They are drawn for the same value X = 0.5. The TTT diagrams confirm once more the significantly slower decomposition in Mg–Y alloys as compared with Mg–Nd ones. The TTT diagram of the Mg–Nd alloy is of the simple C-form with the highest decomposition rate at 300–350°C. The TTT diagram of the Mg–Y alloy is of the more complicated form with two temperature maxima of the decomposition rate. One of the maxima is observed at about 225°C, and the other at about 375°C. At ageing temperatures 250–300°C there is the above mentioned delay of solid solution decomposition.

[†] Temperature/time/transformation diagrams.

Reversion After Ageing in Mg–Re Alloys

Reversion after ageing was observed at first in the investigation of solid solution decomposition in the Mg–8.7 mass % Y alloy [254]. In these experiments the heat treatment regime 260°C, 1 h resulted in a decrease of the hardness produced after ageing at a lower temperature, 200°C. Additional ageing at 200°C restored the hardness of the alloy actually to the same value. Indication of the reversion was obtained also in [255] where the same alloy Mg–8.7 mass % Y was firstly aged at 260°C for 192 h and then annealed at 315°C for 2 h. The visible precipitates of the Y-rich phase after ageing at 260°C for 192 h then disappeared. In [288, 292, 293] the reversion process was studied in more detail in Mg–Y alloy and, moreover, Mg–Gd was also studied for comparison.

The main features of the reversion process are indicated using the results of the later works [288, 292, 293]. Investigation [293] was carried out on the alloys Mg–12 mass % Y and Mg–24 mass % Gd whose compositions were close to the maximum solubility of each rare earth metal in solid Mg at eutectic temperature. The alloys were in the form of hot extruded round rods of about 11 mm in diameter. They were cut into samples which were annealed at 540°C for 2 h (Mg–Y alloy) or at 520°C for 2 h (Mg–Gd alloy) followed by quenching in cold water aiming to obtain the Mg supersaturated solid solution. After quenching the samples were first aged at 200°C for 100 h. This regime was used for both alloys and corresponded to the hardness maximum after solid solution decomposition. The first ageing was followed by a second ageing at higher temperatures in order to achieve reversion. The second ageing was performed isothermally at 225, 250, 275, and 300°C. Its duration was in the range 15 min to 64 h. The reversion process was checked by electrical resistivity measurements. Such a method, unlike hardness measurements, gave direct information about possible dissolution of the RE-rich precipitates back into Mg solid solution and the decomposition of Mg solid solution to form RE-rich precipitates.

Figure 92 shows the results of the resistivity measurements during the second ageing of both alloys, Mg–12 mass % Y (Figure 92a) and Mg–24 mass % Gd (Figure 92b). As one can see for the alloy Mg–12 mass % Y (Figure 92a), at all temperatures of the second ageing, 225,250, 275, and 300°C, an increase of the resistivity is observed indicating dissolution of the precipitates back into Mg solid solution. The resistivity curves indicate also that the reversion effect increases and reaches its maximum at times that are shorter the higher the second ageing temperature. The resistivity values for the second ageing temperatures 250, 275 and 300°C grow significantly compared with those after the first ageing, and this suggests the possibility of tangible reversion after the second ageing at these temperatures for less time than the least time of 15 min used in the experiments. The resistivity falls gradually beyond the maximum as the second ageing time increases indicating decomposition of the magnesium solid solution again. As the decrease of resistivity at 225°C shows, after the longest exposure of 64 h Mg solid solution becomes even more depleted by Y than after the first ageing at 200°C for 100 h. After reversion the solid solution concentration does not reach the same level as after solution treatment, although at the highest temperature 300°C of the second ageing the former becomes close to the latter.

The resistivity measurements of Mg–24 mass % Gd alloy (Figure 92b) indicate that in this alloy the reversion after ageing also takes place at all temperatures used for the second ageing. The reversion effect increases similarly and reaches a maximum after

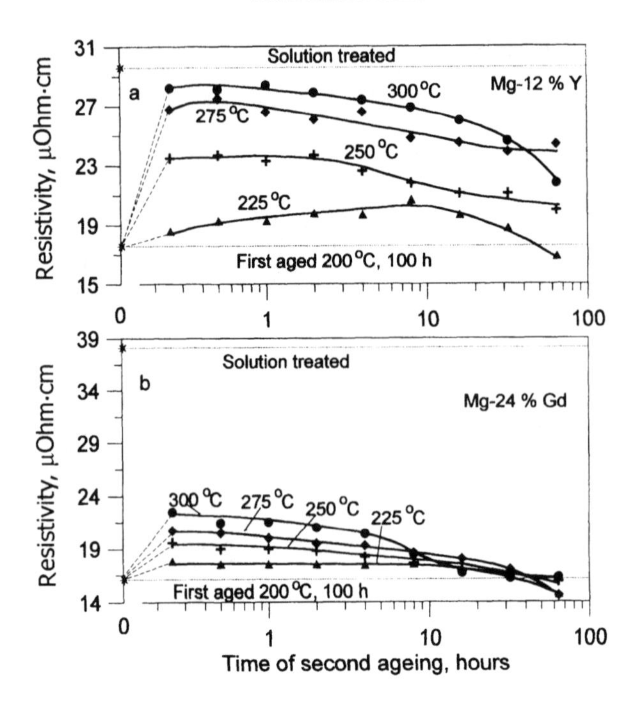

Figure 92. Electrical resistivity of the alloys versus time of second ageing.

shorter times with increasing second ageing temperatures. With increasing second ageing time the resistivity falls, indicating precipitation from Mg solid solution again after reversion at all temperatures. After long times of second ageing the solid solution concentration approximates that of the first aged alloy and may become even lower. The resistivity curves at short ageing times suggests also the possibility of tangible reversion effects after shorter times than the shortest time used. Meanwhile, as electrical resistivity indicates, even after reversion at the highest temperature of 300°C, Mg solid solution in the alloy Mg–24 mass % Gd remains significantly depleted as compared with that after solution treatment. In this aspect the alloy Mg–24 mass % Gd differs notably from the alloy Mg–12 mass % Y.

According to [293], the resistivity method makes it possible to estimate quantitatively the reversion assuming that a measure of it is the ratio $R = (\rho_r - \rho_{ag})/(\rho_{st} - \rho_{ag})$, where ρ_{st}, ρ_{ag}, and ρ_r are the resistivities of the alloy after solution treatment, after ageing at 200°C for 100 h, and after reversion annealing (second ageing), respectively. This ratio

Table 25. Maximum reversion fraction (%) in the alloys for different temperatures of the second ageing.

Alloy	Reversion temperature, °C			
	225	250	275	300
Mg–12 mass % Y	20.8	50.0	80.8	88.8
Mg–24 mass % Gd	6.7	15.8	20.7	26.9

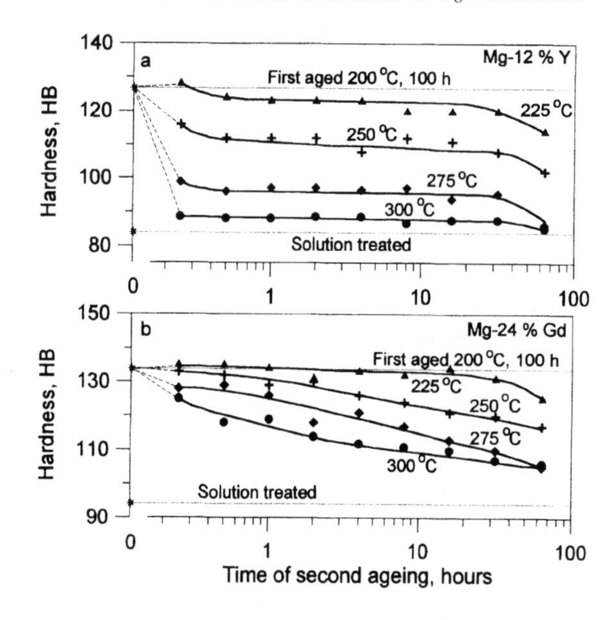

Figure 93. Hardness of the alloys versus time of second ageing.

may be considered as the "reversion fraction". The maximum reversion fraction of Mg–12 mass % Y and Mg–24 mass % Gd at each temperature calculated from the resistivity values is shown in Table 25. The significantly smaller reversion in Mg–24 mass % Gd is evident. In this alloy the reversion fraction reaches only 26.9% at the highest annealing temperature as compared with 88.8% in Mg–12 mass % Y. The smaller reversion effect in the Mg–24 mass % Gd alloy as compared with the Mg–12 mass % Y one is explained in [293] by the higher rate of solid solution decomposition at the temperatures of the second ageing. As a result of this, at the same second ageing temperature solid solution decomposition in Mg–24 mass % Gd alloy superimposes on the reversion process in a shorter time than in the Mg–12 mass % Y alloy.

The hardness measurements of the same alloys (Figure 93) indicate softening at all temperatures of the second ageing as ageing time increases. As the second ageing temperature increases, the hardness becomes significantly lower. For the Mg–12 mass % Y alloy there are "plateaus" on all hardness curves in their middle parts (Figure 93a). In contrast, for the Mg–24 mass % Gd alloy the "plateaus" on the hardness curves are displayed weakly (Figure 93b). Actually they are seen clearly only on the hardness curve for the lowest temperature, 225°C. At 250 and 275°C the "plateaus" are quite short, and at the highest temperature of the second ageing, 300°C, hardness decreases with increasing time without a visible "plateau" for the Mg–24 mass % Gd alloy. The "plateaus" on the hardness curves may therefore be explained by two processes which promote decrease and increase of the alloy hardness simultaneously. They are dissolution of the precipitates into solid Mg and continuation of the precipitation from it. On the other hand, additional ageing after reversion at the same temperature as the first ageing is accompanied by common strengthening [254, 288].

Transmission electron microscopy and X-ray diffraction reveal the same kinds of precipitates in the structure of the alloys both after first ageing and reversion. Reversion is accompanied only by the partial disappearance of the precipitates formed at first ageing [254, 255, 293]. Taking into account these observations the explanation of the reversion in the Mg–Y and Mg–Gd alloys is as follows [293]. Reversion arises as a result of instability of the precipitates formed after first ageing during second ageing at higher temperatures. The instability is connected with the surface energy of the precipitates which is caused by the elastic stresses arising as a result of coherency between them and the matrix. In general, electron microscopy investigations support this viewpoint, suggesting a loss of coherency between the retained precipitates and the matrix after reversion [293].

Comparison of the reversion behaviour of the Mg–Y and Mg–Gd alloys reveals similar features in them. In both alloys the reversion effect continuously increases with elevating temperature of the second ageing within the same temperature range. There is no additional hardening of both alloys when the second ageing continues at all temperatures after the reversion is completed. Both alloys show the same phase transformations during reversion and the subsequent solid solution decomposition. However, there are also differences in the reversion behaviour between the Mg–Y and Mg–Gd alloys, the main one being the lower reversion effect in the alloys of the Mg–Gd system.

Such a relation between the Mg–Y and Mg–Gd system in reversion behaviour conforms to the general tendency in characteristics of Mg–RE alloys, the similarity of the main features of the alloys with various rare earth metals, except Eu and Yb, and the differences in their secondary features with smaller distinctions where the rare earth metals belong to the same subgroup.

The results presented show the extent of reversion after ageing in Mg alloys with Y and Gd and its dependence on the second ageing temperature and time. This process may be expected also in Mg alloys with other rare earth metals of the yttrium subgroup. It can occur quite often in practice at different heating conditions and affect significantly the strength and other important properties of the alloys. No reversion after ageing was observed in Mg alloys with rare earth metals of the cerium subgroup.

Chapter 3

Peculiarities of Plastic Deformation and Recrystallisation in Mg–RE Alloys

Plastic deformation is of great importance for production and properties of commercial magnesium alloys. A great number of commercial magnesium alloys are produced as wrought semiproducts, such as sheets, plates, rods, forgings, and working of them includes plastic deformation. Along with this, plastic deformation can take place in separate parts of Mg alloys when they are used in various machines, and this process is responsible for strength and resistance against failure and distortion of the parts. As a rule, plastic deformation improves strength properties of magnesium alloys.

One of the main peculiarities of Mg alloys is their low plasticity at room and near room temperatures. As a result, significant plastic deformation of magnesium alloys during working is impossible at these temperatures. However, at elevated temperatures the plasticity of Mg alloys becomes sufficiently high, and they can be worked with significant deformation without failure. Therefore, in industrial practice, wrought Mg semiproducts are produced commonly by hot working [294, 295]. Transition from low plasticity at low temperatures to high plasticity at elevated temperatures takes place in a certain temperature range. This temperature range can be rather narrow, so that the transition from small plasticity to high plasticity is clearly revealed in the alloys. Temperatures of the transition depend on the alloy composition, structure of the alloy and deformation conditions.

Increase of the deformation temperature is accompanied by a change of the deformation mechanism in Mg alloys [25, 294–297]. Consequently, the structure of the alloys subjected to plastic deformation at elevated temperatures differs notably from their structure after plastic deformation at near room temperatures. As compared with pure Mg and most of the Mg alloys of other systems, the structural change in Mg–RE alloys after plastic deformation manifests some peculiarities both at near room and elevated temperatures. The main peculiarities are connected with the possibility of solid solution decomposition in most Mg–RE systems and high enough stability of the precipitated phases in the Mg–RE alloys as compared, for example, with the well-known Mg–Al and Mg–Zn alloys, where solid solution decomposition can also proceed. As far as processes of plastic deformation in Mg alloys at near room temperatures (cold deformation) and at elevated temperatures (hot deformation) are different in many aspects, their main features and the properties obtained should reasonably be considered separately.

Behaviour of Alloys During Cold Deformation

Plasticity of Mg–RE Alloys at Near Room Temperatures

One of the peculiar effect of RE additions to Mg and its alloys is commonly an increase of plasticity at near room temperatures. The increase of Mg plasticity resulting from RE is observed already at comparatively small additions of them (even less than 1 mass %) and continues commonly with increasing the RE contents up to certain limits.

As a rule, mechanical tests of Mg–RE alloys with small RE additions and pure Mg, which were treated in the same conditions, show higher characteristics of plasticity for the former than for the latter. So, the plasticity increase was observed during tensile tests of Mg alloys containing various mixtures of rare earth metals of the cerium subgroup enriched in lanthanum, cerium, praseodymium or neodymium [4]. The tested alloys and Mg were then hot extruded and heat treated at various regimes. Increase of plasticity was observed also at small additions of alloying elements in the hot extruded alloys of the systems Mg–Y and Mg–Sc [256, 298]. The effect of the rare earth metals on plasticity at room temperature is illustrated by data on the Mg–Nd alloys presented in Table 26 [239]. They were obtained by tensile tests of the extruded rods quenched in water after annealing at 535°C. As one can see in Table 26, the small addition of 0.16 mass % Nd to Mg results in an increase of the plasticity characteristics, elongation (E) and reduction of area (RA) by several times. Increase of the elongation and reduction of area continues with increasing Nd content up to 4.22 mass %. At higher levels of Nd the plasticity characteristics begin to decrease, and this tendency can be explained by the appearance of the brittle Nd-rich phase in the structure and increase of its quantity.

Table 26 shows the average grain size of Mg–Nd alloys measured in the same samples. The grain size decreases with increasing Nd content, and its substantial diminution takes place already at the lowest Nd concentration of 0.16 mass %. A similar

Table 26. Characteristics of plasticity at room temperature and average grain size of Mg–Nd alloys quenched after annealing at 535°C [239].

Nd, mass %	Elongation, %	Reduction of area, %	Average grain size, mm	Nd, mass %	Elongation, %	Reduction of area, %	Average grain size, mm
0	3.8	2.6	0.86	3.38	32.8	35.7	0.050
0.16	9.6	11.7	0.14	4.22	33.0	37.0	0.030
0.81	17.0	14.8	0.092	6.15	20.7	25.6	0.020
1.49	18.6	16.4	0.15	7.46	21.0	14.2	0.016
2.54	26.5	27.0	0.072				

character of the grain size change with increasing concentration was observed in the Mg binary systems with other rare earth metals [106]. So, increase of plasticity at room temperature with increasing the rare earth contents within concentration range of Mg solid solution may be explained by decrease of its average grain size. On the other hand, rare earth metals added even in small quantities can interact with impurities existing in Mg

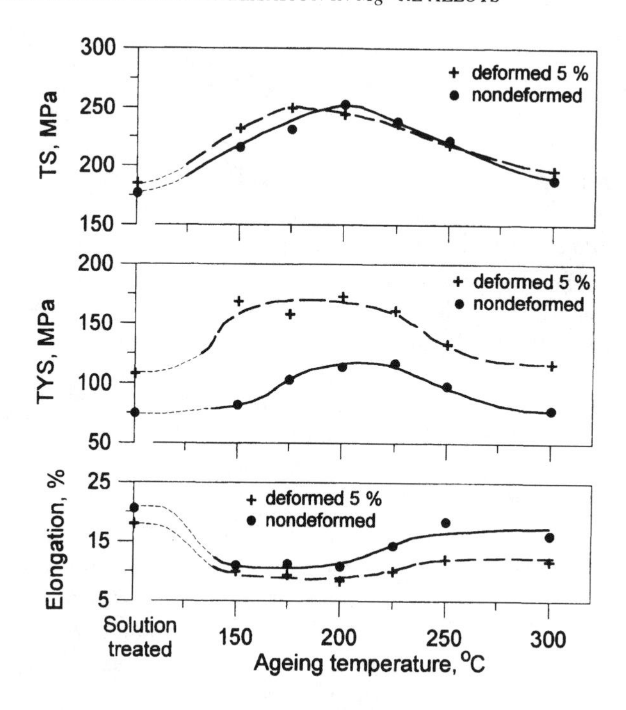

Figure 94. Tensile properties of the alloy Mg–3 mass % Nd versus ageing temperature. Ageing time is 24 h.

including the various gas impurities and form stable compounds with them. The compounds can be removed during melting of the alloys due to their higher density or remain in the structure as quite disperse particles. Such a process must purify the grain boundaries of the Mg solid solution from impurities and thus improve plasticity. Dispersed particles of the compounds between the rare earth metals and impurities in Mg may also result in the grain size diminution in the alloys. The coarse crystals of the Mg–RE compounds in the structure formed during solidification promote, on the contrary, plasticity decrease despite the Mg grain size diminution. Such a dependence can be seen also for the alloys containing more Nd than its solubility in solid Mg at annealing temperature 535°C (4.22–7.46 mass %).

Effect of Cold Deformation on Mechanical Properties of the Mg–RE Alloys

The higher plasticity of Mg alloys with rare earth metals, as compared with Mg alloys of other systems, enables them to be subjected to significant (for Mg alloys) cold deformation without failure. The magnitude of the cold deformation can reach more than 15% by stretching. The cold deformation enables the strength properties of the alloys to be improved, but makes the plasticity worse. The best combination of strength properties and plasticity can be reached by combined action of deformation and heat treatment resulting in solid solution decomposition. Such a treatment was named in Russian practice "low-temperature thermo-mechanical treatment" [239, 299–302].

The character of the tensile properties change as a result of the combined action of cold plastic deformation and ageing is illustrated by the data presented in Figure 94 [239, 299]. The tensile tests were carried out at room temperature using samples of the alloy Mg–3 mass % Nd. The alloy was hot extruded into round rods and then solution treated by quenching from temperature 535°C in cold water. In this condition the alloy structure consisted of only Mg solid solution. One part of the samples was deformed by 5% stretching. Deformed and nondeformed samples were aged simultaneously at various temperatures for the same time of 24 h. As one can see from Figure 94, dependence of tensile stress (TS), tensile yield stress for 0.2% offset (TYS) and elongation (E) on the ageing temperature are similar for deformed and nondeformed samples. The strength characteristics change with increasing ageing temperature passing through a maximum, and elongation changes with increasing ageing temperature passing through a minimum. Plastic deformation of 5% results in some shift of maximum on the tensile stress and tensile yield stress curves to lower ageing temperatures suggesting acceleration of solid solution decomposition.

The maximum values of tensile stress are approximately equal for both deformed and nondeformed samples. As a result of the decomposition acceleration, the TS values of the deformed samples are higher than those of the nondeformed ones at low ageing temperatures before maximum for the nondeformed samples. At higher ageing temperatures the tensile stress values for both deformed and nondeformed samples are close and even some decrease of TS can result from deformation.

Tensile yield stress increases as a result of deformation at all ageing temperatures, although after maximum strengthening by ageing, the difference between TYS values determined with deformed and nondeformed samples becomes less than that before maximum strengthening. Elongation becomes less after deformation at all ageing temperatures.

Table 27 shows the effect of deformation rate and test temperature on the short-run tensile properties of the Mg–3 mass % Nd alloy cold deformed between solution treatment

Table 27. Tensile properties of the alloy Mg–3 mass % Nd cold deformed between solution treatment and ageing at 200°C for 24 hours.

Test temperature, °C	Deformation rate, %	Tensile stress, MPa	Tensile yield stress, MPa	Elongation, %
	0	247	102	18.2
20	5	245	161	14.3
	10	247	203	9.1
	18	257	231	5.4
	0	139	80	23.0
250	5	138	108	23.6
	10	152	128	15.4
	18	167	153	12.2
	0	84	67	28.2
300	5	101	88	35.2
	10	100	83	35.8
	18	112	97	25.4

and ageing. Deformation was performed by stretching. Ageing at 200°C for 24 h corresponded to maximum strengthening. The results of the tests demonstrate increase of strength properties TS and TYS with increasing deformation rate at all test temperatures. Elongation decreases with increasing deformation rate at room temperature and 250°C. At 300°C elongation also retains this tendency in general, but it becomes weaker and less evident because of high level of elongation. The strengthening effect resulting from cold deformation is quite thermally stable and remains at elevated temperatures 250 and 300°C. According to [239], it remains at least up to 400°C in short-run tests. It was observed also at long-run tests at elevated temperatures. So, during long-run tensile tests at constant load the solution treated and aged alloy Mg–3 mass % Nd broke in 100 h at 200 and 250°C for stresses 110 and 60 MPa, respectively, whereas the same alloy cold deformed by 5% between solution treatment and ageing broke during long-term tensile test at constant load in 100 h for stress 120 at 200°C and for stress 65 MPa at 250°C [239].

The similar effect of cold deformation performed before ageing was observed in the Mg–Nd alloys containing additional alloying elements [239, 303] and in the alloys containing yttrium [289, 304, 305]. In all cases there was a notable strengthening effect from cold deformation which was added by the strengthening effect from ageing. For example, in Table 28 results of the tensile experiments of the alloy Mg–10.5Y–4.2Sc–0.79Mn (in mass %) are reproduced [305]. The alloy was hot extruded at about 450°C into round rods and in this condition had Mg solid solution supersaturated by yttrium. The rods were aged at 200°C for 100 h, cold deformed by stretching 5%, or cold deformed by stretching 5% followed by ageing at 200°C for 100 h. The ageing regime provided the maximum strength from the solid solution decomposition. The data presented in Table 28 confirm the highest values of TS and TYS of the alloy at room and at elevated temperatures 250 and 300°C in the case of the combined cold deformation and ageing. Cold deformation and ageing separately decrease plasticity of the alloy Mg–10.5Y–4.2Sc–0.79Mn. After the combined action of them plasticity at room temperature decreases also, but remains close to that after ageing alone. As compared with "as hot extruded" condition, at 250°C plasticity of the alloy does not change actually after each treatment and at 300°C it decreases only insignificantly after each treatment taking into consideration the high level of the elongation (E) values.

Table 28. Tensile properties of the alloy Mg – 10.5% Y – 4.2% Sc – 0.79% Mn (in mass %).

Treatment	Test temperature								
	20°C			250°C			300°C		
	TS, MPa	TYS, MPa	E, %	TS, MPa	TYS, MPa	E, %	TS, MPa	TYS, MPa	E, %
Hot extruded	370	280	8.5	300	240	14.5	235	190	41.0
Aged 200°C, 100 h	385	325	5.0	325	265	12.0	265	225	29.0
Cold deformed 5%	395	375	7.5	340	295	14.5	275	235	26.9
Cold deformed 5% + aged 200°C, 100 h	420	385	4.5	350	295	14.0	275	235	27.0

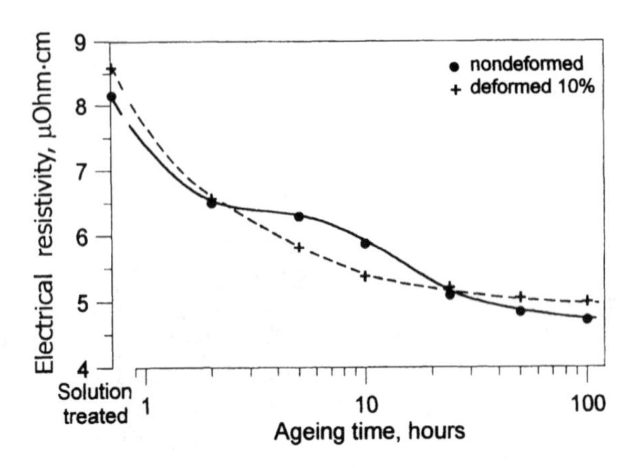

Figure 95. Electrical resistivity of the alloy Mg–2.5 mass % Nd versus ageing time at 200°C.

Effect of Cold Deformation on Kinetics of Solid Solution Decomposition and Accompanying Transformation in Structure

Plastic deformation accelerates decomposition of Mg solid solution. In addition to the noted shift of the strength property maxima to the lower ageing temperatures after cold deformation (Figure 94) there is the direct proof of this effect obtained by electrical resistivity measurements [239]. They are illustrated by data presented in Figure 95. Samples of the alloy Mg–2.5 mass % Nd were solution treated by quenching from 535°C and then aged at 200°C. Before ageing one of the samples was cold deformed 10% by stretching. The other sample was aged without preliminary deformation. As one can see, the cold deformation of the solution treated sample results in increase of electrical resistivity caused by formation of imperfections in the Mg solid solution lattice. Ageing of both samples is accompanied by electrical resistivity decrease showing depletion of Mg solid solution. In the deformed sample, electrical resistivity decreases more quickly and becomes less than electrical resistivity of the nondeformed sample after ageing time 2 h. This fact confirms the quicker solid solution decomposition after deformation. When depletion of Mg solid solution is completed, electrical resistivity of the deformed sample again becomes more than that of the nondeformed sample showing retention of imperfections in the Mg solid solution lattice. Acceleration of the solid solution decomposition after cold deformation is typical of the metal alloys and was observed in other systems [306].

Cold deformation produces imperfections in the Mg lattice which are quite various. They are described in detail in [25, 307–315]. The favoured mechanism of deformation in Mg crystals is the basal slip corresponding to shifts on the (001) planes of the lattice along directions ⟨ 110⟩. Another important mechanism of cold deformation in free Mg crystals is twinning on the {102} planes. Twinning is realised in the lattice when the crystals are oriented unfavourably for basal slip. In polycrystalline aggregate the grain boundaries restrain free plastic flow of the crystals and additional mechanisms of the lattice deformation arise. Cold deformation of the homogeneous solid solution in the Mg–Nd alloys is accompanied by the same changes in the structure as in pure polycrystalline Mg [239, 316]. The various changes in the structure of the Mg–Nd alloy after cold deformation

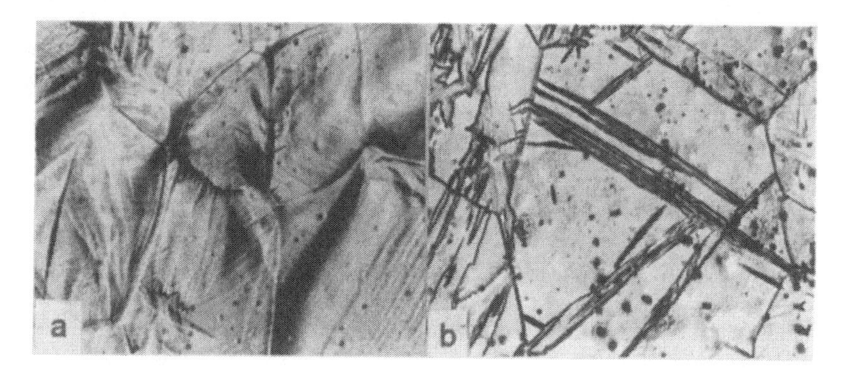

Figure 96. Light micrographs of the alloy Mg–3.4 mass % Nd. a – Solution treated, cold deformed 10%, polished before deformation, × 250, b – Solution treated, cold deformed 18%, repolished after deformation, × 250.

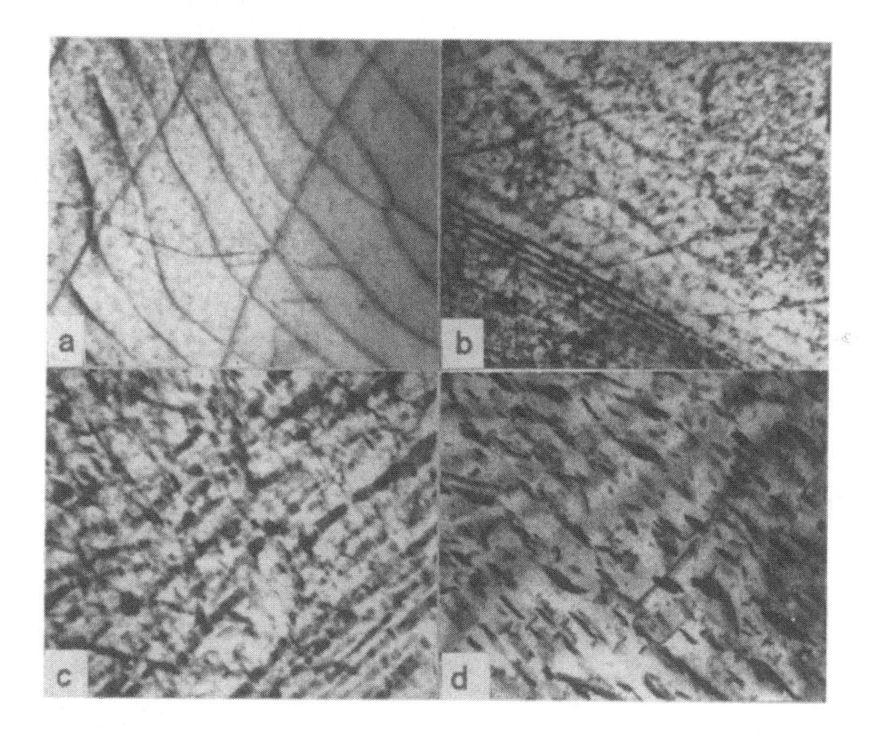

Figure 97. Electron micrographs of the alloy Mg–2.8 mass % Nd solution treated and cold deformed ~1%. a – without ageing, × 12000, b – aged at 250°C for 1 h, × 18000, c – aged at 300°C for 1 h, × 12000, d – aged at 325°C for 1 h, × 12000.

are demonstrated in Figure 96a where a micrograph of the relief arising on the polished surface of the sample after deformation is shown. The straight lines corresponding to basal slip, twins, bend planes, buckling and accommodation folds can be seen. After repolishing almost all signs of the deformation disappear and only twins can be seen (Figure 96b). They

Figure 98. Light micrographs of the alloy Mg–3.4 mass % Nd. a – Solution treated, aged at 350°C for 24 h, × 250, b – Solution treated, cold deformed 10%, aged at 400°C for 24 h, × 250.

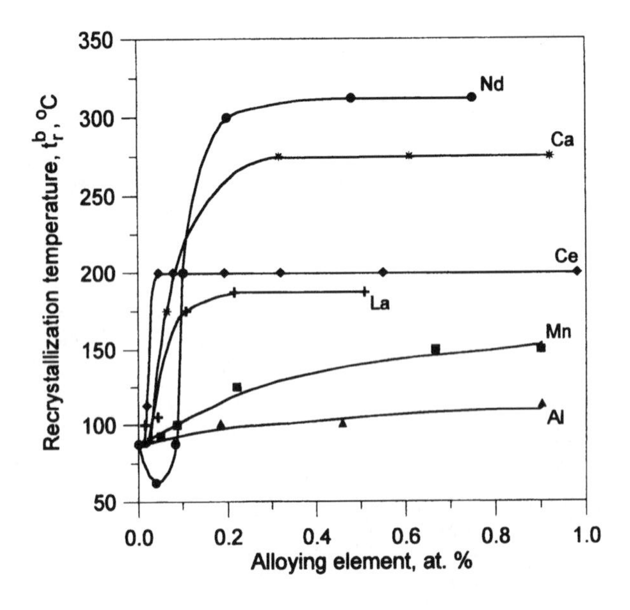

Figure 99. Effect of the alloying elements on temperature of the beginning of recrystallisation of Mg.

are distinguished as bands crossing a whole grain or its parts. The twin boundaries divide the lattice areas with significantly different orientation and, therefore, behave themselves during etching as the common grain boundaries.

The solid solution in Mg–Nd alloys contains comparatively small contents of alloying element and can be considered as dilute. In alloys with rare earth metals of the yttrium subgroup which are more soluble in solid Mg some differences in the deformation mechanism as compared with pure Mg can be believed to be present. This may be suggested taking into consideration results of the investigation [317]. In this study deformation of the alloy Mg–9 mass % Y by compression resulted in only insignificant number of the common twins on the lattice planes {102} unlike pure Mg. Simultaneously, many twins on the planes {111} were formed. In the opinion of the authors [317], this phenomenon is responsible for the smaller difference between tensile yield stress and compressive yield stress of Mg–9 mass % alloy as compared with pure Mg and some Mg-base alloys.

Created by cold deformation the lattice imperfections are favourable places for precipitation from supersaturated solid solution. Such places are, for example, dislocations. Figure 97 shows electron micrographs of the alloy Mg–2.8 mass % Nd solution treated and deformed ~1% by bending. The arisen dislocation net can be seen (Figure 97a). Ageing after deformation of the alloy is accompanied by the preferable precipitation from the Mg solid solution on dislocations and Figures 97b,c,d show successive stages of this process. The growth of the Nd-rich particle size and some degeneration of the dislocation net are observed. Other preferable places for precipitation arisen after deformation are the twin boundaries. Precipitation on them can be seen in light microscope along with precipitation on the grain boundaries (Figure 98a,b). Creation of many new places for preferable precipitation must promote acceleration of the above mentioned solid solution decomposition in Mg–RE alloys after cold deformation.

Behaviour of the Alloys During Annealing After Cold Deformation

Recrystallisation after Heavy Cold Deformation

The effect of some rare earth metals on recrystallisation of Mg after heavy cold deformation was studied in [318–322]. The alloys of the binary systems Mg–La, Mg–Ce, Mg–Nd and some ternary systems were used. Deformation was performed by rolling or squeezing up to 60% either at room or at slightly elevated temperatures. The slightly elevated temperatures were chosen so that no signs of recrystallisation were observed after deformation. Before deformation the alloys were annealed at sufficiently high temperatures (350–500°C) to provide a structure consisting of only equiaxial recrystallised grains. The used annealing temperatures and air cooling of the samples promoted some supersaturation of Mg solid solution before deformation, although the special solution treatment with high temperature annealing followed by quenching was not carried out.

The representative results of the investigations are presented in Figure 99 [319, 320]. They show the temperature of the beginning of recrystallisation for the alloys Mg–La, Mg–Ce, Mg–Nd versus the binary alloy concentration. Recrystallisation temperature was determined for annealing time of 1 h. Deformation rate was 60%. Along with this, temperature of the beginning of recrystallisation determined in the same conditions for other systems of the binary alloys, Mg–Ca, Mg–Al and Mg–Mn is presented in Figure 99 for comparison. As one can see in Figure 99, addition of La, Ce and Nd raise the temperature of the beginning of recrystallisation. The change of it with increasing the rare earth metal content is, in general, of the same character. Temperature of the beginning of recrystallisation increases sharply with comparatively small additions of La, Ce and Nd (at 0.1–0.2 at. %). After this, temperature of the beginning of recrystallisation does not actually change with increasing the rare earth metal content. The highest temperature of the beginning of recrystallisation (at "plateau") increases for alloys in order: La, Ce, Nd, corresponding to successive increase of the rare earth metal solubility in solid Mg. The first small additions of the rare earth metals do not change recrystallisation temperature before its sharp increase. For the Mg–Nd alloys a small decrease of recrystallisation temperature is observed at the first insignificant addition of the rare earth metal (0.039 at. %). Actually, dependence of temperature of the end of recrystallisation on the rare earth metal concentration was observed to be similar to that of the beginning of recrystallisation.

It is natural to connect the character of the recrystallisation temperature change with increasing the rare earth metal concentration in the binary systems with ability of the alloys for solid solution decomposition and, as a result, with formation of the disperse particles of the precipitates in structure. Such particles must hinder nucleation and growth of the recrystallised grains. The proposed hypothesis is supported by the sharp rise of recrystallisation temperature in alloys for the concentration areas where the most formation of the precipitates from Mg solid solution is expected. It is supported also by the successive increase of recrystallisation temperature with increasing solubility of the rare earth metals in solid Mg in order, La, Ce, Nd. Absence of significant change of the recrystallisation temperature at the first small additions of the rare earth metals and even some fall of it for the alloys Mg–Nd may be explained by interaction of the first additions of the rare earth metals with the inevitable impurities in the initial magnesium used making them inactive for recrystallisation.

In the alloys Mg–Mn and Mg–Al presented for comparison recrystallisation temperature grows gradually with increase in the alloying concentration. Meanwhile, up to the most used concentration of about 0.9 at. % recrystallisation temperature of the Mg–Mn and Mg–Al alloys remains substantially lower than that of the alloys with La, Ce and Nd. Another behaviour is observed for the Mg–Ca alloys used for comparison. For these alloys dependence of recrystallisation temperature on the concentration is of similar character as in the case of Mg–RE alloys. Besides, the highest recrystallisation temperature for the Mg–Ca alloys turns out to be more than that of the Mg–La and Mg–Ce alloys, although is lower than that of the Mg–Nd ones. The similarity between Mg–Ca and Mg–RE systems in the dependence of the recrystallisation temperature on concentration may be explained by the similar type of phase diagrams on the Mg side. As in the Mg–La, Mg–Ce and Mg–Nd systems, in the Mg–Ca one there is a small solubility of the alloying element in solid Mg, decreasing when temperature becomes lower [139]. Also, calcium, as with rare earth metals, is quite an active element and can interact with inevitable impurities in the Mg used tying them in the disperse particles.

Addition of rare earth metals promotes an increase of activation energy of recrystallisation in Mg. For cold deformation, 60% activation energy of recrystallisation was determined for Mg to be 17.5 kcal/g-atom as compared with 32, 22.5 and 29 kcal/g-atom in the alloys containing 0.1 at. % La, Ce and Nd, respectively [319, 320].

Recovery and Recrystallisation after Cold Deformation of Solution Treated Alloys

The decisive role of the precipitates from the Mg–RE solid solutions in the annealing processes was confirmed in the investigations where the alloys were specially solution treated before cold deformation [323, 324]. In [323] the alloy Mg–3.4 mass % Nd was solution treated by annealing at 535°C for 4 h followed by quenching in cold water, cold deformed 10% by stretching and then annealed isothermally in the range 200–500°C for

24 h at each temperature. The processes of recovery and recrystallisation in the Mg solid solution lattice were studied by observation of the X-ray patterns for diffraction at high Bragg angles ϑ.

After solution treatment the structure of the alloy consisted of perfect grains of Mg solid solution which in the X-ray patterns gave separate points corresponding to the reflection from them at certain Bragg angles. Cold deformation resulted in appearance of continuous diffuse lines in the X-ray patterns instead of separate points. Such change of the X-ray diffraction confirmed many imperfections in the Mg solid solution lattice created by deformation. Annealing at the lowest temperature 200°C for 24 h was near the ageing regime corresponding to the maximum hardening due to the solid solution decomposition. It was not accompanied by any change of the X-ray pattern. It continued to consist of continuous diffuse lines. After annealing at the next higher temperature 250°C for 24 h, the lines in the X-ray pattern changed notably. They consisted still of continuous lines, but their width became significantly less. As compared with the deformed samples without annealing and annealed at 200°C, after annealing at 250°C for 24 h the doublet of α lines turned out to be evidently split. Such character of the X-ray pattern indicated substantial recovery in the lattice, but without recrystallisation. Mg solid solution's consistency became to include enough perfect areas, but with significantly less size than the grains after solution treatment. Such areas could be twins formed during cold deformation. They could

Table 29. Results of X-ray investigation of the alloy Mg – 3.4 mass % Nd after solution treatment, cold deformation 10% and annealing [323].

Treatment	X-ray pattern	Estimated structure of Mg solid solution
Solution treatment	Separate points	Recrystallized grains of perfect lattice
Cold deformation 10%	Continuous diffuse lines	Lattice with a lot of imperfections and twins
Annealing at 200°C for 24 hours	Continuous diffuse lines	The same
Annealing at 250°C for 24 hours	Continuous thin lines	Areas of perfect lattice (blocks, twins) having smaller dimensions than primary grains
Annealing at 300°C for 24 hours	The same	The same
Annealing at 350°C for 24 hours	The same	The same
Annealing at 400°C for 24 hours	The same	The same
Annealing at 450°C for 24 hours	Separate points on the continuous thin lines	Beginning of recrystallisation
Annealing at 500°C for 24 hours	Separate points only	Completion of recrystallisation

also be blocks within the primary grains of Mg solid solution. Rise of the annealing temperature up to 400°C for 24 h did not result in any visible change of the X-ray patterns. Only after annealing at 450°C for 24 h the separate points appeared on the continuous lines of

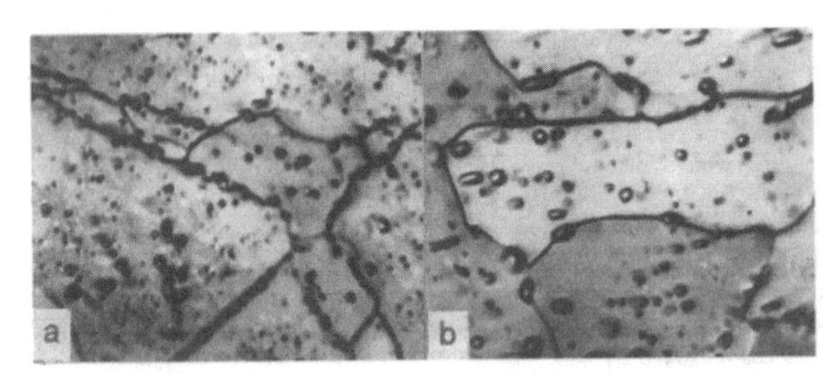

Figure 100. Light micrographs of the alloy Mg–2.85 mass % Nd. a – Solution treated, cold rolled 13%, annealed at 450°C for 6 h, × 1500, b – Solution treated, cold rolled 13%, annealed at 475°C for 6 h, × 1500.

the X-ray pattern showing the beginning of recrystallisation. After annealing at 500°C for 24 h the X-ray pattern consisted of only separate points showing recrystallisation to be completed.

The results of these X-ray experiments are summarised in Table 29. They enable the following conclusions to be drawn taking into consideration previously described observations of the solid solution decomposition in the Mg–Nd alloys.

- The recovery process begins when the precipitation from the supersaturated solid solution in general is completed.
- The structure obtained after recovery consists of the practically perfect areas of Mg solid solution with significantly less size than that of the primary Mg solid solution grains (twins and blocks).
- The structure obtained after recovery is quite stable and remains in place up to high annealing temperatures.
- Recrystallisation begins only at high annealing temperature. At this temperature the solubility of the rare earth metal in Mg solid solution increases significantly, and the particles of the precipitates coalesce significantly in size. As a result, distances between particles of the RE-rich phase in the structure increase.
- Recrystallisation is completed at a temperature near the annealing temperature for solution treatment.

These features of the processes of recovery and recrystallisation were observed in other experiments with Mg–Nd alloys. Their connection with the structure becomes understandable if the restraining action of the precipitates from Mg solid solution is recognised. Such action of the precipitate particles is reasonable to assert because they are obstacles for dislocation movement. The microstructure of the alloys confirms the restraining action of the precipitate particles. As a rule, the particles can be seen on the boundaries of the recrystallised grains. The typical microstructures for the beginning of recrystallisation (a) and completed recrystallisation (b) in the decomposed Mg solid solution are presented in Figure 100. As is common for the cold deformed material, the recrystallised nuclei arise in the grain joints.

In [324] the Mg–Nd alloys with various Nd contents were cold deformed with a significant rate of 60%. Cold deformation was carried out after solution treatment by quenching from 535°C, or preliminary annealing at lower temperatures providing full solid solution decomposition. In order to achieve full solid solution decomposition, the alloys were at first annealed at 400°C for 6 h and then successively at 350, 300, 250, 200, 150 and 125°C with successively increasing annealing time from 8 to 24 h. Deformed samples were annealed anew and their structure was studied. Investigation showed that the temperatures of the beginning and completion of recrystallisation in the sample deformed after solution treatment were significantly higher (by 125–250°C) than those deformed after annealing, provided that full solid solution decomposition took place. This fact can be also considered as a proof of the decisive role of the precipitates from Mg solid solution for the recrystallisation process in Mg–RE alloys.

Behaviour of the Alloys During Hot Deformation and Following Annealing

Effect of Rare Earth Addition on the Hot Deformation Mechanism of Mg at Small Deformations

As deformation temperature rises, the atom mobility increases. Therefore, during deformation at elevated temperatures, in the structure of the alloys the processes of annihilation and redistribution of the lattice imperfections take place simultaneously with their formation and compete with each other. The whole number of lattice imperfections decreases. As a result, the deformation mechanism changes with rising temperature, and the structure obtained differs from that observed after deformation at room and near room temperatures. The changes in the structure of Mg and some Mg alloys after deformation at elevated temperatures are reported in [325–330].

Investigations showed that deformation proceeded without twinning, beginning with a rise to a certain temperature. Also, increase of the deformation temperature promoted block formation. The latter process consisted of dividing of the grains into a few large blocks within a grain and simultaneously of a great number of fine blocks formed along the grain boundaries. After the fine block formation, the grain boundaries became of serrated appearance. Another process which became more important with temperature increase was sliding along the grain boundaries. This process was recognised to be a cyclic one with alternate grain boundary migration in every cycle. Decrease of the deformation rate promoted the high temperature mechanisms of deformation to act at lower temperatures. This fact can be considered as a proof that the high temperature mechanisms of deformation are caused by the processes of thermally activated atom movements.

The peculiarities of the high temperature deformation in Mg alloys containing rare earth metals are connected with the ability of some of them for solid solution decomposition with formation of hard and thermally stable RE-rich particles. Solid solution decomposition can proceed during the deformation process and act on its mechanisms by inhibition of dislocation movement and processes connected with it, such as sliding along grain boundaries, migration of grain boundaries, movement of block walls, and others. On the other hand, imperfections in the grain lattices originating during hot deformation can promote redistribution of the RE-rich particles precipitated from Mg solid solution.

The authors [313] investigated the deformation mechanisms of the alloy Mg–3 mass % Nd during short running stretching by a few percent. The stretching rate was about 5%/min. Previously, the samples were solution treated with quenching and then ageing up to the hardness maximum. A part of the samples were cold stretched between quenching and ageing. For comparison, the deformation mechanisms of pure Mg were studied in the same conditions. In addition, long running creep tests were conducted for the alloy Mg–3 mass % Nd at 250 and 300°C. The creep tests were performed by stretching samples at constant stress corresponding to failure in about 100 h. The study showed transition from cold deformation mechanisms to hot deformation mechanisms in the alloy Mg–3 mass % Nd at higher temperatures than in pure Mg. So, in the alloy with Nd, twins formed up to temperature 350°C as compared with only 250°C for pure Mg. The evident migration of the grain boundaries and formation of the serrated grain boundaries, which were typical of hot deformation of pure Mg, were not observed in the alloy with Nd even at 400°C. Meanwhile, in pure Mg the serrated grain boundaries and migration of the grain boundaries were observed beginning with 200–250°C. These observations confirm the restraining action of the RE-rich particles precipitated from Mg solid solution on realisation of hot deformation mechanisms in the structure.

One of the remarkable features of hot deformation in the Mg–3 mass % Nd alloy was sliding along the basal planes by separated strata of the same thickness. The stratum thickness increased with rising temperature. As a result of this sliding by strata, the grain boundaries acquired a stairs-like appearance. This deformation mechanism became more evident at long-run experiments at elevated temperatures and is illustrated in Figure 101 by the micrographs taken from the sample tested in such conditions. The micrograph in Figure 101a taken at low magnification and heavy etching shows clearly the sliding bands within grains and the connection between them and the “stairs” on the grain boundaries. The micrograph in Figure 101b taken at higher magnification and soft etching shows precipitation of Nd-rich particles along the grain boundary following the “stairs” formed, and precipitation within grains with certain decoration of the sliding planes in them. The previous cold deformation between solution treatment and ageing of the Mg–3 mass % Nd alloy prevented sliding along the baal planes by strata and formation of the stairs-like grain boundaries by this during hot deformation [325].

Effect of Rare Earth Addition on the Transition from Brittle to Ductile Behaviour of Mg with Rising Temperature

The effect of rare earth metals on transition from brittle to ductile behaviour of Mg with rising temperature was studied on the binary Mg–RE alloys using the higher deformation rate of 40% performed by axial compression [106, 331]. Experiments were carried out using round cylindrical specimens produced from hot extruded rods. Compression stress during plastic deformation was applied along the axis of every specimen. Before deformation the specimens were solution treated by annealing at temperatures near the temperatures of the maximum solubility of every rare earth metal in solid Mg followed by quenching in cold water. Solution treatment provided also a structure of Mg solid solution consisting of the full recrystallised equiaxial grains. For comparison, pure Mg and the alloy Mg–9.0 mass % Al were used for the same experiments. Before compression, pure Mg was

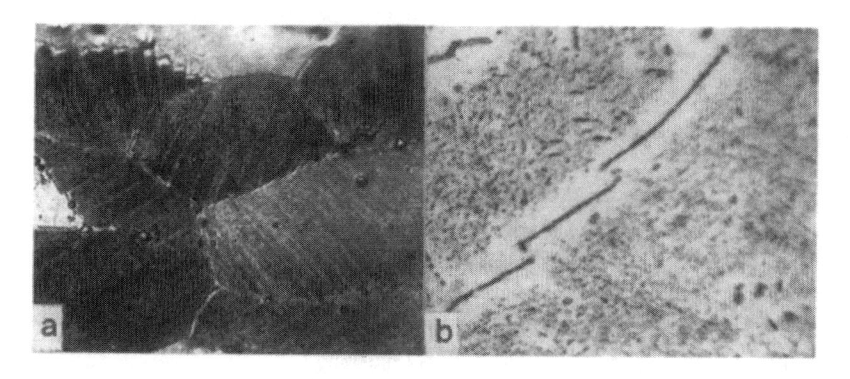

Figure 101. Light micrographs of the alloy Mg–3 mass % Nd. a – test at 250°C, heavy etching, × 250, b – test at 300°C, soft etching, × 1000.

annealed at 435°C for 1 h aiming to obtain the full recrystallised structure consisting of equiaxial grains. The alloy Mg–9.0 mass % Al was solution treated before compression in the same way as the alloys with rare earth metals. The compression temperatures were increased by 50°C steps and sometimes by every 25°C. At low squeeze temperatures the specimens broke by cracking along planes inclined by 45° to the axis. As the temperature was increased the ductility of the alloy increased abruptly and the respective specimens deformed without breaking. So, the temperatures of the transition from brittle to ductile behaviour of the alloys could be determined. Simultaneously temperatures of the beginning of recrystallisation and its completion for the alloys during hot deformation were determined by microscopy observations.

The results of the investigation are summarised in Table 30. For each alloy, the temperature range of the transition from brittle to ductile behaviour is presented. The

Table 30. Results of compression tests of Mg and some binary Mg alloys.

Content of alloying element, mass %	Solution treatment (annealing) temperature, °C	Average grain size before deformation, mm	Transition temperature from brittle to ductile behaviour, °C	Temperature of recrystallisation beginning, °C	Temperature of recrystallisation completion, °C
–	435	0.081	150–175	200–250	400–450
3.2 Nd	500	0.068	300–325	400–450	>500
5.6 Sm	520	0.107	300–325	400–450	500–525
1.4 Eu	540	0.068	250–300	350–400	500–525
1.0 Gd	540	0.124	100–150	350–400	400–450
11.1 Gd	520	0.039	300–325	350–400	450–500
21.5 Gd	520	0.048	450–475	450–500	500–525
19.1 Dy	540	0.036	300–325	400–450	450–500
25.4 Er	545	0.028	300–325	450–500	450–500
2.2 Y	400	0.030	<150	350–400	450–500
12.7 Y	550	0.044	300–350	400–450	>500
9.0 Al	400	0.033	175–200	–	–

lowest value in the range corresponds to the highest deformation temperature when specimen failure was still observed, and the highest value in the range corresponds to the lowest deformation temperature when specimen failure was not already observed. The temperature ranges shown in Table 30 for the beginning and completion of recrystallisation were obtained in the same way. The data presented in Table 30 on the transition temperature from brittle to ductile behaviour show the following main features.

• Alloys including the rare earth metals of the cerium subgroup, Nd and Sm, which are highly soluble in solid Mg, have substantially higher transition temperatures from brittle to ductile behaviour as compared with pure Mg.

• The rare earth metal Eu which is insoluble in solid Mg causes an increase of the transition temperature of Mg, but less than Nd and Sm.

• The rare earth metals of the yttrium subgroup, Gd and Y, added in small quantities which are less than their solubility in solid Mg, decrease the transition temperature of Mg from brittle to ductile behaviour. This fact can be seen from the data for the alloys with 1.0 mass % Gd and 2.2 mass % Y and those for pure Mg.

• Increase of the rare earth metal content where the rare earth metal belongs to the yttrium subgroup promotes an increase of the transition temperature which becomes more than transition temperature of pure Mg. The highest transition temperature is observed in the alloy with 21.5 mass % Gd. It is higher than that of pure Mg by 300°C.

• Aluminium increases the transition temperature from brittle to ductile behaviour of Mg but insignificantly if its effect is compared with effects of the rare earth metals in high enough contents.

Generalisations of these observations lead to the conclusion that RE-rich particles precipitated from Mg solid solution promote a significant increase of the transition temperature from brittle to ductile behaviour of Mg. The precipitation of the RE-rich particles can be expected during deformation because temperatures are sufficiently high. This viewpoint is supported by the greatest increase of transition temperature in alloys with Nd, Sm, Y, Gd, Dy, Er, where a large quantity of precipitates can be expected taking into account the respective phase diagrams. The highest transition temperature is established for the alloy with 21.5 mass % Gd where the largest quantity of the RE-rich precipitates can be expected during solid solution decomposition. Meanwhile, the nature of the RE-rich precipitates and the change of the solubility in solid Mg seem to be important. This conclusion can be made taking into consideration the insignificant increase of the transition temperature in the alloy with 9.0 mass % Al as compared with the alloys with rare earth metals. The coarse particles of the RE-rich phases formed during crystallisation can promote increase of the transition temperature from brittle to ductile behaviour, but their action is quite weak as compared with the RE-rich particles precipitated from Mg solid solution. This is seen from the comparison of Mg–Eu alloy with alloys with other rare earth metals where decomposition of Mg solid solution can take place. Finally, small quantities of rare earth metals within the Mg solid solution can increase ductility and decrease transition temperature. This effect can be ascribed to the refining action of the small quantities of rare earth metals which can supposedly tie impurities in Mg in compounds and thus purify the grain boundaries.

Addition of the rare earth metals to Mg raises the temperatures of the recrystallisation beginning and recrystallisation completion in conditions of hot deformation. This fact is evident in Table 30. Besides, in alloys where solid solution decomposition can take place, temperatures of the recrystallisation beginning and completion turn out to be higher than those in alloys where the alloying elements are present in Mg solid solution, but their contents are too small for supersaturation (the alloys Mg–1.0 mass % Gd and Mg–2.2 mass % Y). Increase of the Mg solid solution supersaturation results in an increase of the recrystallisation temperature. This is seen by comparing the alloys with 11.1 and 21.5 mass % Gd. In the alloys where the solid solution decomposition can happen during hot deformation, recrystallisation is observed at temperatures at which the precipitated particles turn out to be significantly coagulated and present in less quantity because of the increased RE solubility in solid Mg at higher temperatures. The noted features of the effect of rare earth metals on recrystallisation of Mg during hot deformation confirm the restraining action of the RE-rich particles precipitated from the supersaturated solid solution on this process as in the case of the considered recrystallisation during annealing of cold deformed alloys.

Structural Peculiarities of Recovery and Recrystallisation during Hot Deformation

Some peculiarities of recovery and recrystallisation during hot deformation have been revealed by electron and light microscopy observations [106, 331]. At the lowest temperatures of ductile behaviour (only slightly above transition temperature) the most remarkable feature is the formation of serrated grain boundaries. According to [328], such grain boundaries arise from formation of blocks along them during hot deformation. Some blocks become recrystallisation nuclei and they also can be seen. The typical structure of the alloy in this state observed by light microscope is shown in Figure 102a. It is worth noting that most of the recrystallised nuclei have a well outlined boundary at one side and a weakly outlined boundary on the opposite side. This fact confirms the origin of the nuclei from the blocks formed along the primary grain boundaries during hot deformation. The well outlined boundary of the nucleus conforms to contact with the foreign primary grain for the initial block, and the weakly outlined one conforms to contact with the parent primary grain for the initial block. The next micrograph presented in Figure 102b shows the tendency to recrystallisation with increasing deformation temperature when the structure is observed by light microscope. Fine equiaxial recrystallised grains arise along the boundaries of the coarse primary deformed grains. The recrystallised grains become well outlined and arranged by chains around the primary grains. With further increase in temperature of deformation, the new recrystallised grains become more coarse and more numerous. They occupy a greater part of the alloy volume surrounding separate remnants of the primary grains (Figure 102c). At the highest deformation temperatures, the full recrystallised structure is obtained. It consists of only new equiaxial grains (Figure 102d). As with cold deformation, hot deformation is accompanied by damage to the crystal lattice within grains of Mg solid solution. As imperfections arise, they destroy the arrangement of the precipitates formed during solid solution decomposition. The places within grains where the imperfections are concentrated become favoured for precipitates. The next two micrographs show typical examples of solid solution decomposition in hot deformed Mg alloys. Figure 102e demonstrates preferable precipitation along the sliding bands within the primary grain of Mg solid solution along with grain boundaries. The sliding bands form after hot deformation and divide the grain into few pieces. Figure 102f demonstrates Gd-rich particles precipitated in hot deformed Mg–Gd alloy. Directions of the plate-like particles change within the grain following misorientation of the separate grain parts

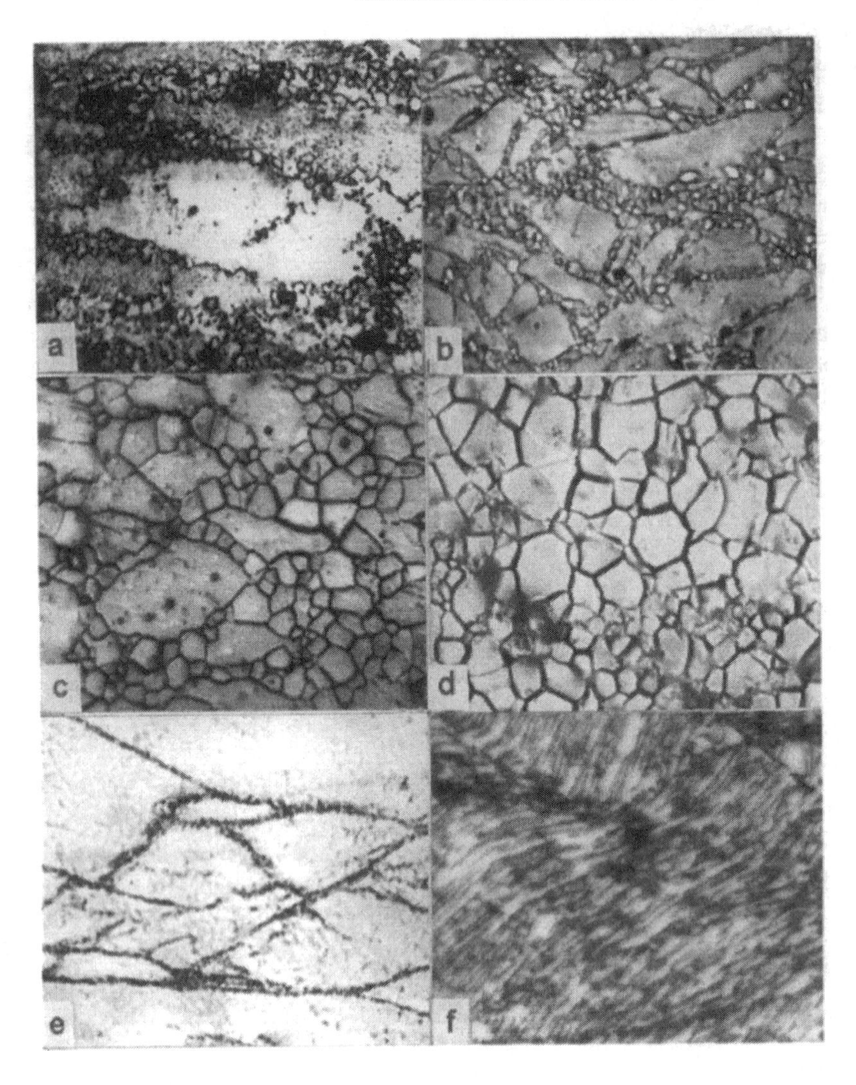

Figure 102. Light micrographs of the Mg–RE alloys. a – Mg–3 mass % Nd, deformed 80% at 350°C, × 1000, b – Mg–13.6 mass % Gd, deformed 40% at 400°C, × 240, c – Mg–2.2 mass % Y, deformed 40% at 450°C, × 240, d – Mg–19.1 mass % Dy, deformed 40% at 500°C, × 250, e – Mg–19.1 mass % Dy, deformed 40% at 350°C, × 1000, f – Mg–21.5 mass % Gd, deformed at 400°C, × 1000.

formed as a result of hot deformation.

Additional information on the structure of the hot deformed Mg–RE alloys was given by transmission electron microscopy. Some results of the investigations by this method are illustrated in Figure 103. Transmission electron microscopy clearly revealed the blocks within grains of Mg solid solution formed during hot deformation. The typical microstructure with blocks is shown in Figure 103b. As compared with the heavy cold deformed structure presented in Figure 103a, the blocks in Figure 103b are distinguished by strict boundaries and some misorientation between each other. The inner constitution of the

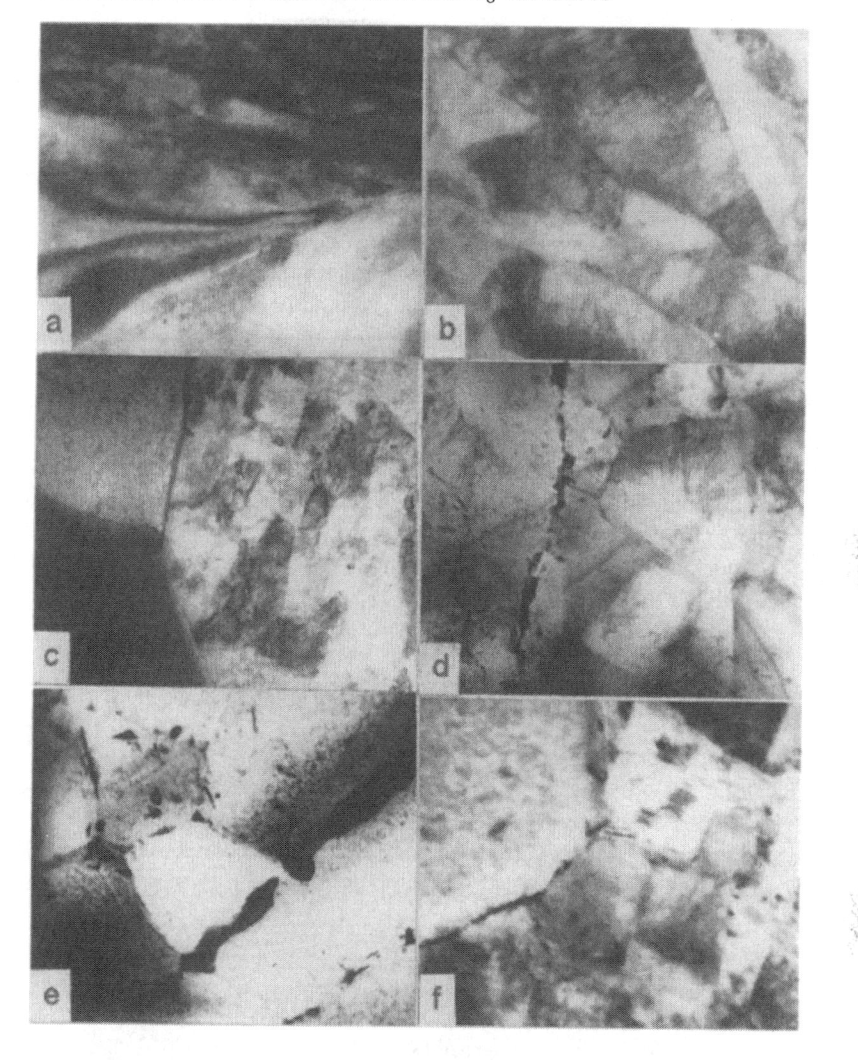

Figure 103. Electron micrographs of the Mg–RE alloys. a – Mg–2.8 mass % Nd, cold deformed 13%, × 12000, b – Mg–1.0 mass % Gd, deformed 40% at 350°C, × 15000, c – Mg–1.0 mass % MM (mishmetal), deformed ~80% at ~350°C, × 3500, d – Mg–12.7 mass % Y, deformed 40% at 450°C, × 3000, e – Mg–3.2 mass % Nd, deformed 40% at 450°C, × 6000, f – Mg–2.8 mass % Sm, deformed 88% at ~420°C, × 14000.

blocks is visibly inhomogeneous suggesting the existence of imperfections including dislocations. As compared with the microstructure presented in Figure 103b, with reducing deformation temperature the block boundaries become more diffuse. By contrast, with increasing deformation temperature the block boundaries become better outlined, and their inner constitution becomes more homogeneous. Figure 103c demonstrates the microstructure of the hot rolled alloy. It consists of recrystallised and deformed grains and may be characterised as "partially recrystallised". The recrystallised grains are in the left part of the micrograph and show inner homogeneous constitution unlike the hot deformed grain in the right part of the micrograph consisting of blocks. The difference in inner constitution between the recrystallised grains and the hot deformed one is evident. The above considered electron micrographs were taken from alloys where the solid solution decomposition can not take place because of low deformation temperature (Figure 103a), low RE concentration as compared with its solubility in solid Mg (Figure 103b) or low RE concentration caused by small solubility of the mishmetal in solid Mg (Figure 103c). The next three electron micrographs illustrate the peculiarity of microstructure in alloys where solid solution decomposition can take place during hot deformation in larger volume. Figures 103d shows blocks in the alloy where the solid solution decomposition proceeded. The precipitates tend to be disposed along the primary grain boundaries and around blocks. The precipitates around the block boundaries are certainly finer than those along the primary grain boundaries. The next electron micrograph in Figure 103e demonstrates the same tendency with more shaped blocks at greater magnification. The last electron micrograph in Figure 103f is taken from the hot extruded alloy with partially recrystallised microstructure. It shows the Sm-rich precipitates along boundary of the recrystallised grain (it is in the left upper corner) and along boundaries of the blocks. Disposition of the RE-rich precipitates along boundaries of the blocks and new recrystallised grains suggest the restraining action of the precipitates on the recovery and recrystallisation processes.

Effect of Deformation Temperature on Mechanical Properties of the Alloys

The effect of hot deformation on the mechanical properties of Mg–Nd and Mg–Y alloys was studied in [332–336]. In general, increase of the hot deformation temperature promoted decrease of the strength properties and increase of plasticity. The change of tensile properties with increasing deformation temperature may be derived from the data obtained on the alloy Mg–3 mass % Nd (Table 31). In this case pieces of hot extruded rods

Table 31. Tensile properties of the Mg–3 mass % Nd rods after various treatments.

Treatment	TS, MPa	TYS, MPa	E, %
Solution treatment from 535°C + ageing at 200°C for 8 hours	235–255	115–155	10–15
Annealing at 535°C for 1–1.5 h, cooling down to extrusion temperature, extrusion			
at 260°C	335–345	275–315	1.5–2.5
at 300°C	330–340	305–330	0.5–2.5
at 350°C	245–285	175–225	8–16
at 420°C	205–225	120–145	10–15

of the alloy were at first annealed at 535°C for 1–1.5 h in order to obtain all Nd in Mg solid solution, then promptly cooled down to temperatures 260, 300, 350 and 420°C and hot extruded additionally at these temperatures into round rods with an area reduction of 80%. The rods obtained were used for tensile tests without ageing because ageing of them was established not to be accompanied by any strengthening. For comparison, Table 31 includes the results of tensile tests carried out on rods of the same alloy obtained by

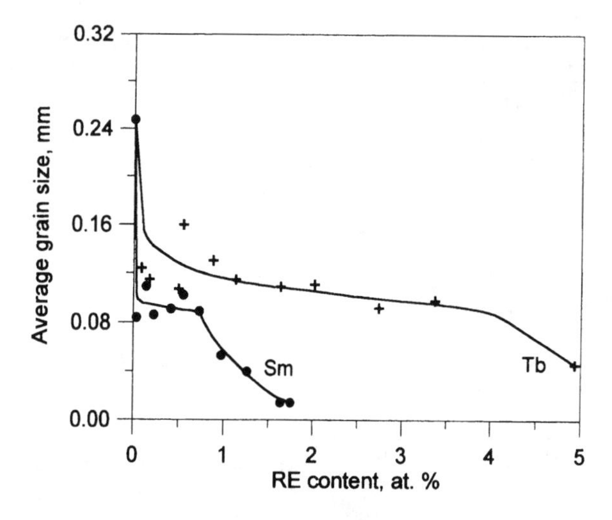

Figure 104. Average grain size of the alloys Mg–Sm and Mg–Tb annealed at 500°C for 5 h after hot extrusion.

common hot extrusion. The latter rods were solution treated by annealing at 535°C followed by quenching in cold water and aged at 200°C for 8 h. The chosen ageing regime conformed approximately to the achievement of the hardness maximum.

The data presented in Table 31 show the highest strength properties at the lowest extrusion temperatures 260–300°C. However, at these deformation temperatures an inadmissibly low plasticity is obtained. With rising extrusion temperature the strength properties successively decrease with increasing elongation. Meanwhile, the difference between elongation values at extrusion temperatures 350 and 420°C is insignificant. The change of mechanical properties with rising extrusion temperature correlates with progress of recrystallisation in the structure of the alloy. The optimum combination of strength and plasticity is achieved at extrusion temperature 350°C. The structure of the alloy at this extrusion temperature may be characterised as "partially recrystallised" and consists of both recrystallised and hot deformed grains of Mg solid solution. The strength properties after hot extrusion at 350°C are significantly higher than those after solution treatment and ageing with actually the same plasticity.

Recrystallisation of Hot Deformed Alloys during Annealing

Additional annealing of the hot deformed alloys results in completion of recrystallisation and grain growth. The grain size formed depends on the RE content and the RE solubility in solid Mg at annealing temperature. In addition to the above considered grain size of Mg–Nd alloys, in Figure 104 two typical dependencies of the average grain size on contents in the binary Mg–RE systems are shown [106]. The opted systems are representatives of those with small or large solubility of the rare earth metal in solid magnesium, Mg–Sm and Mg–Tb, respectively. Alloys of both systems were annealed at the same temperature 500°C for 5 h after hot extrusion with area reduction 90%, and had in this state the fully recrystallised structure.

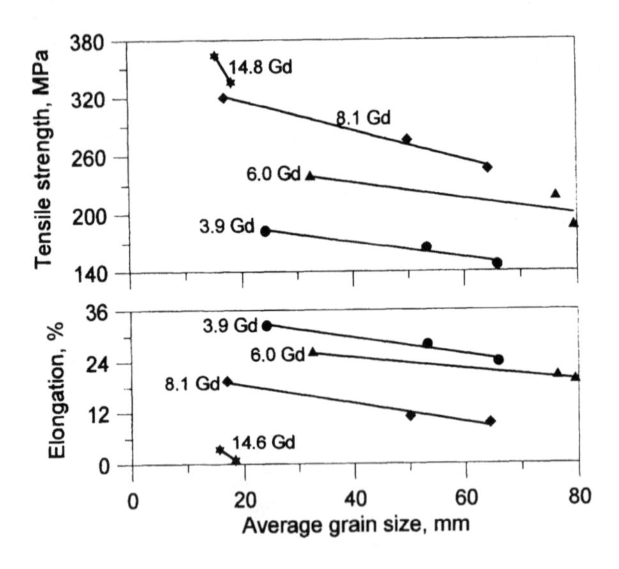

Figure 105. Mechanical properties of the Mg–Gd alloys versus grain size. Gd content is in mass %.

As in the above mentioned Mg–Nd system, the curves in Figure 104 show abrupt decrease of the average grain size at the first small additions of the rare earth metals in both systems. It can be explained by interaction of the rare earth metals with small impurities in the initial Mg resulting in formation of insoluble compounds. The dispersed particles of the compounds must effectively restrain the grain growth during annealing.

Within concentration areas corresponding to solid solution formation, the average grain size continues to decrease with increasing RE content, but more gradually than it takes place at the first small additions of the rare earth metals. The most reasonable explanation of grain size decrease within the limits of Mg solid solution formation can be suggested by assuming preferable disposition of rare earth atoms on the grain boundaries and thus diminution of their surface energy. As a result, the grain growth becomes less energy profitable and decelerates. Such an assumption can be made taking into account the quite different atomic radii of the rare earth metals as compared with the atomic radius of Mg. The supposed factor can act on the grain size also at small contents of rare earth metals as an additional one. Transition to the two-phase area in the phase diagram is accompanied by a steeper decrease of the average grain size. It starts at lower concentrations for alloys with less solubility of the rare earth metal in solid Mg. In this case the particles of the RE-rich phase formed during crystallisation restrain the grain growth during annealing. Their restraining action seems to be more effective than the presence of the rare earth metal in Mg solid solution.

The grain size is of a great importance for mechanical properties of Mg alloys with rare earth metals. With increasing grain size, both strength properties and plasticity decrease. The scale of the change of these characteristics corresponding to grain size changes was established experimentally in Mg–Gd alloys [106]. The alloys were hot extruded into round rods, which were solution treated and then aged at 200°C for 24 h to achieve maximum strengthening. The different average grain size in the alloys was obtained by annealing during solution treatment at different temperatures in the limits of the Mg solid solution area in the phase diagram. The results of the tensile tests of the alloys are presented in Figure 105. They confirm decrease of strength and plasticity with increasing grain size. Meanwhile, the data presented in Figure 105 show evidently clearly the greater effect of Gd content on the properties of the alloys than that of the grain size.

Chapter 4

Effect of Rare Earth Metals on Mechanical and Some Other Properties of Magnesium

The main advantage of magnesium alloys with rare earth metals is their high strength properties at ambient and especially at elevated temperatures. These characteristics of the alloys were, therefore, investigated in most detail. The high strength level of magnesium alloys with rare earth metals conditioned the main applications of them in practice as light structural materials.

Mechanical Properties of the Binary Mg–RE Alloys

Investigation of mechanical properties of the binary Mg–RE alloys made possible the assessment of every separate rare earth metal and the choice of the most favourable of them for development of commercial materials. Almost all binary Mg–RE systems were studied with determination of mechanical properties of the alloys, but in different volume. Most attention was paid to systems with attractive characteristics of the alloys and the less expensive rare earth metals, keeping in mind the possibility to realise commercial production of the alloys in industry.

Binary Magnesium Alloys with Lanthanum, Cerium, Praseodymium and Neodymium

In general, most attention was paid to the alloys with the first four elements of the lanthanum series. Lanthanum, cerium, praseodymium and neodymium are the cheapest of the rare earth metals and are the main constituents of mishmetal which is widely used in practice. The first and most detailed investigations of the binary magnesium alloys with La, Ce, Pr and Nd were carried out by T. Leontis [3, 4]. They included determination of mechanical properties of the cast and hot extruded alloys with or without the performance of additional heat treatment. At that time the rare earth metals used were not pure enough and contained a great deal of other rare earth metals of the same four elements. Only the La and Ce used were strictly individual rare earth metals. Their contents amongst the rare earth metals in the used mixtures were 98.5 and 97.3 mass %, respectively. Meanwhile, the Nd used ("didymium") contained in the mixture of the rare earth metals 80.5% Nd, 9.8% La, 8.8% Pr and 0.9% Ce (in mass %). The Pr used was represented by mixtures of the rare

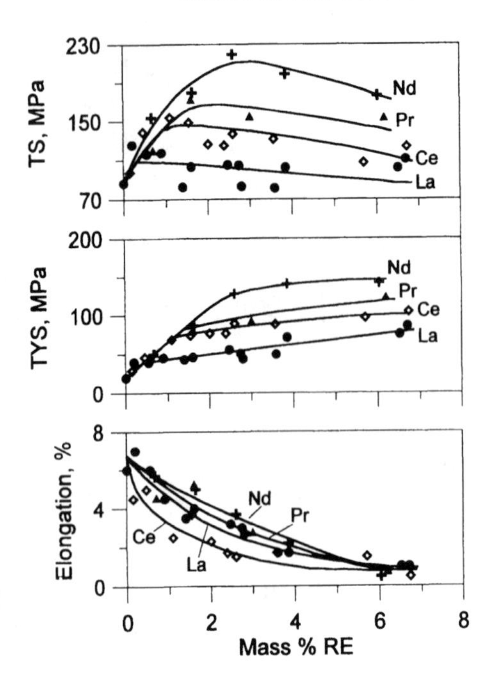

Figure 106. Tensile properties of the cast binary Mg alloys with La, Ce, Pr and Nd at room temperature. T6 condition.

earth metals consisting of 68.9% Pr, 27.2% La, 3.0% Nd and 0.8% Ce (in mass %). The rare earth metals used contained also other impurities, the main one being iron. Along with the above mentioned individual rare earth metals T. Leontis [3, 4] used also the common mishmetal containing in the sum of the rare earth metals 51.7% Ce, 23.1% La, 18.6% Nd, 6.5% Pr (in mass %). Despite the Nd and Pr used, being actually mixtures of rare earth metals enriched by the main component, the general results obtained in [3, 4] became quite convincing and were supported in many other subsequent works.

Heat treatment of cast alloys with La, Ce, Pr and Nd used in [3] consisted of solution treatment with or without subsequent ageing at 204°C for 16 h for maximum strengthening. In the alloys with La, heat treatment resulted in practically no change of strength properties. In the cast alloys with the other rare earth metals, Ce, Pr and Nd, solution treatment with ageing provided the highest strength properties. The data on mechanical properties of the cast binary alloys with La, Ce, Pr and Nd obtained by tensile tests at room temperature are presented in Figure 106. It shows increase of tensile strength (TS, σ_B) and tensile yield strength (TYS, $\sigma_{0.2}$) as all the four rare earth metals are added to Mg. Nevertheless, the strengthening effect from the separate alloying elements is different. The lowest level of strength properties is shown by alloys Mg–La. The strength properties of the other alloys are higher and their levels grow successively for the systems Mg–Ce, Mg–Pr and Mg–Nd. So, there is growth of the level of strength properties of the alloys with increasing atomic number of the rare earth elements and their solubility in solid Mg.

It is worth noting the kinks on the curves characterising dependence of the strength properties on the rare earth content in the alloys. After the kinks the strength properties grow more slowly or fall with increasing RE concentrations. Disposition of the kinks on

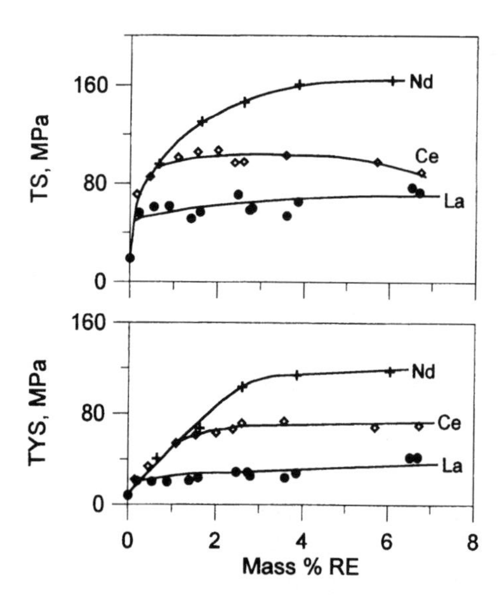

Figure 107. Tensile strength properties of the cast binary Mg alloys with La, Ce and Nd at 260°C. T6 condition.

the curves corresponds approximately to the solubility of the rare earth metals in solid Mg at temperatures of solution treatment and, consequently, close to their maximum solubility. It is remarkable that for TYS the kinks on the curves are manifested especially clearly as compared with those for the TS. Plasticity of all alloys decreases with increasing RE contents, and at concentrations of more than 4 mass % the elongation values (E, δ) become less than 2%. There is no significant difference between plasticity of the alloys of different systems at the same RE concentrations, but the alloy Mg–Ce shows the lowest elongation values.

The main regularities noted for mechanical properties of the alloys with La, Ce, Pr and Nd at room temperature were observed also during tensile tests at elevated temperatures [3]. Some of the results obtained are shown in Figure 107 for testing temperature 260°C. As one can see, in general the rare earth metals used, La, Ce, Nd, increase the strength properties of Mg. The Mg–La alloys demonstrate the lowest level of the TS and TYS values, whereas the Mg–Nd alloys demonstrate the highest level of the TS and TYS values and the Mg–Ce the intermediate level of them. As in the case of tensile tests at room temperature, the curves characterising dependence of the strength properties at 260°C on the RE contents show kinks corresponding to the transition from steep growth to slow increase and even to decrease as it takes place for tensile strength of the Mg–Ce alloys. Disposition of the kinks conforms approximately to the solubility of the rare earth metals in solid Mg at the solution treatment temperatures. The kinks are manifested also more clearly for tensile yield strength than for tensile strength.

The character of the plasticity change at elevated temperatures with increasing RE content was observed by [3] to be mainly the same as at room temperature. However, the absolute values of the plasticity (elongation) increased and ceased to be critical. For example, at testing temperature 260°C elongation for the alloys with small alloying additions was about 20% and for the alloys with lowest plasticity containing more than 6%

RE, it was within the limits 3–12%. At elevated temperatures plasticity of alloys with high content of La became notably more than that of the alloys with high content of other rare earth metals. So, at 204°C elongation was found to be 7.5% for the alloy with 6.70 mass % La and 3.0, 3.0 and 3.2% for alloys with 6.73 mass % Ce, 6.19 mass % Pr and 6.03 mass % Nd, respectively. At 260°C elongation was obtained to be 10.5% for the alloy with 6.70 mass % La and 3.2 and 6.5% for the alloys with 6.73% Ce and 6.19% Nd, respectively [3].

Mechanical properties of the alloys containing about 3 mass % mishmetal (MM) coincided practically with those of the alloy with 3 mass % Ce. However, at higher concentrations, about 6 mass %, mechanical properties of the cast alloys with mishmetal turned out to be rather higher than those of alloys with Ce. So, after solution treatment and ageing, the tensile properties at room temperature were obtained as follows for the alloy with 6.33 mass % MM, TS = 200 MPa, TYS = 167 MPa, E = 1.0%, and for the alloy with 6.73 mass % Ce, TS = 177 MPa, TYS = 149 MPa, E = 0.5%. At 260°C the tensile properties were obtained as follows for the alloy with 6.33 mass % MM, TS = 124 MPa, TYS = 92 MPa, E = 4.2%, and for the alloy with 6.73 mass % Ce, TS = 102 MPa, TYS = 71 MPa, E = 3.0%. The alloys with mishmetal showed also more creep resistance as compared with the alloys with Ce. The stress resulting in elongation of 0.2% during creep at 260°C was determined to be 47 MPa for the alloy with 2.85% MM, and 33 MPa for the alloy with 3.57 mass % Ce [3].

Investigation of the hot extruded alloys of Mg with lanthanum, cerium and neodymium also showed an increase of strength properties with increasing alloying elements at various test temperatures [4]. In this case the strengthening effect from the rare earth metals depended on their atomic number in the same way, in general, as in the cast alloys, but with some deviations. The alloys were tested directly after hot working and after additional heat treatment consisting of either direct ageing at 204°C for 16 h (T5 condition), solution treatment only (T4 condition) or solution treatment + ageing at 204°C for 16 h (T6 condition). Solution treatment was carried out at 510°C for the alloys containing Nd or at 566°C for the alloys with the rest of the rare earth metals. In "as extruded" condition the strength properties at room temperature turned out to be the highest for the alloys with lanthanum. At La content of 5.7% the strength properties were determined to be TS = 295 MPa, TYS = 265 MPa, CYS† = 205 MPa. The higher strength properties of alloys with La also showed after solution treatment only (T4 condition).

Table 32. Mechanical properties of the hot extruded magnesium alloys with lanthanum, cerium and neodymium at room temperature [4].

Treatment	Mg–2.5 mass % La		Mg–2.3 mass % Ce		Mg–2.7 mass % Nd	
	TS, MPa	E, %	TS, MPa	E, %	TS, MPa	E, %
Hot extruded	273	9.2	220	13.2	214	19.4
Hot extruded + ageing at 204°C for 16 h	274	9.8	234	11.4	293	12.6
Hot extruded + solution treatment + ageing at 204°C for 16 h	245	9.0	240	8.8	289	10.6

† CYS is compressive yield strength.

After ageing, alloys with Ce and Nd strengthened more than alloys with La. As a result, the strength properties at room temperature of the Mg–Ce and Mg–Nd alloys approached those of the Mg–La alloys and even surpassed them. The scale of the change of mechanical properties of the hot extruded alloys after heat treatment can be assessed from the data presented in Table 32. These data show that the tensile strength of the alloy with Ce remains lower than that of the alloy with the same La content after heat treatment, despite the higher strengthening effect due to ageing after both hot extrusion and solution treatment. The alloy with Nd in the "as extruded" condition shows lower tensile strength than that of the alloy with La, but after ageing the tensile strength of the alloy with Nd becomes higher than that of the alloy with La due to the large strengthening effect. In the case of ageing directly after extrusion (T5 condition) tensile yield strength of the alloy with Nd was lower than that of the alloy with La, but in the case of ageing after solution treatment (T6 condition) tensile strengths of the alloys with Nd and La were actually the same [4].

Figure 108 presents the mechanical properties of the alloys for one of the conditions of heat treatment, solution treatment + ageing at 204°C for 16 h, extracted from [4]. The left part of Figure 108 illustrates the main features of the changes in properties with increasing the RE content at room temperature. Along with the mentioned growth of the strength properties with increasing the RE content, there is also some growth of plasticity at first additions of rare earth metals. However with further increasing the RE content plasticity decreases. This general tendency can be explained, as was above mentioned, by decrease of the grain size of the extruded alloys when a small quantity of rare earth metals is added, and formation of coarse brittle crystals of RE compounds in the structure when much more rare earth metals are added to Mg. The effect of plasticity growth at small RE additions increases in the order La, Ce, Nd. This fact can be explained by increase in the same order of the refining action of the rare earth metals when they tie impurities in initial Mg. Besides, cerium and neodymium can be dissolved in solid Mg up to higher concentrations and, consequently, form fewer coarse crystals of RE compounds in the structure at the same concentrations than lanthanum. The second factor seems to be responsible for the higher plasticity of the alloys with cerium and neodymium as compared with alloys with lanthanum at high concentrations.

The right side of Figure 108 illustrates the mechanical properties of the extruded alloys at one of the elevated temperatures used, 260°C, for the same T6 condition. As one can see, at elevated temperature regular change of the strength properties of the alloys with increasing the rare earth atomic number is revealed more clearly and is similar to that in the cast alloys. In the order, La, Ce, Nd, tensile strength and tensile yield strength successively increase so that they are the highest for the Mg–Nd alloys and the lowest for the Mg–La alloys with intermediate values for the Mg–Ce alloys. On the concentration curves kinks may be conceived as in the cast alloys, but they are less evident because of the small number of experimental points. Disposition of the kinks correlates with solubility of the rare earth metals in solid Mg at temperature of solution treatment, although quite approximately, also because of the small number of the alloys used. Nevertheless, the general tendency is revealed quite clearly. It consists of more gradual growth of the strength properties after kinks with increasing RE content. In the Mg–La alloys the strength properties grow weak actually without kinks.

Plasticity of the hot extruded alloys at elevated temperatures after solution treatment and ageing turned out to be the least for the system Mg–Nd and the highest for the system

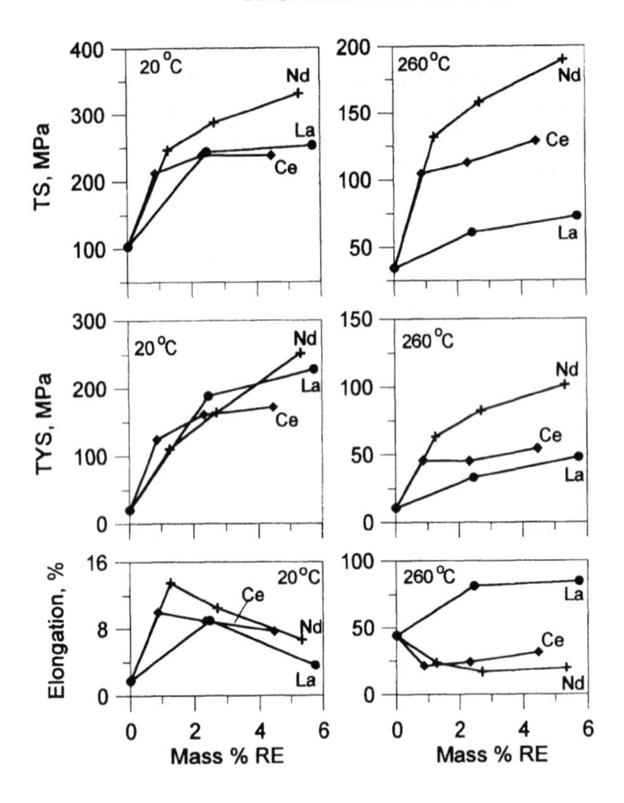

Figure 108. Tensile properties of the extruded Mg–La, Mg–Ce and Mg–Nd alloys at 20°C (left) and 260°C (right). T6 condition.

Mg–La. Incidentally, the Mg–La alloys showed quite a high level of plasticity at elevated temperatures. At 260°C elongation of them achieved values as high as more than 80%.

The main results of the work described on binary magnesium alloys with lanthanum, cerium and neodymium [4] were confirmed, in general, in the work of the Russian researchers [337] where wrought magnesium alloys with lanthanum, cerium and neodymium containing additionally about 2 mass % Mn were studied.

The above works [3, 4] where mechanical properties of magnesium alloys with the four first metals of the lanthanum series, La, Ce, Pr and Nd were studied, confirm the importance of the rare earth metal's ability to form solid solutions in Mg and their decomposition with precipitation of the RE-rich disperse phases. In accordance with increasing the solid solubility in Mg the strength properties tend to grow in order, La, Ce, Pr and Nd. This tendency becomes especially evident at elevated temperatures and for yield strength. The kinks on the curves characterising change of the strength properties with increasing the RE content confirm also the decisive role of the RE solid solubility and connected with it the ability of the alloy for solid solution decomposition. Most growth of the strength properties with increasing rare earth metal addition takes place within the limits of the Mg solid solution formation. When the solubility limits are surpassed, the strength properties can continue to grow, but significantly more slowly. In general, the effect of the RE solubility on mechanical properties of cast and hot extruded alloys is similar, although there are also certain peculiarities in the behaviour of them in each condition.

The role of coarse RE-rich crystals formed in the structure of the alloys when the solubility limits are surpassed is worth considering separately. Their effect is more simple for tensile yield strength and less simple for tensile strength. Tensile yield strength, in general, continues to grow with increasing the coarse crystal quantity, at least in the limits of the studied compositions of the alloys (less than 7 mass %). This tendency is observed in both cast and hot extruded alloys at room and elevated temperatures. Tensile strength grows with increase in the coarse crystal quantity in hot extruded alloys at room and elevated temperatures. In cast alloys the growth of this strength property is insignificant, but it can also fall as the coarse crystal quantity increases. The latter behaviour is revealed more clearly at room temperature, and the former behaviour occurs more often with increase in the test temperature. The different effect of the coarse crystals on the strength properties of cast and hot extruded alloys is understandable. In cast alloys the coarse crystals are bigger and disposed between dendrite branches of Mg solid solution. Meanwhile, hot extrusion breaks up the coarse crystals in the cast structure into ones which are significantly smaller and isolated. The coarse RE-rich crystals promote a decrease of plasticity at room temperature. However, at elevated temperatures, their presence in the structure can promote an increase of plasticity. An example of this can be seen in Figure 108 for the elongation values of Mg–La alloys obtained during tests at 260°C. The Mg–La alloys can be considered as those where the action of the coarse RE-rich crystals in the structure is manifested within a matrix of almost pure magnesium. The more characteristic features of these alloys after hot extrusion are the achievement of high strength properties at room temperature with low plasticity, but achievement of quite low strength properties with high plasticity at elevated temperatures.

The effect of the coarse RE-rich crystals on the creep resistance turned out to be different in alloys of various systems. So in Mg–La alloys coarse crystals resulted in an increase of the creep resistance [4], but in Mg–Ce and Mg–Nd alloys their presence in the structure promoted a decrease of the creep resistance [4, 155]. According to [155], the negative influence of the coarse crystals in alloys with cerium on the creep resistance is conditioned by two factors. One is the decrease of the strengthening effect of disperse particles precipitated from Mg solid solution when the coarse crystals are present, and the other factor is the decrease of the Mg solid solution grain size when the coarse crystals are present.

The coarse crystals of the RE-rich phases in the structure of magnesium alloys can promote strength due to their greater hardness at room and elevated temperatures as compared with a matrix consisting of the decomposed solid solution. This viewpoint is supported by results of microhardness measurements of the phases at room and elevated temperatures presented in Table 33 [338–340]. For comparison, the results of the microhardness measurements of the second phases in other widespread systems of magnesium alloys with Al, Zn, Ca, Th and those in the alloys with Y, Sm, Dy are also shown. Table 33 also contains microhardness values obtained for three solid solutions in alloys Mg–1 mass % Ce, Mg–4 mass % Th and Mg–10 mass % Al. The alloys were solution treated by annealing at temperatures near the eutectic ones and then quenched. They could decompose during tests at elevated temperatures. The microhardness measurements presented were carried out with the same indentation duration of 30 seconds. The data of Table 33 show some softening of the second phases in the Mg–Ce and Mg–Nd alloys with increasing temperature, but always their microhardness remains substantially more than the microhardness of all the solid solutions, even the most heat resistant of them in the alloy Mg–4 mass % Th. Softening of the second phases in Mg–Ce and Mg–Nd alloys with increasing temperature is notably less than that of the second phases in the Mg–Al, Mg–Zn and Mg–Ca alloys. This fact correlates with the lower strength properties

Table 33. Microhardness of phases (HV) in Mg alloys at various temperatures [338–340].*

Phase	Temperature, °C					
	20	150	200	250	300	350
Compounds:						
$Mg_{17}Al_{12}$	183	168	158	125	84	–
MgZn	271	234	193	157	97	–
Mg_2Ca	156	139	131	91	47	–
$Mg_{12}Ce$	168	152	147	134	125	–
$Mg_{12}Nd$	186	169	165	139	104	–
Mg_5Th	256	258	228	222	196	–
$Mg_{24}Y_5$	218	214	201	196	173	132
$Mg_{41}Sm_5$	232	217	188	164	132	698
$Mg_{24}Dy_5$	223	216	178	146	114	90
Solid solutions in:						
Mg – 4% Th	42.6	48.7	37.6	38.8	23.1	–
Mg – 1% Ce	43.6	38.3	30.6	21.8	21.2	–
Mg – 10% Al	55.0	58.4	41.4	37.6	31.2	–

*Comments: In the original article [338] for the compounds in the systems Mg–Ce and Mg–Nd the formulas Mg_9Ce and Mg_xNd_y were written. In the original article [340] for the compound in the Mg–Sm system the formula $Mg_{6.2}Sm$ was written.

of the Mg–Al, Mg–Zn and Mg–Ca alloys at elevated temperatures. On the other hand, microhardness of $Mg_{12}Ce$ and $Mg_{12}Nd$ is less and remains less with rising temperature than that of the phases Mg_5Th, $Mg_{24}Y_5$, $Mg_{41}Sm_5$ and $Mg_{24}Dy_5$.

Binary Alloys Mg–Sm

The mechanical properties of the binary alloys Mg–Sm were studied in [270, 341]. The alloys were prepared by melting in steel crucibles under a flux consisting of chlorides and fluorides of alkaline and alkaline earth metals. The initial materials were magnesium of 99.95 mass % purity and samarium of 99.92 mass % purity. The melts were cast into round ingots which were then homogenised at 500°C for 12 h. After homogenisation the ingots were hot extruded into round rods of 10.5 mm in diameter with area reduction of about 90%. The rods prepared were tested in two conditions, T5 and T6. T5 condition corresponded to ageing at 200°C for 12 h immediately after hot extrusion (with cooling in air), and T6 condition corresponded to the same ageing at 200°C for 12 h after preliminary solution treatment consisting of annealing at 500°C for 2 h followed by quenching in cold water. As the special experiments revealed, ageing immediately after hot extrusion is accompanied by a significant strengthening effect showing that high Sm contents remain in Mg solid solution. The opted ageing regime was chosen to result in maximum hardness.

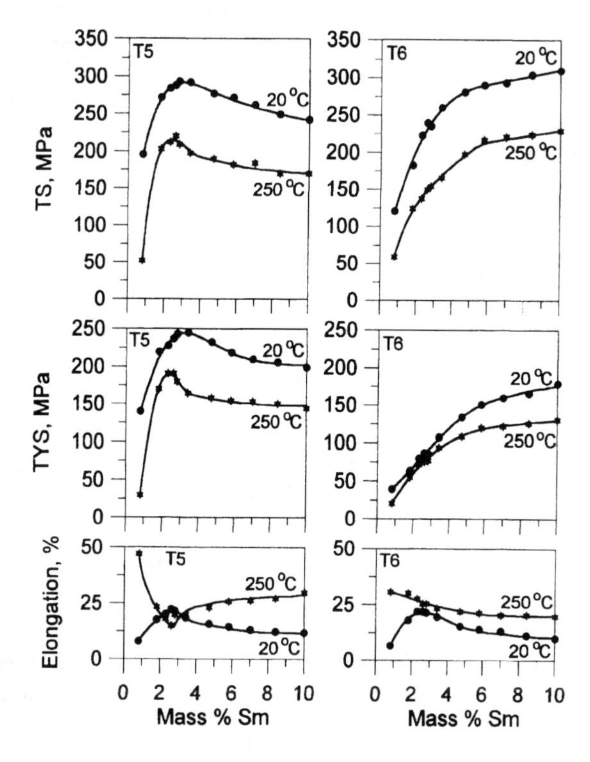

Figure 109. Mechanical properties of the extruded Mg–Sm alloys at room temperature and 250°C in T5 (left) and T6 (right) conditions.

The results of tensile tests of the alloys are shown in Figure 109, with the left part for T5 condition and the right part for T6 condition. As the data presented show, the addition of samarium promotes increase of the strength properties of the alloys both at room temperature and at 250°C for both cases of heat treatment, T5 and T6. Nevertheless, the character of the change in strength properties with increasing Sm content is different for T5 and T6 conditions. For T5 condition (ageing immediately after extrusion) the strength properties increase more steeply with the first Sm additions and reach a maximum at 2–4 mass %. These contents correspond approximately to the solubility of Sm in solid magnesium at the temperature of hot extrusion. After the maximum the strength properties of the alloys in T5 conditions decrease gradually with further increasing Sm content. Such a character of the change of the strength properties with increasing Sm content takes place at both testing temperatures, 20 and 250°C. For T6 condition (solution treatment + ageing) increase of the strength properties of the alloys with increasing Sm content is less at first addition, but continues up to the most used Sm addition of 10 mass %. However, after Sm content corresponding approximately to the solubility in solid magnesium at the solution treatment temperature (~5 mass %), the strength properties change markedly less with increasing alloying element than before. So, the solid solution existence with its ability to decompose during ageing is confirmed as the most important thing for the strengthening effect from Sm addition. In this aspect the Mg–Sm alloys in the T6 condition are similar to the Mg–Ce, Mg–Pr and Mg–Nd alloys considered above and follow the general tendency of the change of properties with increasing the RE content.

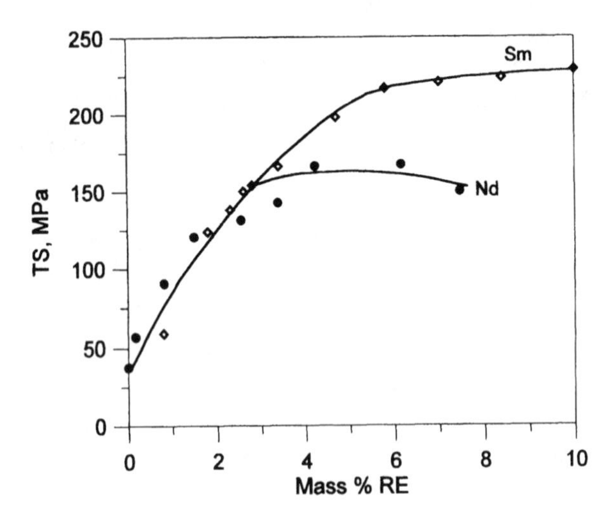

Figure 110. Tensile strength at 250°C of the extruded Mg–Nd and Mg–Sm alloys. T6 condition.

As the solubility of Sm in solid magnesium is higher than that of Nd, it is reasonable to expect higher strength properties of Mg–Sm alloys after the solubility content is reached, if the alloys of both systems are compared. Such a comparison is shown in Figure 110 and confirms this fact [342]. The alloys of the systems Mg–Sm and Mg–Nd were obtained and tested in the similar conditions. They were used as hot extruded round rods which had undergone solution treatment and then ageing to maximum hardening (T6 condition). As Figure 110 shows, the tensile strength values at 250°C for alloys with Sm and Nd coincide actually within the contents corresponding to the possibility of solid solution formation simultaneously in both systems at the solution treatment temperature, but turn out to be different within the contents behind the solid solution formation. Within the contents areas behind the solid solution formation tensile strength values of the Mg–Sm alloys are evidently higher than those of the Mg–Nd alloys as a result of the higher solubility of Sm in Mg solid solution.

According to the microstructure investigations carried out in [270] the specific dependence of the strength properties of the Mg–Sm alloys on Sm content for the T5 condition can be explained by different recrystallisation of the alloys during hot extrusion. At low Sm the grains of magnesium solid solution become finer, and the structure turns out to be only partially recrystallised with increasing Sm content. This factor promotes steeper increase of the strength properties superimposed on simultaneous increase of solid solution concentration. The maximum strength properties in T5 condition are reached at about 3 mass % Sm when in the structure of the alloy, coarse crystals of Sm-rich phase appear in accordance with the lower solubility of Sm in solid magnesium at lower temperature of hot extrusion (420–450°C) as compared with the temperature of solution treatment (500°C). Appearance of the coarse crystals of Sm-rich phase is accompanied by formation of the completely recrystallysed structure and grain growth. As a result the strength properties of the Mg–Sm alloys decrease with Sm content more than about 3 mass % in T5 condition. The above mentioned change of the recrystallisation process with increasing alloying element in the alloys of the systems with limited solid solutions was described in [343].

The plasticity of the Mg–Sm alloys at room temperature both in T5 and T6 condition changes with increasing Sm content following the elongation curves with a maximum of about 3 mass %. Increase of elongation at small Sm content can be explained by reduction of the grain size, and the decrease of elongation after its maximum can be explained by more precipitates being formed after ageing and appearance of Sm-rich coarse crystals in the structure. In general, plasticity of the alloys in both conditions turns out to be high enough at room temperature with elongation values being more than 10% at 10 mass % Sm. Plasticity at 250°C changes with increasing Sm content in T5 and T6 conditions differently. For T5 condition it follows the elongation curve with a minimum, and for T6 condition it decreases successively with increasing Sm content. In both conditions, plasticity of the Mg–Sm alloys at 250°C is high enough not to be critical for their application as structural materials. The different character of the elongation curves for 250°C as compared with those for room temperature is connected with different behaviour of magnesium alloys during plastic deformation at room and elevated temperatures.

Comparison of the strength properties of the alloys in T5 and T6 conditions leads to the conclusion that the former condition is preferable for structural materials. The maximum values of tensile strength (TS) in T5 and T6 are quite close, but the maximum values of tensile yield strength (TYS) in T5 are significantly higher than those in T6. Besides, the maximum values of TS and TYS are achieved in T5 condition at lower contents of the quite expensive samarium. This fact also makes usage of the Mg–Sm alloys in T5 condition preferable to the T6 condition.

Binary Alloys Mg–Y

Investigations of the binary Mg–Y alloys created grounds for development of a number of commercial alloys with high strength properties. The investigations were performed using rods produced by hot extrusion and sheet produced by hot rolling with and without various additional treatment [8, 256, 305, 344, 345]. In the first investigation [8] the strength properties in T5 condition (ageing after hot rolling without special solution treatment) were established to be significantly higher than the strength properties in T6 condition (ageing after special solution treatment at 524°C). So, the alloy with 8.2 mass % Y showed at room temperature in T5 condition TS = 351 MPa, as compared with TS = 274 MPa in T6 condition. The alloy with 9.0 mass % Y showed in T5 condition TS = 373 MPa, as compared with TS = 327 MPa in T6 condition. In [256] lower strength properties of the Mg–Y alloys were reported for T6 condition as compared with the "as extruded" one. The representative results of the tensile tests of the Mg–Y alloys are shown in Table 34 [256]. In experiments [256] additional ageing of the hot extruded alloys at 200°C for 36 h resulted

Table 34. Mechanical properties of the Mg–Y alloys.

Mass % Y	Testing temperature, °C	As extruded			Solut. treat. + aged at 200°C/36 h		
		TS, MPa	TYS, MPa	E, %	TS, MPa	TYS, MPa	E, %
6.14	20	272	212	8.4	229	176	9.8
	250	187	125	20.3	151	80	27.5
12.12	20	356	290	9.0	293	229	15.4
	250	284	229	15.4	284	227	13.6

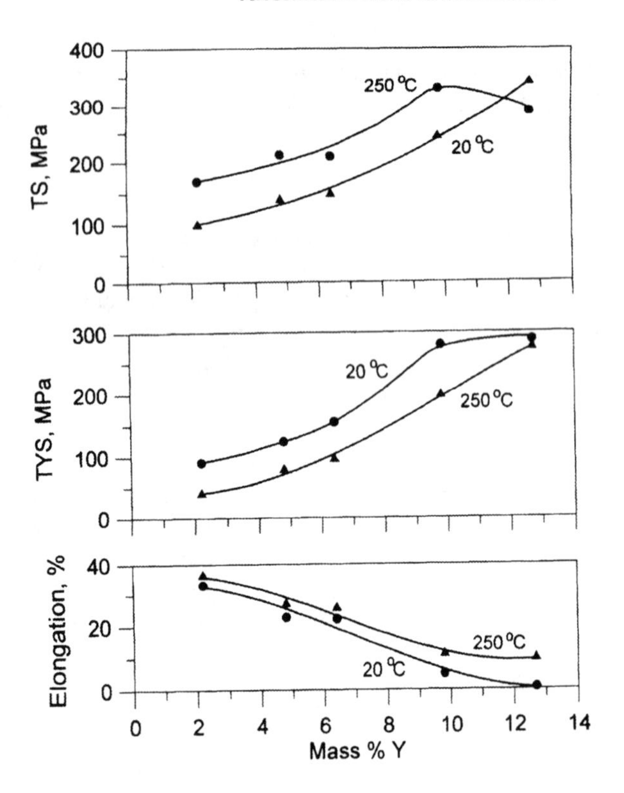

Figure 111. Tensile properties of the extruded Mg–Y alloys at 20°C and 250°C. T6 condition.

in increase of strength properties at room temperature and at 250°C by about 50 MPa. At 300 and 350°C strength properties of the alloys in the "as extruded" condition and after additional ageing were close together.

Tensile properties of the Mg–Y alloys in T6 conditions are presented in Figure 111. They were determined in the experiments described in [106, 345]. Although the strength properties in T6 condition are lower than those in T5 condition, they reveal more certainly the influence of the Y contents in Mg solid solution, solid solution decomposition and the presence in the structure of coarse crystals of Y-rich phase formed during crystallisation. The alloys were solution treated after hot extrusion by annealing at different temperatures, 400°C for the alloys with 2.2 and 4.8 mass % Y, 425°C for 6.4 mass % Y, 500°C for 9.8 mass % Y and 550°C for 12.7 mass % Y, and then quenched in cold water. The different temperatures for solution treatment were chosen in order to obtain in all alloys a structure with similar grain size of about 1500 grains per 1 square mm on the sample plane. All alloys, except that with 12.7 mass % Y, were quenched from the area of the homogeneous solid solution in the phase diagram. The alloy with 12.7 mass % Y was quenched from the two-phase area, and in its structure the coarse crystals of the phase $Mg_{24}Y_5$ remained. After solution treatment the alloys were aged at 200°C for 100 h for maximum strengthening.

Figure 111 shows continuous increase of strength properties, TS and TYS, of the alloys at room temperature with increasing Y content up to 9.8 mass %, and simultaneous decrease of elongation E down to 3.5%. Such a change of tensile properties results from

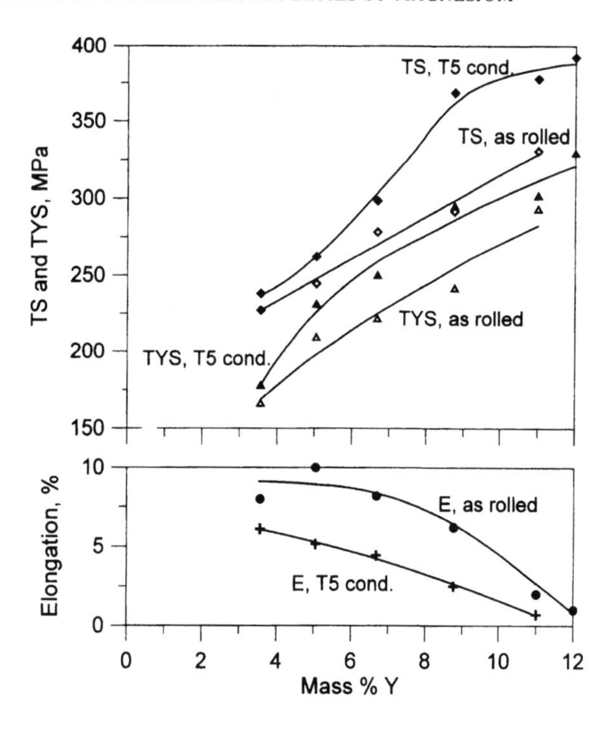

Figure 112. Tensile properties of the rolled Mg–Y alloys at room temperature.

increase of Y content in Mg solid solution only. The remarkable feature of the strength property curves in these parts is the weak kinks on them near 5 mass % Y. After this Y content the solid solution decomposition becomes significant, and this results in a steeper increase of the strength properties with increasing Y. Further increase of the Y content up to 12.7 mass % demonstrates the effect of the appearance of coarse $Mg_{24}Y_5$ crystals in the structure. It is accompanied by decrease of elongation down to 0.2%. At such low elongation value TYS remains actually the same as compared with the previous alloy with 9.8 mass % Y, and TS even decreases.

At 250°C the strength properties (Figure 111) increase continuously with increasing Y content within the whole studied composition range with simultaneous decrease of plasticity (elongation). The rise of TS and TYS continues also when in the structure of the alloys coarse $Mg_{24}Y_5$ crystals appear between 9.8 and 12.7 mass % Y. For the two-phase alloy with 12.7 mass % Y the value of TYS at 250°C turns out to be near that at room temperature, and the value of TS at 250 ° is even higher than that at room temperature. Elongation at 250°C is higher than that at room temperature for all studied alloys.

Figure 112 shows the tensile properties of Mg–Y sheets with various Y content [344]. The sheets of 2 mm in thickness were produced by hot rolling flat ingots having a thickness of 23 mm. Heating temperature of ingots before rolling was 490°C. The sheets were tensile tested at room temperature both in "as rolled" condition and after additional ageing (in T5 condition). The data presented in Figure 112 demonstrate significant increase of strength properties TS and TYS and decrease of elongation E after ageing. The strength properties increased and elongation becomes lower continuously with increasing Y content in the alloys up to 12 mass %, at which TS reached a value of about 400 MPa and E fell to less than 2%. The strengthening effect caused by ageing (difference between TS and TYS values, respectively, before and after ageing) grows with increasing Y content. For sheets of Mg–Y alloys, insignificant anisotropy of strength properties was observed. In the transverse direction they were commonly less only by 3–5% than in the longitudinal one. However the difference between elongation values of the sheets in transverse and longitudinal direction was 20–40%. In general, the dependence of tensile properties on Y content for sheets is similar to that observed for extruded rods. The only exception is the behaviour at the highest Y content of 12 mass % corresponding to appearance of the coarse $Mg_{24}Y_5$ crystals in structure. The strength properties of the sheets continue to increase at this high content whereas in the case of the extruded rods considered, they either did not change (TYS) or decreased (TS). The possible reason for this discrepancy may be that the tests of the sheets and rods were done in different conditions. The sheets were tested without high temperature annealing such that the finer grain size and the finer $Mg_{24}Y_5$ crystals remained in the structure.

In [8] close values of tensile yield strength (TYS) and compressive yield strength (CYS) in sheets of Mg–Y alloys were observed. In the case of tests along the longitudinal direction the sheets of the alloy with 8.2 mass % Y hot rolled and aged at 200°C showed TYS = 357 MPa, CYS = 358 MPa, and the sheets of the alloy with 10.6 mass % Y in the same condition showed TYS = 379 MPa, CYS = 310 MPa. Such behaviour of the Mg–Y alloys during tests of mechanical properties was unexpected. In most known magnesium alloys in rolled condition, CYS is commonly within the limits 0.5 to 0.67 of TYS. According to the investigation [317], addition of Y to Mg results in suppression of twinning along the lattice planes {102}. In this investigation the twins of this type were found in great numbers in deformed pure Mg, but occurred quite rarely in the deformed alloy Mg–9 mass % Y. Ability of the {102} twin formation is commonly connected with significantly different TYS and CYS in Mg and most known Mg alloys, because the twins along the planes {102} form when compression is directed parallel to the basis planes, but do not form when tension is parallel to them. As far as significantly different TYS and CYS are connected with ability of Mg lattice to be deformed by twinning along the planes {102}, the closeness of TYS and CYS in the sheets of Mg–Y alloys was explained by suppression of this deformation mechanism in them [317]. Taking into consideration the above mentioned constitution of the Mg–Y solid solution, it is reasonable to suppose that suppression of the {102} twin formation in it is caused by the existence of short ordering or metastable phases formed in the lattice during solid solution decomposition.

Binary Magnesium Alloys with Gd, Tb, Dy, Ho and Er

The mechanical properties of these binary alloys show many similar features, although some differences between them exist, as well. Also, there is the expected change of the mechanical properties with increasing atomic number of the rare earth metal. There is also a certain similarity between the binary alloys with Gd, Tb, Dy, Ho, Er and the Mg–Y alloys. Mechanical properties of alloys with Gd, Tb, Dy, Ho and Er were mainly studied using round rods obtained by hot extrusion of the ingots with an area reduction of about 90% [104, 106, 260, 280, 342, 346, 347]. The rods were tested at room and elevated temperatures "as extruded", in T5 and T6 conditions. Apart from the alloys with erbium, the ageing regimes corresponded to maximum strengthening during solid solution

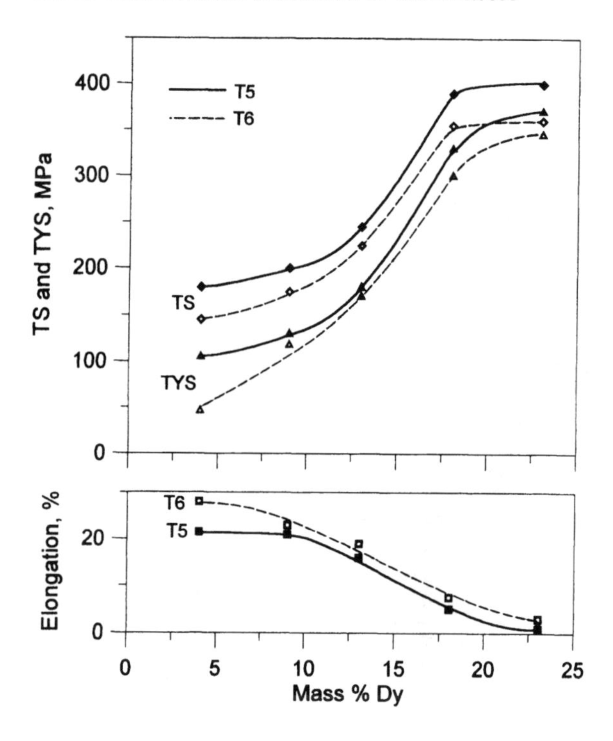

Figure 113. Tensile properties at room temperature of the Mg–Dy alloys in T5 and T6 conditions.

decomposition established previously. For alloys with erbium the ageing regimes were chosen arbitrarily based mainly on the results of investigations into the Mg–Y alloys and, as it was proved afterwards, did not correspond to maximum strengthening, which requires about 400 h and more at 200°C according to investigations described in Chapter 2. Meanwhile, the ageing regimes used for the Mg–Er alloys were 200°C/50–200 h and 300°C/100 h [106, 347].

Except for the Mg–Er alloys, ageing was established to achieve significant increase of strength and decrease of plasticity. Thus, the alloy Mg–18.8 mass % Gd showed in T5 condition (ageing at 200°C for 24 h) at room temperature TS = 413 MPa, TYS = 393 MPa, E = 2.0%, as compared with TS = 343 MPa, TYS = 304 MPa, E = 2.0% in the "as extruded" condition. The alloy Mg–20 mass % Tb showed at room temperature tensile properties in T5 condition (ageing at 200°C for 50 h) TS = 468 MPa, TYS = 383 MPa, E = 6.4%, as compared with TS = 343 MPa, TYS = 259 MPa, E = 15.2% in the "as extruded" condition. The alloy Mg–18.6 mass % Dy showed at room temperature in T5 condition (ageing 200°C for 50 h) TS = 354 MPa, TYS = 308 MPa, E = 6.0%, as compared with TS = 322 MPa, TYS = 282 MPa, E = 10.5% in "as extruded" condition. The alloy Mg–28.3 mass % Ho showed at room temperature in T5 condition (ageing at 200°C for 50 h) TS = 374 MPa, 318 MPa, E = 1%, as compared with TS = 314 MPa, TYS = 292 MPa, E = 4% in the "as extruded" condition [106].

The strength properties in T5 condition were higher than those in T6 condition for all the systems considered where the strengthening effect from ageing was significant. This

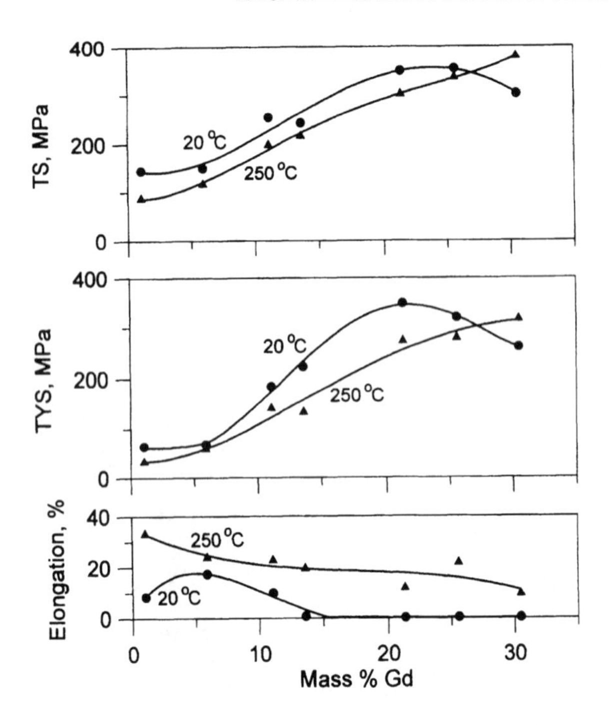

Figure 114. Tensile properties of the Mg–Gd alloys at 20°C and 250°C. T6 condition.

peculiarity is illustrated by the tensile properties of the Mg–Dy alloys [346] in Figure 113. As can be seen in Figure 113, the character of the tensile strength variation with increasing RE content is similar in general for T5 and T6 conditions, but for all compositions tensile strength (TS) and tensile yield strength (TYS) of the alloys remain higher in T5 condition. The evident cause of this is the smaller grain size and the remains of the nonrecrystalised structure in T5 condition as compared with T6. Another remarkable feature of the strength property curves is the clear kinks at 10–15 mass % Dy. Beginning with such Dy contents the strengthening effect resulting from ageing is expected to become tangible according to the Mg–Dy phase diagram. On the other hand, at the highest Dy near its solubility limit in solid Mg, the growth of the strength properties with increasing Dy content becomes slower (TYS) or even ceases (TS). Such a character of the strength properties curves is similar to that noted above for Mg–Y alloys.

Another similarity with the Mg–Y alloys is demonstrated in Figure 114 where tensile properties of the Mg–Gd alloys at room and an elevated temperature are compared. The alloys were tested in T6 condition (quenching from 520°C for 2 h + ageing at 200°C for 24 h). At room temperature the growth of TS and TYS continues up to a Gd content corresponding approximately to the solubility limit at the temperature of solution treatment. After this the strength properties begin to decrease with increasing Gd. By contrast, at 250°C the strength properties continue to grow with increasing Gd content in the two-phase area, so that in the alloy with 30.5 mass % Gd TS and TYS at 250°C they become more than those at room temperature.

Plasticity of the alloys with Gd, Tb, Dy, Ho and Er tended to decrease with increasing RE content and at room temperature it became quite low at high RE with elongation values

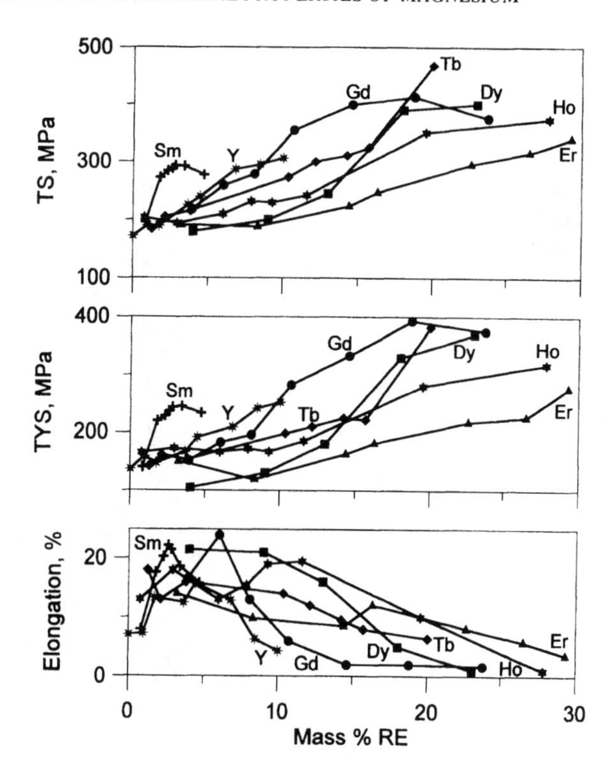

Figure 115. Comparison of the tensile properties of the Mg–RE alloys at room temperature. T5 condition.

being lower than 2%. The above mentioned decrease of the strength properties of the alloys at high RE contents seems to be connected with such low plasticity. At elevated temperatures, plasticity of the alloys increased significantly, and this factor promoted increase of the strength properties even when coarse brittle crystals of RE-rich phases appeared in structure.

It is reasonable to compare tensile properties of the alloys with the relevant rare earth metals. Such a comparison is demonstrated in Figure 115 for tests at room temperature and in Table 35 for tests at room and elevated temperatures. Tensile properties in Figure 115 are presented for all systems in T5 condition providing the highest level of strength for extruded alloys. The data of the alloys with Gd, Tb, Dy, Ho and Er in it are added by the relevant data of the Mg–Y and Mg–Sm alloys in T5 condition [304, 341] prepared and treated in exactly the same way.[†] The most remarkable peculiarity of the data presented is the tendency of the strength property curves to be shifted in the direction of greater contents of alloying elements with increasing atomic number of the rare earth metals in order from Sm to Gd, Tb, Dy, Ho and Er. The order corresponds to the respective increase of solubility of the rare earth metals in solid magnesium, and, in accordance with this, the position of

[†] The used ageing regimes were: 200°C/12 h for Mg–Sm, 200°C/100 h for Mg–Y, 200°C/24 h for Mg–Gd, 200°C/50 h for Mg–Tb, 200°C/50 h for Mg–Dy, 200°C/50 h for Mg–Ho and 300°C/100 h for Mg–Er.

Table 35. Mechanical properties of the binary extruded Mg alloys with various rare earth metals [106, 270, 346].

Mass % RE	Condition	Test temperature, °C	TS, MPa	TYS, MPa	Elongation, %
3.0 Nd	Sol. treat.+ageing 200°C/24 h	20	274	127	10.4
		250	173	116	15.4
		300	114	94	19.8
3.4 Sm	Ageing 200°C/12 h	20	292	245	18.6
		250	198	108	19.8
5.8 Sm	Sol. treat.+ageing 200°C/12 h	20	290	151	14
		250	217	120	21.2
18.8 Gd	Ageing 200°C/24 h	20	413	393	2.0
18.8 Gd	Sol. treat.+ageing 200°C/24 h	20	343	304	2.0
25.6 Gd	Sol. treat.+ageing 200°C/24 h	20	353	320	0.4
		250	334	213	21.5
		300	283	227	18.6
20.0 Tb	as extruded	20	343	259	15.2
		300	277	243	25.6
20.0 Tb	Ageing 200°C/50 h	20	468	383	6.4
18.6 Dy	as extruded	20	322	282	10.5
		300	150	114	42.0
18 Dy	Ageing 200°C/50 h	20	390	370	6
		250	300	330	9.6
23 Dy	Ageing 200°C/50 h	20	400	370	1.0
		250	380	340	3.0
18.8 Ho	As extruded	20	297	226	13.8
		300	123	110	37.6
28.3 Ho	As extruded	20	314	292	4.0
		300	128	105	49.0
28.3 Ho	Ageing 200°C/50 h	20	374	314	1
22.6 Er	Ageing 300°C/100 h	20	295	218	8.1
		250	212	170	35.5
29.3 Er	As extruded	20	351	281	6.2
29.3 Er	Ageing 300°C/100 h	20	341	277	3.6
32.1 Er	As extruded	20	348	283	1.0
32.1 Er	Ageing 300°C/100 h	250	249	185	26.5
9.8 Y	Sol. treat.+ageing 200°C/100 h	20	330	284	3.5
		250	245	198	12.5
		300	154	147	44.2
12.7 Y	Sol. treat.+ageing 200°C/100 h	20	292	292*	0.1
		250	341	281	11.2
		300	192	161	28

*For deformation corresponding to failure (less than 0.2%).

the alloys Mg–Y correlates with the intermediate solubility of yttrium between those of samarium and gadolinium.

Another important point concerns the maximum values of TS and TYS. The highest of them both at room temperature and elevated temperatures were observed in the Mg–Tb alloys, although in the Mg–Gd and Mg–Dy alloys almost as high values were achieved. In these alloys the TS values of more than 400 MPa at room temperature and more than 250 MPa at 300°C were achieved. In the range from Sm to Y and then to Gd, Tb and Dy there is, in general, a successive increase of the maximum values of the strength properties with increasing solubility of the rare earth metals in solid magnesium. However, in the systems with Ho and Er staying after Dy in the lanthanum series and having more solubility in solid Mg, the maximum strength properties tend to be certainly less than those in systems with Gd, Tb and Dy. For the Mg–Er system the lower maximum strength properties can be explained by the actual absence of strengthening during ageing at the ageing regime used. Perhaps, the ageing regime used was responsible for the low strength properties of the Mg–Ho alloys, too, but only partially, because a strengthening effect after ageing of these alloys at 200°C for 50 h was observed. Anyway, the highest strength properties were achieved in the systems Mg–Gd, Mg–Tb, Mg–Dy, and with further increase of atomic number of the rare earth metals decrease of the strength properties was observed. The high strength properties of the alloys Mg–20 mass % Gd and Mg–20 mass % Tb were established also in [277].

The highest strength properties in the alloys were achieved at quite low plasticity. The respective curves of the elongation tend also to shift to more RE content with increasing solubility of the rare earth metals in solid Mg, but this effect is revealed to a less extent and mainly at high RE contents.

Although the alloys with Gd, Tb and Dy show quite high strength properties, they are achieved only at high RE contents, about 20 mass % and more. This fact poses two questions. One of them is how much is the increase of density of Mg alloys containing so much heavy rare earth metals. Calculations show that at the heavy RE content of 5 mass %, the density of Mg alloys becomes about 1.80 g/cm^3 as compared with 1.74 g/cm^3 for pure Mg. At 10 mass % heavy rare earth metals the calculated density of Mg alloys becomes 1.90 g/cm^3, and at 20 mass % heavy rare earth metals it becomes 2.10 g/cm^3 [106]. So, Mg alloys containing 20 mass % RE remain significantly lighter than Al alloys used in industry. The second question concerns the high cost of the heavy rare earth metals. Increase of their contents as compared with lanthanum, cerium, neodymium and yttrium makes their cost much higher. Therefore, alloys with high contents of heavy rare earth metals are expected to be attractive only in areas where the weight saving or, may be, some other specific characteristics of these materials will be especially important.

Binary Magnesium Alloys with Europium and Ytterbium

Mechanical properties of the binary alloys Mg–Eu and Mg–Yb were studied using hot extruded round rods of about 11 mm in diameter produced in a similar manner to the majority of the binary alloys described above (with temperature of ingots of about 400°C and reduction area of about 90%) [98, 106]. The results of tensile tests of the rods without additional heat treatment are shown in Table 36.

As one can see in Table 36, in general, the strength properties at room temperature increase and elongation decreases with increasing Eu content. At 6.7–8.2 mass % Eu the strength properties were determined to be TS = 283–297 MPa, TYS = 234–276 MPa at quite low elongation E = 1.2–2.1%. At 300°C the Mg–Eu alloys showed very low strength properties and very high elongation. In this respect the Mg–Eu alloys are similar to the Mg–La ones, and the reason for this is the very low solubility of both rare earth metals in solid magnesium. Strengthening by addition of Eu at room temperature is connected solely

Table 36. Tensile properties of the hot extruded Mg–Eu and Mg–Yb alloys.

RE, mass %	Room temperature			300°C		
	TS, MPa	TYS, MPa	E, %	TS, MPa	TYS, MPa	E, %
0	223	–	6.9	33	21	36.0
0.9 Eu	220	170	6.2	43	31	42.6
2.0 Eu	261	235	11.7	39	27	74.1
5.9 Eu	302	257	4.4	48	31	89.2
6.7 Eu	294	256	1.9	44	31	92.0
7.2 Eu	297	234	1.2	51	36	79.8
8.2 Eu	283	276	2.1	45	33	102.6
0.2 Yb	194	124	14.1	39	26	50.2
1.2 Yb	231	196	11.5	77	56	35.4
3.2 Yb	249	221	10.9	69	55	38.2
5.8 Yb	215	180	14.5	56	47	10.6
6.2 Yb	226	179	14.3	63	53	62.2
11.3 Yb	216	182	11.2	51	40	79.6
12.9 Yb	257	234	9.4	38	26	67.0

with appearance of coarse Eu-rich crystals in the structure of the alloys during their crystallisation. As Eu is not soluble practically in solid magnesium, heat treatment including ageing did not apply to the Mg–Eu alloys. The alloys were, however, annealed at 400°C for 1 h. Such an annealing promoted some increase of elongation and decrease of the strength properties. The above mentioned alloys with 6.7–8.2 mass % Eu showed after annealing at 400°C for 1 h tensile properties at room temperature as follows, TS = 175–204 MPa, TYS = 166–196 MPa at E = 5–8% [106].

In magnesium alloys with ytterbium the strength properties at room temperature in the "as extruded" condition were somewhat lower and plasticity was somewhat higher than those of the Mg–Eu alloys. By contrast, at 300°C the strength properties of the alloys with ytterbium seemed to be higher than the strength properties of alloys with europium and this fact can be explained by existence of tangible solubility of ytterbium in solid magnesium unlike europium whose solubility in solid magnesium is practically nil. Ageing of the hot extruded Mg–Yb alloys resulted in some increase of tensile strength properties and decrease of elongation. After ageing at 200°C for 8 h corresponding to the hardness maximum the alloys with 5.8–11.3 mass % Yb showed at room temperature TS = 231–242 MPa, TYS = 192–216 MPa, E = 6.0–9.5% [98, 106].

In general, both the Mg–Eu and Mg–Yb alloys show good low strength properties, especially if they are compared with the alloys with neodymium, samarium, yttrium and other rare earth metals of the yttrium subgroup.

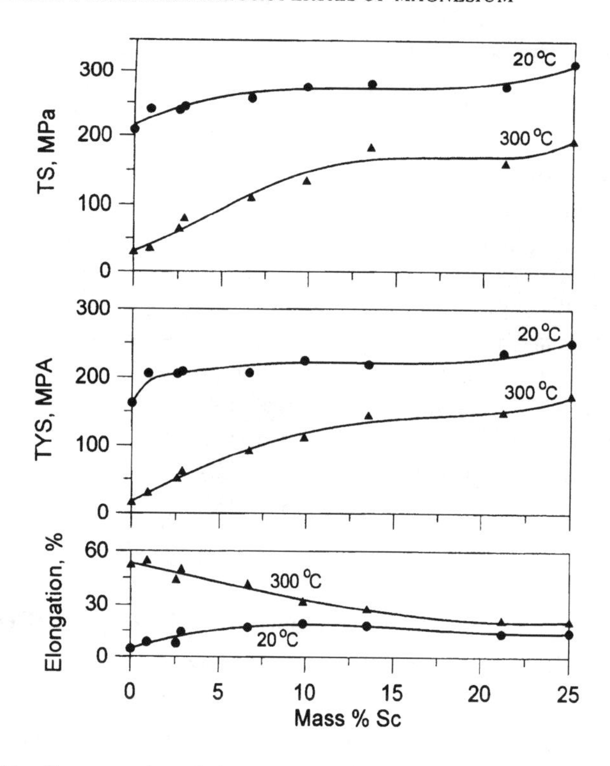

Figure 116. Tensile properties of the extruded Mg–Sc alloys at room temperature and 300°C.

The Alloys Mg–Sc

The alloys Mg–Sc are interesting because of the significant solubility of Sc in solid Mg. Besides, Sc has a low density (2.989 g/cm^3) unlike other rare earth metals, such that even with large contents in Mg alloys the density increase is not substantial. Tensile properties of the binary Mg–Sc alloys were determined using round rods of about 11 mm in diameter produced by hot extrusion of ingots at 440–460°C with reduction of area of about 90% [298]. Results of the tensile tests of them are shown in Figure 116.

At room temperature the tests show a steep increase of the strength properties with the first small additions of scandium (up to about 3 mass %). With further increase of the Sc content the increase of the strength properties at room temperature continues, but it becomes sluggish. Eventually at the highest Sc contents near 25 mass %, the strength properties begin to rise steeply again. The latter Sc concentration corresponds to about its solubility in solid Mg and can be considered to result in the appearance in the structure of the Sc-rich phase. Plasticity of the Mg–Sc alloys at room temperature changes with increasing Sc concentration following the curve with a maximum typical of binary alloys with other rare earth metals. However, unlike alloys with other rare earth metals, elongation remains high (more than 10%) at high concentrations as a result of the quite extended solid solution area in the Mg–Sc system.

At elevated temperature (300°C) the increase in strength properties is more significant and continues to be steep in the higher concentration range, up to about 15 mass % Sc. At the highest Sc concentration the strength properties begin to raise steeply again as at room temperature. Plasticity of the Mg–Sc alloys at 300°C decreases successive with increasing Sc content. The high level of the strength properties of the Mg–Sc alloys at elevated temperatures is noteworthy, especially at Sc contents of more than 20%. The tests of such alloys showed at 250°C TS ≈ 250 MPa, TYS ≈ 200 MPa, at 300°C TS ≈ 160–190 MPa, TYS ≈ 150–170 MPa, at 350 ≈ 60–90 MPa, TYS ≈ 50–80 MPa [106, 298].

Increase of Sc content in Mg–Sc alloys is accompanied by increase of the elastic modulus [298]. It becomes successively more with increasing Sc concentration within both the Mg solid solution area and when the Sc-rich phase appears in the structure. As high as 20 mass % Sc resulted in increase of elastic modules by about 7% [298]. The elastic modulus change with increasing Sc concentration testifies to the increase of interatomic bonds in Mg solid solution lattice as Sc dissolves. This feature of the Mg–Sc alloys seems to be responsible for the rise of the strength properties with increasing Sc content.

The high strength properties of magnesium alloys containing significant contents of scandium (up to 30 mass %) were reported also in [9].

Effect of Additional Alloying Elements on Mechanical Properties of Magnesium Alloys With Rare Earth Metals

Additional alloying elements in magnesium alloys with rare earth metals can be rationally divided into two groups. The first of them includes other rare earth metals added to the alloys containing a rare earth metal as a base alloying element. The second one includes all metals which do not belong to the rare earth metals and are usually added to magnesium alloys for improvement of their properties.

Effect of Additional Rare Earth Metals on Properties of Mg–RE Alloys

This case is realised commonly in practice. Actually, rare earth metals are used in commercial magnesium alloys jointly, because it is significantly more expensive and not justified to separate them to a high purity for alloying. Typical example of this is mishmetal consisting of Ce (the greatest constituent), La, Nd and Pr. Another example is didymium, named "technical neodymium" in Russia, consisting substantially of Nd (not less than 70 mass %) and Pr.

The influence of individual rare earth metals, La, Ce, Pr and Nd, added together on mechanical properties of both cast and hot extruded magnesium alloys can be accurately assessed from the results of T. Leontis's works [3, 4]. As was mentioned above, in the works [3, 4] a number of the mixtures including La, Ce, Pr and Nd were used for alloying of magnesium. Compositions of the mixtures along with the "pure" rare earth metals Ce and La used were as follows (in mass %).

Didymium, 0.8 Ce, 72.3 Nd, 7.9 Pr, 8.8 La, joint solubility ~1.83 mass %.
Ce-free mishmetal, 0.2 Ce, 35.2 Nd, 12.2 Pr, 46.0 La, joint solubility ~0.91 mass %.
Pr/La mixture, 0.8 Ce, 2.9 Nd, 65.6 Pr, 25.9 La, joint solubility ~0.49 mass %.
Mishmetal, 50.6 Ce, 18.2 Nd, 6.4 Pr, 22.6 La, joint solubility ~0.59 mass %.
Cerium, 92.2 Ce, 0.9 Nd, 0.3 Pr, 1.4 La, joint solubility ~0.28 mass %.

Lanthanum, 0.5 Ce, 0.7 Nd, 0.2 Pr, 95.1 La, joint solubility ~0.048 mass %.

In general, the mixtures are arranged in the following order by promotion of the increase of the strength properties over those of magnesium at elevated temperatures, didymium, Ce-free mishmetal, Pr/La mixture, mishmetal, cerium and lanthanum. Taking into account the solubility of the separate rare earth metals in solid magnesium and assuming the approximately linear dependence of it on the content of every element when they are present in a magnesium alloy together, the joint RE solubility for the mixture may be calculated. The solubility values calculated in this way for 500°C are shown above immediately after the mixture compositions. As one can see, the calculated solubility values correlate with the order of the mixtures by their promotion of the strength of the alloys at elevated temperatures. The more joint solubility of the rare earth metals in solid magnesium, the higher strength at elevated temperatures results from the mixture addition. The only disturbance in this regularity is between Pr/La mixture and mishmetal. The calculated solubility for the latter is some more than that for the former, although the Pr/La mixture is disposed in the order of the mixture strength promotion before mishmetal. This disturbance is, however, insignificant, because the strength properties of alloys with Pr/La mixture and mishmetal were, in general, close, and the difference in joint solubility of them, taking into account the approximate calculation, is quite small. The estimation made shows certainly that in the case of rare earth metals of the cerium subgroup with small solubility in solid magnesium, the effect of individual rare earth metals in their mixtures on mechanical properties of magnesium alloys is caused at most by the respective influence of every one of them on the joint RE solubility in Mg solid solution.

The conclusion made conforms to the results of the investigation [348] where decrease of the strength properties of the extruded Mg alloys containing Nd was observed when Nd was replaced partially by Ce or La.

In the case when at least one of the rare earth metals has significant solubility in solid magnesium, the influence of the individual rare earth metals on the strength properties of the alloys is more complicated and depends on both the ratio between every rare earth metal and the sum of them. Such an example is the system Mg–Nd–Y. It was studied in [349]

Table 37. Tensile properties of the hot extruded Mg–Nd–Y alloys.

Alloy composition, mass %	Test temperature, °C	Hot extruded			Aged after extrusion*			Reference
		TS, MPa	TYS, MPa	E, %	TS, MPa	TYS, MPa	E, %	
3.2 Nd	room	220	170	20.4	220	139	15.0	[349]
	300	123	76	27	90	74	50	
3.2 Nd, 6.0 Y	room	260	204	10.1	290	238	14.0	[349]
	300	180	149	49	173	141	39	
0.9 Nd, 9.8 Y	room	306	249	7.2	409	331	11.0	[349]
	300	226	157	48	223	160	47	
9.1 Y	room	266	207	13.0	323	204	6.4	[350]
	300	166	140	19	177	147	32	
0.6 Nd, 8.3 Y	room	291	205	13.8	319	258	7.4	[350]
	300	183	140	22.6	183	183	41	

*Aged regimes were 210°C/36 h in [349] and 200°C/100 h in [350].

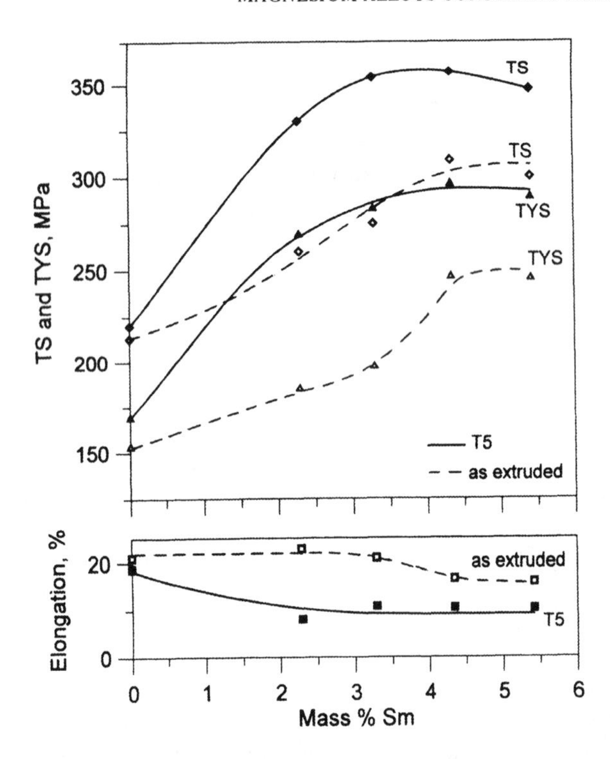

Figure 117. Effect of Sm on tensile properties of the alloy Mg–5.5 mass % Gd at room temperature.

using hot extruded rods produced with a reduction area of about 90%. The rods were tested in the "as extruded" condition and after additional ageing at 210°C for 36 h (T5 condition). The results of the tests at room and elevated temperatures are presented in Table 37. They show clearly an increase of the strength properties with increasing Y content even when Nd content decreases. In another investigation [350] carried out with rods prepared in the same conditions, small addition of 0.6 mass % Nd resulted in notable increase of the strength properties of Mg alloys containing ~8–9 mass % Y. The rods were tested also in "as extruded" and in T5 conditions, but ageing was at 200°C for 100 h. Results of the latter investigation are also included in Table 37. The data presented show that in the system Mg–Nd–Y both rare earth metals supplement each other in increase of the strength properties of the alloys.

Another example of the studied systems with two rare earth metals having substantial different solubility in solid magnesium and belonging to different RE subgroups is the system Mg–Sm–Gd [282]. In this investigation the alloy Mg–5.5 mass % Gd was chosen for additional alloying by samarium within the limits of Sm solubility in solid magnesium in the Mg–Sm binary system. The alloys were prepared by melting with casting into round ingots which were then homogenised at 500°C for 12 h. After homogenisation the ingots were hot extruded into round rods of 10 mm in diameter with heating up to 450°C and reduction area of 87%. The used high temperature homogenisation and extrusion at high temperature followed by cooling in air provided the supersaturated Mg solid solution in the alloys. Taking this into account, some of the specimens of the alloys were aged at 200°C

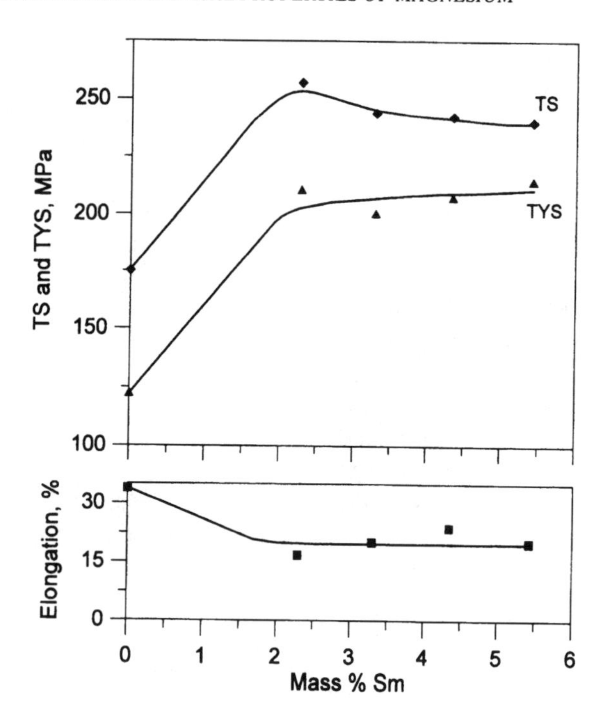

Figure 118. Effect of Sm on tensile properties of the alloy Mg–5.5 mass % Gd at 250°C. T5 condition.

for 12 h which corresponded to the hardening maximum for the Mg–Sm alloys. The opted Gd content was significant less than its solubility in solid magnesium in the Mg–Gd binary system so that Gd action on the strength properties might be connected only with its presence in Mg solid solution or dissolution in the Sm-rich precipitates formed during solid solution decomposition.

The results of tensile tests of the alloys at room temperature in "as extruded" and T5 condition are presented in Figure 117. They show increase of the strength properties with increasing Sm content up to about 4 mass %. After this concentration the strength properties tend to remain at the same level and even to decrease a little. Plasticity falls at first with increasing Sm content and remains then approximately constant. There is a substantial strengthening effect from ageing. It increases with increasing Sm content up to the same Sm concentration of about 4 mass % as the rise of the strength properties. Plasticity falls after ageing with its greatest decrease at the same Sm concentration as the highest increase of the strength properties. In the aged condition T5 the strength properties are increased at most up to about 2 mass % Sm and change more slowly after this concentration. Plasticity in the aged condition T5 falls substantially only up to 2 mass % Sm. In general, the effect of Sm on mechanical properties at room temperature with the presence of Gd is similar to that in the binary Mg–Sm system. Meanwhile, Gd promotes an increase of the strengthening effect from ageing after hot extrusion and raises the general level of the strength properties. They reach TS ~360 MPa, TYS ~310 MPa at E ~10% [282].

In Figure 118 tensile properties of the Mg–Sm–Gd alloys at 250°C are presented. The alloys were aged after hot extrusion by the same regime 200°C, 12 h [282]. At 250°C addition of Sm to the alloy Mg–5.5 mass % Gd also results in an increase of the strength properties and a fall of plasticity, but only up to about 2 mass % Sm. After the steep increase of tensile strength with tensile yield strength, and the reduction of elongation to about 2 mass % Sm, they do not practically change unlike tensile properties at room temperature when the strength properties and plasticity continue to change up to about 4 mass %. The steep increase of strength properties and steep reduction of plasticity up to about 2 mass % Sm seem to be connected with increase of the Sm-rich precipitates in the structure of the alloys and with the appearance of primary crystals of the Sm-rich phase at concentrations of Sm more than about 2 mass %. Following this some difference in the behaviour of the alloys at room temperature and at 250°C with increasing Sm content may be explained by the different effect of the Sm-rich primary crystals on tensile properties of the alloys at every temperature.

The joint effect of two rare earth metals with significant solubility in solid magnesium was established in [351]. In this investigation the effect of 6 mass % Y addition to Mg-base alloy containing 10 mass % Gd was studied. The alloys contained also about 0.6 mass % Mn, but, as will be specifically considered later, Mn does not interfere in the behaviour of rare earth metals in Mg alloys. The used materials were hot extruded round rods prepared substantially in the same conditions as the above considered Mg–Sm–Gd ones. After extrusion the rods were aged at 200°C for 24 h and tensile tested at room temperature and 300°C. Addition of 6 mass % Y resulted in significant increase of strength properties of the alloy with 10 mass % Gd and 0.6 mass % Mn both at room temperature and 300°C. The increase of the strength properties was accompanied by a decrease of plasticity. Results of the tests are presented in Table 38. The increase of strength properties and fall of plasticity

Table 38. Tensile properties of the hot extruded Mg alloys. Ageing after extrusion at 200°C for 24 h.

Alloy composition, mass %	room temperature			300°C		
	TS, MPa	TYS, MPa	E, %	TS, MPa	TYS, MPa	E, %
10 Gd, 0.6 Mn	340	280	10	170	140	26
10 Gd, 0.6 Mn, 6 Y	440	390	5	230	200	15

with Y addition to the alloy with 10 mass % Gd can be reasonably explained by increase of the total content of the rare earth metals in the alloy.

In [352] mechanical properties of Mg alloys containing both Y and Sc were studied. In this system both rare earth metals have significant solubility in solid Mg, although the solubility of Y is notably less than that of Sc. The alloys studied contained 1–11 mass % Y and 1–11 mass % Sc. At high Y contents decomposition of Mg solid solution could be expected with precipitation of Y-rich phases. By contrast, the Sc contents were significantly less than its solubility in solid Mg in the binary Mg–Sc system, so that Sc could be present in the alloys either in Mg solid solution or in the Y-rich precipitates. The alloys were investigated in the form of hot extruded round rods prepared in conditions similar to the above considered Mg–Sm–Gd [282] and Mg–Gd–Y [351] alloys. The rods were tested at room temperature and 300°C in "as extruded" and T5 condition with ageing at 200°C for 100 h which corresponded to the hardness maximum for the binary Mg–Y alloys. All alloys contained 0.6 mass % Mn which is known not to influence the behaviour of the rare earth metals in Mg alloys. Results of the tests are presented in Table 39. They show increase of the strength properties with increasing Y content both at room temperature and 300°C. Such a dependence on the Y content is observed both for the "as extruded" and for T5 condition. Elongation of the alloys remains high enough with increasing Y content, except in the case of the richest by additives alloy with 11 mass % Sc and 8% Y with E = 8% in "as extruded" rods and E = 4% after ageing. Ageing of the alloys with high Sc content did not change significantly the strength properties of the alloys. This fact can be explained by the high joint solubility of Y and Sc in solid Mg at high Sc content and, as a result, insignificant supersaturation of Mg solid solution in the alloys after hot extrusion. Tensile properties of the last alloy in Table 39 with 11 mass % Y and 1 mass %

Table 39. Tensile properties of the hot extruded Mg–Y–Sc–Mn alloys. Ageing after extrusion at 200°C for 100 h.

Alloy composition, mass %	Condition	Room temperature			300°C		
		TS, MPa	TYS, MPa	E, %	TS, MPa	TYS, MPa	E, %
11 Sc, 1 Y, 0.6 Mn	extrud.	310	235	11.5	185	150	37.0
	T5	295	235	14.0	185	155	36.5
11 Sc, 4 Y, 0.6 Mn	extrud.	355	280	13.0	245	210	32.0
	T5	345	275	13.0	215	185	32.0
11 Sc, 6 Y, 0.6 Mn	extrud.	365	295	11.0	235	205	37.0
	T5	355	285	12.0	265	225	30.0
11 Sc, 8 Y, 0.6 Mn	extrud.	375	325	8.0	265	220	32.0
	T5	360	325	4.0	245	205	26.0
1 Sc, 11 Y, 0.6 Mn	extrud.	335	275	5.0	195	165	57.5
	T5	375	325	1.5	165	135	55.0

Sc confirms this explanation. Its strength properties at room temperature are notably higher and elongation is lower after ageing than after only extrusion in accordance with the more Y/Sc ratio as compared with the previous alloys containing 11 mass % Sc and 1–8 mass % Y. At higher Y/Sc ratio in the last alloy the joint solubility of Y and Sc becomes less, so that a supersaturation of Mg solid solution becomes possible and, as a result, the notable strengthening effect after ageing was observed.

In [289] the effect of 0.05–1.0 mass % La, 0.1–1.0 mass % Ce and 0.05–1.0 mass % Gd on mechanical properties of Mg–Y alloys with 9–10 mass % Y was reported. The alloys were used as hot extruded rods. Mechanical properties were determined in the "as extruded" condition and aged at 200°C for 100 h (T5 condition). The addition of La resulted in increase of strength properties at room temperature in the "as extruded" condition and their decrease in T5 condition. At 300°C, addition of La did not increase the strength properties. On the contrary, some alloys with La showed less strength than the binary Mg–Y alloy prepared and tested in the same conditions. Addition of Ce did not result in any increase of the strength properties of the Mg–Y alloys, but some decrease of the properties took place. Addition of Gd resulted in increase of the strength properties at room temperature and 300°C in accordance with [351].

In summary, the effect of joint alloying by rare earth metals on mechanical properties of Mg can be different. Its character depends mainly on the joint solubility of the rare earth metals in solid Mg as compared with the solubility of individual elements. This rule is applicable the most clearly when the solubility of the rare earth metals is small, as in the case of the elements of the cerium subgroup. If at least one of the rare earth metals has significant solubility in solid Mg, as one of the elements of the yttrium subgroup, the strengthening effect from individual rare earth metals can be superimposed on each other so that the joint alloying will be accompanied by a higher strengthening effect than that from each of the rare earth metals alone.

Effect of Non-Rare Earth Metals on Mechanical Properties of the Mg–RE Alloys

The investigation into how these alloying elements act on the strength properties of Mg alloys with various rare earth metals is of great importance. Additional alloying can be expected to result in some improvement of mechanical and other significant properties. Besides, industrial production of commercial alloys can be accompanied by accidental contamination of them by alloying elements which are present in other magnesium alloys. So, the limits of their permissible contents should be established.

The effect of additional alloying on mechanical properties of Mg–RE alloys was studied in [238, 349, 351, 353–368]. Results of the works in this direction are also presented in a large volume in the books [106, 239, 289, 294] and patents. Most investigations were into alloys containing rare earth metals which used to be used as alloying additives for magnesium alloys, such as mishmetal, cerium, neodymium and yttrium. For the additional alloying elements, mostly used were the metals well known as common alloying elements in commercial magnesium alloys. In general, the results of the investigations showed a similar character of the properties change if the same additional alloying element was used for the alloys with different rare earth metals. Such behaviour of the Mg–RE systems as the additional alloying is applied correlates with the similar constitution of the respective ternary phase diagrams with different rare earth metals.

Effect of manganese

Manganese is commonly used as an alloying additive for Mg alloys. It is known to improve the corrosion resistance of Mg alloys. For Mg alloys with rare earth metals the addition of Mn is used also for some improvement of their strength properties.

In [353, 354] increase of the strength properties with addition of Mn was established for alloys with mishmetal consisting of about 50 mass % Ce. The cast and hot extruded materials were tested at room temperature and 316°C. Hot extrusion gave the rods which were solution treated and then aged at 204°C for 16 h. For tests at 316°C the rods were annealed additionally at 343°C for 24 h. Addition of 0.81–1.90 mass % Mn to the alloys with 2 mass % mishmetal resulted in an increase of TS from 195 MPa to 205–230 MPa, TYS from 120 to 140–195 MPa at room temperature and TS from 50 MPa to 59–69 MPa, TYS from 29 MPa to 41–51 MPa at 316°C.

The small increase of the strength properties from Mn addition was confirmed for Mg–Nd alloys in [106, 238, 239, 362]. The alloys were used for tests as rods produced by hot extrusion. The rods were solution treated and then aged for T6 condition. At room temperature the highest strengthening effect was observed at the highest Mn content of 1.60

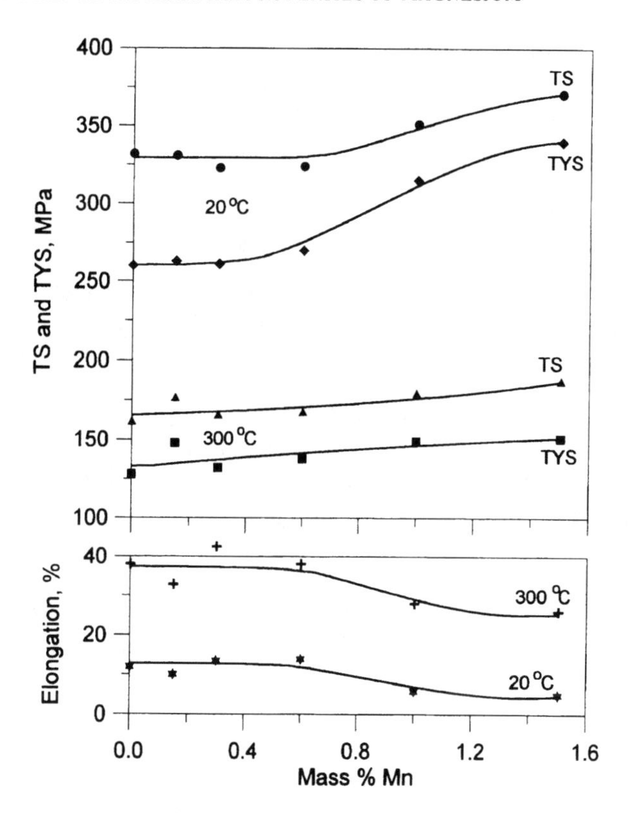

Figure 119. Effect of Mn on tensile properties of the alloy Mg–10 mass % Gd. T5 condition.

mass %. Used for comparison the binary Mg–Nd alloy contained about 2 mass % Nd. They gave at room temperature TS = 242 MPa, TYS = 74 MPa, whereas the alloy with the same Nd content and 1.6 mass % Mn gave TS = 271 MPa, TYS = 126 MPa. At 250°C the binary Mg–Nd alloy gave TS = 137 MPa, TYS = 73 MPa, whereas the Mg–Nd–Mn alloy gave TS = 142 MPa, TYS = 105 MPa. So, TYS increased more than TS after addition of Mn as in the case of the alloys with mishmetal [353, 354]. Manganese improved also to some extent the strength of the Mg–Nd alloys during long-loaded tests at elevated temperatures [106, 239]. Plasticity of the Mg–Nd alloys was practically unchanged when Mn was added [106, 239].

Similar results were obtained with Mn addition to Mg–Gd alloys [351]. The investigation was also carried out using hot extruded rods. The alloy Mg–10 mass % Gd was used as the base for additional alloying by Mn up to 1.5 mass %. After extrusion the rods were aged at 200°C for 24 h for T5 condition. Results of the tensile tests are shown in Figure 119. As in the cases of magnesium alloys with mishmetal and neodymium, addition of Mn resulted in increase of the strength properties with TYS increasing more than TS. However, unlike the alloys with mishmetal and neodymium, addition of Mn to the alloy with Gd was accompanied by significant decrease of plasticity.

According to the ternary phase diagrams of the Mg–RE–Mn systems studied there are no compounds formed between RE and Mn in Mg-rich alloys. Manganese does not actually change the solubility of the rare earth metals in Mg solid solution, either. These facts lead to the conclusion that there is actually no interaction between Mn and RE in Mg solid solution and, as a consequence of this, Mn does not change behaviour of Mg–RE alloys during ageing. Mg–RE alloys retain its ability to strengthen during solid solution decomposition when Mn is present.

Effect of zirconium

Zirconium is known as an alloying element which results in refinement of the grain size of cast Mg and some Mg alloys. This phenomenon is connected with precipitation of practically pure Zr during crystallisation of the Mg–Zr alloys and closeness between lattices of the precipitated Zr and Mg solid solution. Some alloying elements in Mg alloys hinder Zr action, but Mg alloys with rare earth metals are those where the refinement action of Zr takes place. Due to the finer grain the strength properties and plasticity are expected to increase together, and results of experimental investigations in general confirm this supposition. Solubility of Zr in molten Mg at temperatures common for melting Mg alloys is not more than 0.8–1 mass % [139, 370–372]. Therefore, its content in the alloys is limited by this quantity. The typical concentration of Zr in Mg alloys is 0.4–0.6 mass %, although the refinement effect could be observed sometimes even at 0.2 mass % Zr.

The scale of the increase of the mechanical properties due to Zr addition is illustrated by data obtained on separately sand cast specimens of alloys containing mishmetal (MM) [355]. In this work specimens of the alloy Mg–3.10 MM–0.46 Zr (in mass %) showed TS = 130 MPa, TYS($\sigma_{0.1}$) = 82.5 MPa, E = 3%, as compared with TS = 89.5 MPa, TYS($\sigma_{0.1}$) = 52.5 MPa, E = 1% of the binary alloy Mg–3.01 MM (in mass %).

The effect of Zr on the cast alloys with mishmetal, cerium and didymium was studied also in [355]. Didymium contained (in mass %) 75 Nd, 7.9 Pr, 8.8 La, 0.8 Ce, remainder Fe and some other elements. The character of the change of mechanical properties with the addition of Zr for all three kinds of rare earth mixtures turned out to be similar. In the case of the alloys with 3 mass % didymium, the addition of 0.57 mass % Zr resulted in an increase of TS from about 210 MPa to 280 MPa at room temperature and from about 85 MPa to 120 MPa at 204°C. TYS increased from about 140 MPa to 160 MPa at room temperature and from about 125 MPa to 150 MPa at 204°C. Meanwhile, tests at a higher temperature, 316°C, showed a certain decrease of the strength properties with the addition of Zr to the alloy with 3 mass % didymium. As 0.57 mass % Zr was added TS decreased at 316°C from about 60 MPa to 52 MPa and TYS from about 90 MPa to 80 MPa. The alloys were tested after quenching from 538°C + ageing at 204°C for 16 h. Decrease of the strength properties at the highest test temperature of 316°C with addition of Zr could be expected because the fine grain size was known to be a factor which promoted, in general, softening of metal materials at high temperatures.

Plasticity of the alloys with mishmetal, cerium and didymium increases with the addition of Zr. In the case of the cast alloys with 3–4 mass % didymium, which were quenched and aged, elongation at room temperature increased with addition of Zr from 3–5% to 6–9% [358].

Addition of Zr improves the castability of Mg alloys with rare earth metals [355].

In wrought Mg alloys with rare earth metals of the cerium subgroup, addition of Zr did not result in significant improvement of their mechanical properties [356]. This fact could be explained by the absence of any significant difference in grain size between alloys with and without Zr after hot working. Nevertheless addition of Zr to the wrought Mg alloys with the rare earth metals is still justified because Zr improves castability of the alloys during casting of ingots and the fine grain size in ingots facilitates their hot working when wrought semiproducts are produced.

Zirconium does not actually change the behaviour of rare earth metals during solid solution decomposition accompanied by the increase of strength.

Influence of nickel

Investigations carried out on wrought magnesium alloys containing mishmetal (MM) or neodymium together with manganese showed that their strength properties could be improved at elevated temperatures by small Ni additives (up to 1 mass %) [337, 353, 362, 363]. At room temperature, no improvement of the strength properties of the alloys Mg–MM–Mn and Mg–Nd–Mn by Ni addition was observed. Data showing how properties change when Ni is added are presented in Tables 40 and 41. They were obtained

Table 40. Tensile properties of the Mg–MM(mishmetal)–Mn–Ni alloys [353]. T6 condition.

Alloy composition, mass %	Room temperature			300°C		
	TS, MPa	TYS, MPa	E, %	TS, MPa	TYS, MPa	E, %
2.41MM, 1.90 Mn	228	187	7.2	69	51	116
2.33 MM, 1.90 Mn, 0.34 Ni	230	181	7.0	86–87	60–65	84–132

by testing hot extruded rods which had undergone solution treatment with quenching and ageing. Addition of Ni lowers the solidus temperature of the alloys. Therefore, the alloys containing Ni were solution treated at lower temperatures as compared with respective alloys without Ni. So, the alloy of the system Mg–MM–Mn–Ni was solution treated at

Table 41. Tensile properties of the hot extruded Mg–Nd–Mn–Ni alloys [362]. T6 condition.

Alloy composition, mass %	Room temperature			250°C			300°C		
	TS, MPa	TYS, MPa	E, %	TS, MPa	TYS, MPa	E, %	TS, MPa	TYS, MPa	E, %
2.68Nd, 1.48 Mn	275	154	16	182	104	22	86	67	40
2.72 Nd, 1.28 Mn, 0.13 Ni	274	132	16	196	103	30	121	93	35
2.76 Nd, 1.30 Mn, 0.45 Ni	275	98.1	18	188	104	20	124	77	29

538°C whereas the alloy Mg–MM–Mn used for comparison was solution treated at 560°C. The alloys Mg–Nd–Mn–Ni were solution treated at 500°C and their counterpart without Ni of the Mg–Nd–Mn system was solution treated at 535°C. Specimens of the alloys containing mishmetal were aged at 204°C for 16 h for maximum strengthening. Those which were intended for tests at 316°C were additionally annealed at this temperature for 24 h. The alloys containing Nd were aged for maximum strengthening at 175°C for 24 h. The presence of Ni changes the phase constitution of the Mg–RE alloys. It is the cause of the increase of strength properties at elevated temperatures when Ni is added.

Addition of Ni even in small quantities results in an abrupt decrease of corrosion resistance [373]. Besides, as Ni lowers the melting temperature of Mg–RE alloys, it makes it difficult to work them at elevated temperatures. Alloys containing Ni show bad weldability. All these factors hinder the practical use of Mg–RE alloys containing additional Ni.

Influence of silver

Addition of silver can improve the strength characteristics of magnesium alloys containing rare earth metals. In accordance with this, commercial alloys containing neodymium and silver were developed and used in industry [294, 374–376]. In [356] a detailed investigation of the influence of Ag on mechanical properties of the alloys containing mishmetal and neodymium was carried out using both cast and wrought materials. Along with this a number of other investigations on the Mg–RE base alloys containing silver were conducted. Most of the alloys studied contained about 0.6 mass % Zr.

According to [356], tensile strength of the cast alloys with 0.6 and 1.4 mass MM in T6 condition increased continuously with addition of Ag up to 3 mass %. The tensile yield strength of the alloys increased also with addition of Ag, but only up to ~2 mass %. Further increase of Ag contents up to about 3 mass % practically did not change the tensile yield strength of alloys with mishmetal. The plasticity of the alloys hardly changed with addition of Ag up to 3 mass %. The cast alloy with 1.4 mass % MM containing additionally 3.2 mass % Ag showed TS = 250 MPa, TYS($\sigma_{0.1}$) = 138 MPa, E = 1.9% as compared with TS = 212 MPa, TYS($\sigma_{0.1}$) = 105 MPa, E = 1.8%.

The cast alloys with neodymium containing additional silver showed higher values of strength properties than alloys with mishmetal and silver [356]. Optimum concentrations of neodymium and silver added together to the alloy with 0.6 mass % Zr were about 2 mass % Nd and 2 mass % Ag. At such concentrations of the alloying elements, a quite high level of strength properties was achieved. So, the cast alloy Mg–2.50 Ag–1.76 Nd–0.6 Zr (in mass %) quenched from 535°C and aged at 200°C for 16 h showed TS = 282 MPa, TYS($\sigma_{0.1}$) = 188 MPa with E = 7%. Mg alloys with Nd and Ag are the most attractive owing to their high tensile yield strength especially at elevated temperatures. At 250°C the cast alloy with 2.7 mass % Nd and 2.5 mass % Ag showed after quenching and ageing TYS($\sigma_{0.1}$) ≈ 123 MPa, TYS($\sigma_{0.2}$) ≈ 132 MPa with TS ≈ 157 MPa [356]. The alloys also have a high creep resistance up to 200°C. The alloys Mg–Nd–Ag–Zr are characterised by high strengthening effect during ageing with silver promoting increase of the strengthening effect.

The structure investigations [356] led to the conclusion that for normal content silver was present in the Mg–Nd alloys dissolved in Mg solid solution. Following this observation the authors [356] supposed that silver changed the mechanism of solid solution decomposition in the Mg–Nd alloys and this was the cause of higher strength properties of the alloys containing silver.

The Mg–Nd–Ag Zr alloys show good castability. Their mechanical properties determined on the separately cast specimens are commonly close to those established on specimens cut from thick sections of cast parts.

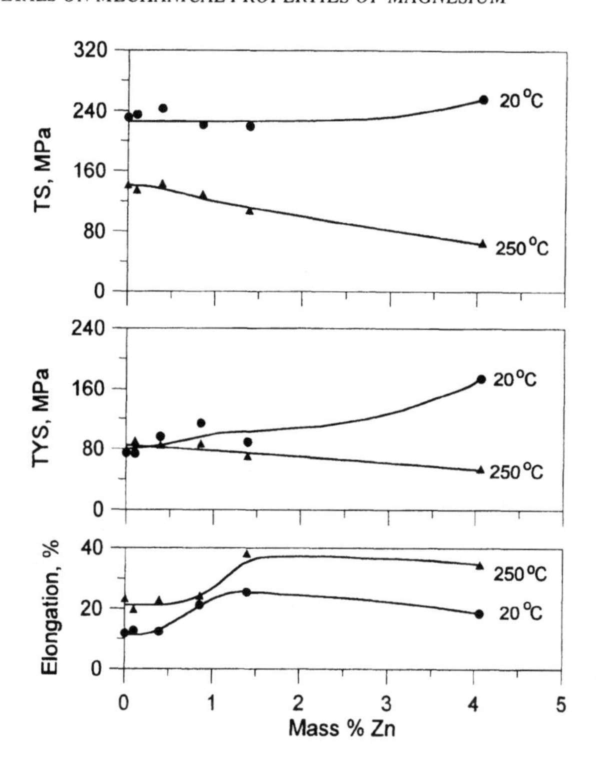

Figure 120. Effect of Zn on tensile properties of the alloy Mg–2 mass % Nd. T6 condition.

Investigation of the wrought alloys containing mishmetal or neodymium with zirconium and silver revealed high strength properties for some of them. For example, the alloy Mg–0.83 MM–1.99 Ag–0.6 Zr (in mass %) which was successively forged, solution treated by quenching from 570°C and aged at 175°C for 24 h showed TS = 308 MPa, TYS($\sigma_{0.1}$) = 271 MPa, E = 4.5%. The alloy Mg–1.36 Nd–2.42 Ag–0.6 Zr was successively forged, solution treated by quenching from 530°C and aged at 200°C for 8 h. This alloy showed TS = 337 MPa, TYS($\sigma_{0.1}$) = 306 MPa, E = 4%. Nevertheless, some alloys showed rather low strength properties. There was also a notable anisotropy of the strength properties in the forgings obtained. The properties in the longitudinal direction were more than twice as high as those in the transverse direction [356].

Improvement of the strength properties at elevated temperatures up to 250°C from silver was also observed in [362] where about 1–4 mass % Ag was used for additional alloying of Mg–Nd–Mn alloys containing about 2.5 mass % Nd and 1.4 mass % Mn. Investigation [362] was carried out on rods obtained by hot extrusion with solution treatment and then ageing for maximum strengthening (T6 condition). With increasing test temperature and at long-term tests the strengthening effect from the Ag additives became, however, less and eventually disappeared.

Influence of zinc

Zinc is a widely used alloying element for commercial Mg alloys. It is used for improvement of the strength properties of both cast and wrought alloys. Investigations have shown that zinc improves the strength properties of magnesium alloys with rare earth metals, too, but only with certain limitations.

In the work [358] the effect of 1.8–3.8 mass % Zn on the properties of the cast alloys Mg–MM–Zr and Mg–Nd–Zr was studied. The properties were determined after heat treatment of ageing or solution treatment with quenching followed by ageing. Addition of Zn resulted in decrease of the strength properties both at room and at elevated temperatures. Meanwhile, the authors [377] reported that addition of 0.3 mass % Zn resulted in increase of the long-term strength (for 100 h) of the Mg–Nd–Zr alloy containing 2.5 mass % Nd at 200 and 250°C.

According to the data [337], where hot extruded alloys were studied, addition of 1–6 mass % Zn to the alloy Mg–1.5 Nd–0.5 Zr (in mass %) resulted in continuous increase of the strength properties and decrease of plasticity at all test temperatures from ambient to 250°C. However, strengthening caused by the Zn addition decreases with rising temperature. The alloy Mg–1.3 Nd–5.5 Zn–0.5 Zr (in mass %) treated according to T5 regime showed the following mechanical properties: at room temperature TS = 365–390 MPa, E = 5–6%, at 250°C TS = 145–165 MPa, E = 25–30%.

The effect of Zn on mechanical properties of the extruded Mg–Nd alloys after solution treatment and ageing (T6 condition) can be characterised by data presented in Figure 120 [238, 362]. The alloys contained about 2 mass % Nd. As one can see, in T6 condition the increase of TS at room temperature with increasing Zn content is, in general, insignificant, but TYS at room temperature increased substantially (more than twice). At 250°C both strength characteristics fell substantially with increasing Zn content. Plasticity of the Mg–Nd in T6 condition remained high with the addition of Zn. It is reasonable to note a significant increase and high values of elongation at ~1.5–4 mass % Zn. Addition of Zn to Mg–Nd alloys necessitated the use of lower temperatures for solution treatment. For the alloy with 4.06 mass % Zn it became as low as 435°C instead of 535°C for the binary Mg–Nd alloy. The lower solution treatment temperature and decrease of the Nd solubility

Table 42. Tensile properties of the hot extruded Mg–Y–Zn alloys at room temperature [289].

Alloy composition, mass %	TS, MPa	TYS, MPa	E, %
4.8 Y	232	164	13.2
3.83 Y, 1.20 Zn	308	258	7.2
4.39 Y, 3.30 Zn	319	265	4.0
4.29 Y, 4.30 Zn	388	364	2.2
8.00 Y	298	220	7.0
8.41 Y, 1.03 Zn	308	271	4.0
8.11 Y, 1.60 Zn	321	239	8.0
8.20 Y, 3.30 Zn	390	348	3.2
8.23 Y, 3.80 Zn	349	272	6.4

in solid Mg with increasing Zn content according to the Mg–Nd–Zn phase diagram resulted first in a decrease of the strengthening effect during ageing and then its disappearance at ~2 mass % Zn [362]. So, the mentioned change of mechanical properties with addition of Zn for the Mg–Nd alloys in T6 condition corresponded to a decrease of the precipitates from Mg solid solution during ageing and increase of the coarse crystals of

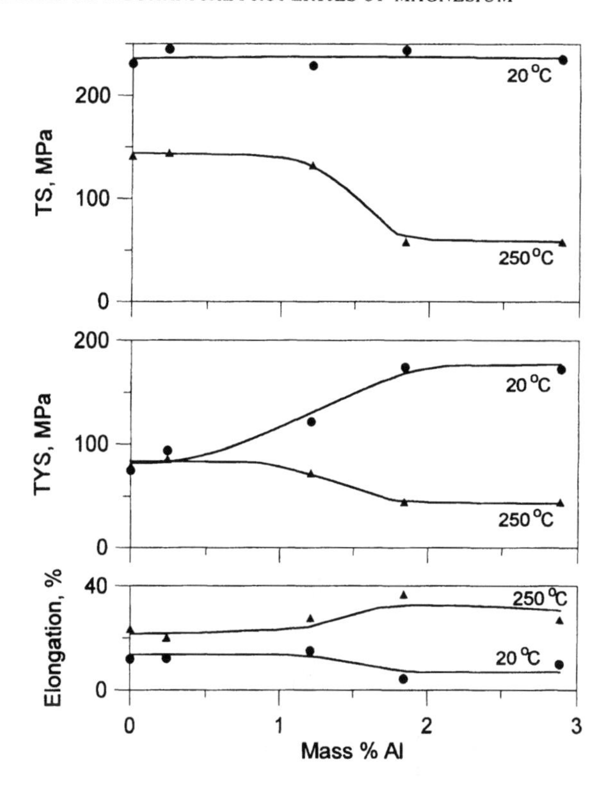

Figure 121. Effect of Al on tensile properties of the alloy Mg–2 mass % Nd. T6 condition.

the second phase in the structure of the alloys. Results of the works [337, 362] on the Zn influence on extruded Mg–Nd alloys agree with results of the work [357].

The effect of the Zn additives on mechanical properties of hot extruded Mg–Y alloys was reported in [289]. The alloys were tested at room temperature in "as extruded" condition. Representative results of the experiments carried out are shown in Table 42. They demonstrate increase of the strength properties with increasing Zn content. Sufficiently high values of TS and TYS are achieved already at 1.2–1.6 mass % Zn, and acceptably high values of elongation are retained.

Influence of aluminium

Aluminium is, as well as zinc, one of the widely used alloying elements for Mg alloys because of its significant strengthening effect. Aluminium is present in many commercial Mg alloys.

The influence of aluminium on mechanical properties of Mg–Nd alloys was studied in [238, 362]. Hot extruded alloys containing about 2 mass % Nd were solution treated and aged for T6 condition. Results of the tests at room temperature and 250°C are presented in Figure 121. They lead to the conclusion that addition of Al up to about 1–3 mass % practically do not change the tensile strength of Mg–Nd alloys at room temperature. Tensile yield strength at room temperature practically does not change at small Al addition (at least up to 0.25 mass %), but increases significantly with addition of about 1–3 mass %

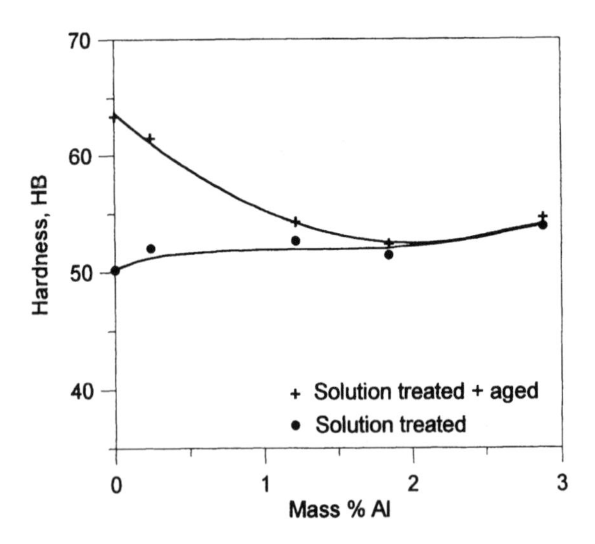

Figure 122. Effect of Al on hardness of the alloy Mg–2 mass % Nd.

Al. At 250°C both tensile strength and tensile yield strength hardly change at small contents of Al, but abruptly reduce beginning with about 1 mass % Al. Elongation tends to decrease a little at room temperature and to increase a little at 250°C with addition of Al. Microscopy investigations showed significant increase of the second phase in the structure of Mg–Nd alloys with addition of Al. On the other hand, hardness measurements showed at first a decrease and then disappearance of the strengthening effect during ageing as Al was added [238, 362]. Results of the hardness measurements of the alloys after solution treatment and after solution treatment + ageing at 175°C for 24 h are shown in Figure 122. Both microscopy observations and the hardness measurements pointed to a decrease of Nd solubility in solid Mg with addition of Al. Actually, addition of Al makes behaviour of the Mg–Nd alloy during tensile tests similar to that of the Mg–La alloys which are distinguished by insignificant solubility of the rare earth metal in solid magnesium. In investigations by the author, Al exerted the same influence on the mechanical properties of magnesium alloys with mishmetal as in the case of the Mg–Nd alloys described above.

In [289] results of tensile tests of the hot extruded Mg–Y alloys containing Al are reported. In Table 43 a representative part of them for room temperature tests in the "as extruded" condition is presented. The data can be summarised as follows. Already at small Al contents of about 0.25–0.3 mass %, the strength properties of the Mg–Y alloys with approximately the same Y content become significantly more. With further increase of Al content the strength properties remain approximately at the same level as with the small amounts of Al content and then tend to decrease. In alloys with about 3–5 mass % Y the strength properties increase at about 7 mass % Al again up to the same level as at 0.3–0.4 mass % Al. Increase of the strength properties of the Mg–Y alloys with addition of Al is accompanied by reduction of plasticity. In other series of tests, hot extruded Mg–Y alloys containing 9–10 mass % Y and 0–1 mass % Al were tested [289]. The tests at room temperature in the "as extruded" condition confirmed that the change of the strength properties with addition of Al followed the curve with a maximum at 0.6 mass % Al, corresponding to the values TS = 324 MPa, TYS = 277 MPa at E = 7.0% as compared with

TS = 283 MPa, TYS = 213 MPa at E = 14% for the alloy without addition of Al. Tests of the alloys with 9–10% Y after ageing at 200°C for 100 h (T5 condition) did not reveal any strengthening effect from the Al addition. Also no improvement of the strength properties was observed in the case of tests at 300°C. Tests of the hot extruded Mg–Y alloys with

Table 43. Tensile properties of the hot extruded Mg–Y–Al alloys at room temperature [289].

Alloy composition, mass %	TS, MPa	TYS, MPa	E, %
4.62 Y	215	136	14.6
3.90 Y, 0.3 Al	260	202	7.0
4.03 Y, 0.42 Al	257	219	4.8
3.99 Y, 0.86 Al	232	172	19.2
4.23 Y, 1.80 Al	206	149	19.8
4.49 Y, 3.98 Al	220	158	14.2
3.04 Y, 4.92 Al	215	158	10.6
5.07 Y, 6.99 Al	262	181	9.8
3.09 Y, 6.30 Al	266	177	9.8
7.73 Y	265	139	14.4
7.70 Y, 0.27 Al	335	275	6.8
7.65 Y, 0.58 Al	352	298	5.0
7.90 Y, 0.97 Al	351	301	7.2
7.30 Y, 1.73 Al	327	275	11.8
7.58 Y, 1.97 Al	332	287	11.0

about 8 mass % Y in T6 condition also showed a change of the strength properties with addition of Al following a curve with a maximum. Contents of Al additives were in the range 0.34–1.06 mass %. Tests were carried out at room temperature and 300°C [366].

Influence of cadmium

According to [239] addition of 1.8–3.0 mass % Cd was accompanied by some strengthening of hot extruded rods of Mg–Nd–Mn alloys. The alloys were tested after solution treatment and ageing (in T6 condition). The strengthening effect from Cd consisted of some increase of tensile yield strength at room and elevated temperatures up to 300°C. At room temperature tensile strength did not change practically with addition of 1.8–3.0 mass % Cd. Tensile strength did not change with addition of Cd at elevated temperatures up to 250°C but did increase at 300°C.

No significant change of hot extruded Mg–Y alloys was noticed when they were alloyed by 0.55–5.8 mass % Cd [289].

Influence of silicon

The influence of silicon was studied for the Mg–Nd alloys [362] and Mg–Y alloys [289]. According to [362], small additives of Si (0.04–0.05 mass %) practically did not change the strength properties of hot extruded Mg–Nd alloys in T6 condition. The same content of 0.05 mass % Si added to the Mg–Y alloy containing about 10 mass % Y resulted in a significant decrease of the strength properties in "as extruded" and T5 condition. With increasing Si content up to 0.47 mass % the strength properties of the Mg–Y alloys continued to fall [289].

Influence of tin, bismuth, calcium, barium, lithium and cobalt

These elements are rarely used as alloying additives to Mg alloys, but they were checked for Mg–Nd alloys [238, 362]. Results of the respective tests are shown in Table 44. The alloys were hot extruded into rods. The rods were then solution treated and aged at 175°C for 24 h for T6 condition. The tests were carried out at room temperature and 250°C by stretching. The main results can be summarised as follows.

Table 44. Tensile properties of the alloy Mg – 2 mass % Nd alloyed additionally by various elements [362].

Additive, mass %	Room temperature			250°C		
	TS, MPa	TYS, MPa	E, %	TS, MPa	TYS, MPa	E, %
0	233	75	12.2	137	77	19.9
0.27 Sn	243	92	14.8	138	89	11.0
0.47 Sn	235	108	18.8	131	92	19.4
0.73 Sn	217	102	19.0	120	70	19.6
1.87 Sn	257	184	2.2	45	39	41.8
4.20 Sn	263	192	3.6	50	39	36.4
0.09 Bi	231	73	14.2	149	80	15.6
0.65 Bi	232	72	14.0	139	77	16.4
1.01 Bi	230	64	17.2	118	69	18.8
1.37 Bi	223	75	16.4	114	67	21.6
2.39 Bi	177	58	24.8	106	51	30.6
0.2 Ca	239	82	12.6	149	77	18.6
0.67 Ca	230	98	7.8	139	80	19.2
0.79 Ca	228	110	7.8	124	75	19.4
0.2 Li	231	75	10.8	125	66	17.1
0.8 Li	224	96	7.0	129	73	16.6
0.16 Ba	218	73	11.0	145	73	27.6
0.23 Ba	248	92	17.6	135	71	26.2
0.75 Ba	241	90	12.0	127	84	25.8
1.40 Ba	245	118	9.8	139	90	22.6
0.98 Co	245	86	9.4	139	92	17.8
1.80 Co	247	101	9.0	139	90	18.0

The effect of tin up to 4.20 mass % is similar, in general, to that of aluminium. This additive results in an increase of the strength properties at room temperature, especially tensile yield strength, and a decrease of the strength properties at 250°C. Simultaneously, plasticity falls at room temperature and becomes significantly more at 250°C with increasing Sn content. Behaviour of Sn, as that of Al, can be explained by diminution of the Nd solubility in Mg solid solution when Sn is added. Such an explanation is supported by decrease of the strengthening effect of the Mg–Nd alloys during ageing with increasing Sn content in them [238, 362].

Addition of bismuth to the Mg–Nd alloys results in a decrease of the strength properties both at room temperature and 250°C. As aluminium and tin, bismuth decreases the strengthening effect of Mg–Nd alloys during ageing. So its influence on the Mg–Nd alloys can be connected with decrease of the Nd solubility in Mg solid solution. Addition of calcium, lithium, barium and cobalt in the range used results in some increase of tensile yield strength at room temperature with tensile strength remaining practically unchanged. At 250°C addition of calcium, lithium, barium and cobalt practically does not change both tensile strength and tensile yield strength of Mg–Nd alloys.

Influence of indium and thallium

Addition of 0.5 mass % In results in an increase of the long-term strength of cast Mg–Nd–Zr alloys with 2.5 mass % Nd at 200 and 250°C [364, 377]. Tests at constant load were accompanied by failure of the specimens in 100 h at stresses of 93 and 44 MPa at 200 and 250°C, respectively, for the alloy with 0.5 mass % In, as compared with 88 and 27 MPa for the Mg–Nd–Zr alloy without In. The authors [377] explained the increase of the long-term strength of the alloys by addition of indium by deceleration of the decomposition of the Mg solid solution and suppression of the precipitate coagulation.

Addition of 3 mass % thallium to the Mg–Nd–Zr alloy with 1.9–2.6 mass % Nd and 0.2–1.0 mass % Zr resulted in an increase of the strength properties and plasticity in the "as cast" condition without additional heat treatment. Tensile strength of the "as cast" alloy with Tl turned out to be close to that of the alloy without Tl after solution treatment and ageing for T6 condition [359].

Mechanical Properties of Mg–RE Alloys Compacted After Rapid Solidification

Rapid solidification is quite an effective way of enhancing the strength properties of metals and it was also applied to Mg alloys including those with various rare earth metals [241–247, 378–380]. The high strength properties of the metal materials after rapid solidification are caused by peculiarities of their structure. One of them is the very fine structure constituents such as grains of metal-base solid solution and particles of the second phases containing alloying elements. At the highest solidification rates it is possible to obtain alloys in amorphous or glassy state. As a rule, sufficiently rapid solidification is achieved by cooling materials in thin sections. One of the widespread processes for this is cooling of melts on a fast rotating disk which is called "spinning". Therefore, alloys with substantially fine structure are commonly obtained in the form of thin ribbons, thin flat pieces or wires. The typical thickness of them is about 20–100 μm. Such thin ribbons and pieces are hardly suitable for practical application as useful parts. However, they can be compacted into bulk forms by pressing and then wrought into various kinds of common semiproducts or finished parts. Such processes are similar to those used in powder metallurgy technologies.

Rapid solidification was applied to Mg alloys of various systems. Investigations showed quite high strength properties for some of them immediately in thin sections in accordance with the expected fine structure. So, according to [379], thin ribbons of several Mg-base alloys containing rare earth metals along with other elements could achieve

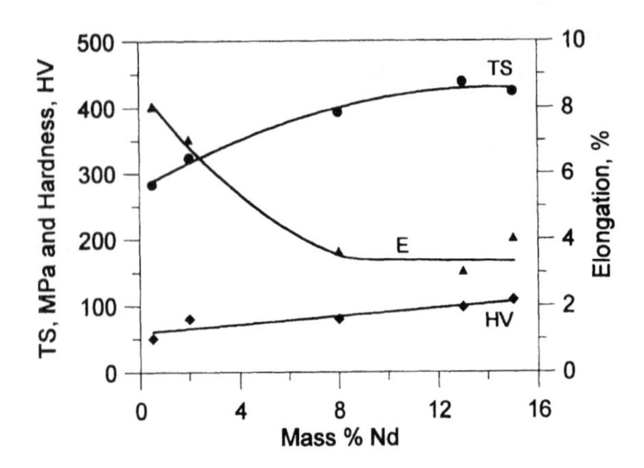

Figure 123. Mechanical properties of the Mg–Nd alloys after rapid solidification and hot extrusion.

tensile strength values of 870–1280 MPa [247]. In [379] Mg-base alloys containing 0–40 at. % Ni or Cu and 0–40 at. % Y were prepared with an amorphous structure using the rapid solidification technique. Tests of them by bending showed stress of failure to be 800–830 MPa.

Compacting and working of thin ribbons and flat pieces of Mg alloys prepared by using rapid solidification require heating for better performance. Therefore, these processes are accompanied by a certain destruction of the fine structure formed in thin ribbons with coarsening the structural constituents. So, the inevitable damage to high strength of the materials obtained immediately after rapid solidification should be expected. Nevertheless, by choosing special compositions of the alloys and using suitable conditions of compacting and working the high strength of the rapidly solidified materials could be retained in bulk parts, at least to a great extent. The relevant investigations of magnesium alloys with rare earth metals are described below.

Mechanical Properties of Binary Mg–RE Alloys after Rapid Solidification, Compacting and Hot Working

In [381] binary Mg–Nd and Mg–Y alloys prepared by using rapid solidification were studied. Rapid solidification was performed by quenching the melts on a cold rapidly rotating disk in He atmosphere. The alloys were obtained in the form of thin ribbons with thickness of 40–90 μm. The cooling rate calculated from the dendrite parameter in the structure of the alloys prepared was determined to be 10^5–10^6 K/s. Microhardness of the Mg–Nd ribbons increased continuously with increasing Nd content from 35 up to 118 by the Vickers scale at 18 mass % Nd. Microhardness of the Mg–Y alloys also increased continuously with increasing Y content up to 107 by the Vickers scale. The thin ribbons of the alloys were cut into small pieces which were compacted by pressing and then hot extruded into rods. Figure 123 demonstrates the results of tensile tests and microhardness measurements of the rods of the Mg–Nd alloys. They show that microhardness of the alloys after hot extrusion increases with increasing Nd content following the same tendency as in the ribbons, but with slightly higher values at low Nd contents and slightly lower

Table 45. Tensile properties of magnesium alloy rods prepared by using the powder metallurgy/rapid solidification (PM/RS) and ingot (IN) technologies [381].

Composition, mass %	TS, MPa	TYS, MPa	E, %
Mg−7 Y (PS/RS)	345	300	9.9
Mg−7 Y (IN)	290	235	8.0
Mg−11.5 Y (PS/RS)	381	367	9.1
Mg−11.5 Y (IN)	330	275	3.0
Mg−0.5 Nd (PS/RS)	300	275	9.7
Mg−0.5 Nd (IN)	180	40	15.0
Mg−2 Nd (PS/RS)	350	300	7.0
Mg−2 Nd (IN)	230	90	10.0

values at higher Nd contents. Tensile strength (TS) of the rods successively increases and elongation (E) successively decreases with increasing Nd content with tensile strength value reaching 430 MPa. In this aspect the changes of the properties with increasing content of the alloying element remind us of those observed in the case of alloys obtained by ingot technology, but without clear kinks on the property curves corresponding to the solubility of Nd in solid Mg for equilibrium conditions. Such a difference between the Mg–Nd alloys obtained by using rapid solidification and ingot technology is understandable because of the enormous increase of the solubility of Nd in solid Mg during rapid solidification [250]. In general, the same tendency was observed for rods of the Mg–Y alloys. The rods of the alloy Mg–17 mass % Y showed TS = 450 MPa, TYS = 400 MPa at E = 4.6%.

As compared with the rods obtained by ingot technology, the rods prepared from ribbons obtained by rapid solidification show higher strength properties. This conclusion can be drawn from data presented in Table 45, where results of tensile tests of rods prepared by both technologies are compared. The rods obtained by ingot technology were additionally heat treated at regimes corresponding to the maximum strength (hot extrusion + ageing). As one can see in Table 45, the rods prepared using rapid solidification show substantially higher strength properties, if alloys of the same composition are compared. The high strength properties of the rods prepared using rapid solidification are accompanied by sufficiently high values of elongation.

As the hardness measurements showed, no hardening effect took place during ageing of the alloys prepared using rapid solidification, compacting and hot extrusion. This fact can be considered as the result of the Mg solid solution decomposition in the rods during hot extrusion. Meanwhile, investigations of the structure showed that the dispersed structure of the rapid solidified alloys remained to a great extent after compacting and hot extrusion [381].

Mechanical Properties of Multicomponent Mg Alloys with Rare Earth Metals after Rapid Solidification, Compacting and Hot Working

Mechanical properties of multicomponent Mg alloys containing a rare earth metal as one of the components were studied in a number of the works [241, 244, 246, 381, 382]. In [244] thin ribbons of the alloys were prepared by spinning in an inert atmosphere. The ribbons were cut into small pieces (500–250 μm) which were compacted and then hot

Table 46. Tensile properties of multicomponent magnesium alloys prepared by using powder metallurgy / rapid solidification technology.

Composition, mass % (remainder Mg)	Density, g/cm^3	TS, MPa	TYS, MPa	E, %	Reference
5.1 Zn, 5.2 Al, 2.7 Ce	1.89	425	359	17.5	[244]
4.9 Zn, 5.1 Al, 5.3 Ce	1.93	487	425	10.1	[244]
5.1 Zn, 5.2 Al, 2.7 Pr	1.89	427	352	15.9	[244]
5.0 Zn, 5.1 Al, 5.3 Pr	1.94	491	447	3.5	[244]
4.9 Zn, 5.1 Al, 0.5 Si, 5.3 Pr	1.94	516	476	5.0	[244]
5.1 Zn, 5.2 Al, 3.4 Y	1.90	496	448	4.3	[244]
4.9 Zn, 5.1 Al, 6.7 Y	1.93	513	456	5.0	[244]
4.9 Zn, 5.1 Al, 5.5 Nd	1.94	475	434	13.8	[244]
4.7 Zn, 5.1 Al, 1.0 Mn, 5.4 Nd	1.96	476	441	14.0	[244]
2.6 Zn, 8.4 Al, 1.1 Si, 2.8 Nd	1.88	490	465	1.6	[244]
2.6 Zn, 8.7 Al	1.85	372	272	9.5	[244]
2.6 Zn, 8.6 Al, 1.1 Si	1.86	418	365	5.3	[244]
2.6 Zn, 8.6 Al, 2.2 Si	1.88	468	448	1.7	[244]
5.2 Zn, 5.3 Al, 2.2 Si	1.80	455	427	5.7	[244]
5 Al, 8 Y	–	441	413	1.4	[382]
7.5 Zn, 11 Y	–	469	467	1.5	[382]
5 Al, 10 Nd	–	406	386	7.2	[382]
9 Al, 10 Nd	–	500	472	2.8	[382]
9 Al, 5 La	–	485	445	9.4	[382]
7 Al	–	380	294	14.0	[382]

extruded at 200–300°C into round or rectangular rods. Results of tensile tests of the alloys after hot extrusion are presented in Table 46. They show quite a high level of strength properties of the alloys with the rare earth metals, Ce, Pr, Nd and Y at room temperature combined with acceptably high plasticity. Especially high strength properties with tensile strength more than 500 MPa were obtained in alloys with neodymium and yttrium. The alloys also showed high corrosion resistance in salt solutions. The high contents of Zn and Al in the chosen alloys resulted in small solubility of the rare earth metals in solid Mg and the presence of them in the structure mainly in the form of quite dispersed second phases. In general, the strength properties of the alloys with rare earth metals surpassed the strength properties of the alloys of the system Mg–Zn–Al–Si without rare earth metals prepared in the same conditions.

Table 46 includes also data obtained on the multicomponent alloys containing rare earth metals in the work of Baikov Institute of Metallurgy [382]. They confirmed the high level of the strength properties established in of the work [244] and other investigations.

Effect of Rare Earth Metals on Acoustical Properties of Magnesium

Acoustical properties of Mg alloys were studied specially because of their application to ultrasonic delay lines [383, 384]. Ultrasonic delay lines are devices used in electronics to delay electrical signals by a period of time for their treatment. Electrical signals in the ultrasonic delay lines are transformed into ultrasound ones which move along the ultrasound guide and are transformed back at the finish into electrical signals. Delay of the signals results from the substantially less velocity of ultrasound (by several orders) as compared with the velocity of electrical signals. One of the main requirements for material for an ultrasound guide is low ultrasound attenuation. Amongst metallic materials magnesium and its alloys are distinguished by especially low ultrasound attenuation. However, a number non-metallic materials show also quite low ultrasound attenuation. They are quartz glass, single crystals of some salts and some others. Magnesium alloys have the advantages of low cost, better mechanical properties and low density [383–390]. Ultrasound attenuation is the most important acoustical characteristic of magnesium alloys which accounts for their application in ultrasonic delay lines. Another important acoustical characteristic is sound velocity. It amounts to about 5700 and 3200 m/s in pure magnesium for longitudinal and transverse waves, respectively, and changes insignificantly with common alloying and transformations in the structure [391].

Physical Nature of the Ultrasonic Attenuation in Metallic Materials

The physical nature of ultrasound attenuation in solids has been studied in many investigations. It is quite complicated and can be different depending on frequency, amplitude and mode of ultrasound waves. Magnesium ultrasonic delay lines are used commonly with longitudinal or transverse ultrasound waves having a frequency in the range of 5–40 Mc and low amplitude of $\sim 10^{-7}$. For such a frequency range, the amplitude level and wave modes, the four main mechanisms of ultrasound attenuation in polycrystalline metallic materials can be substantial [391, 392]. They are

- scattering on the grain boundaries [386, 387, 393–395],
- scattering arising from presence of different phases in structure of the alloys [396, 397],
- oscillation of dislocations in the metal lattice [398],
- heat flows between areas of extensions and compressions in the longitudinal wave [399].

Analysis of the possible mechanisms of the ultrasound attenuation in common magnesium alloys showed that all of them should be taken into consideration, but scattering on the grain boundaries was the most important [384, 400]. Scattering on the grain boundaries of Mg solid solution provides the main part of the whole attenuation. Contributions by the other mechanisms of ultrasound attenuation are significantly less. Especially low attenuation in common magnesium alloys is expected from heat flows between areas of extensions and compressions in longitudinal waves. Actually this mechanism can be ignored at typical frequencies of 10–20 Mc for Mg delay lines. A quite low contribution to the whole attenuation can be expected also from scattering by crystals of the second phases in the Mg solid solution matrix. Calculations showed that it became substantial only with large quantities of the second phases and great size of their crystals, whereas in the common Mg alloys used in practice the quantities of the second phases in the structure are insignificant and the size of their crystals is small [384, 400]. Attenuation connected with oscillation of dislocations is quite sensitive to small quantities of impurities in the lattice decreasing abruptly with their increase [401–404]. Such a behaviour of

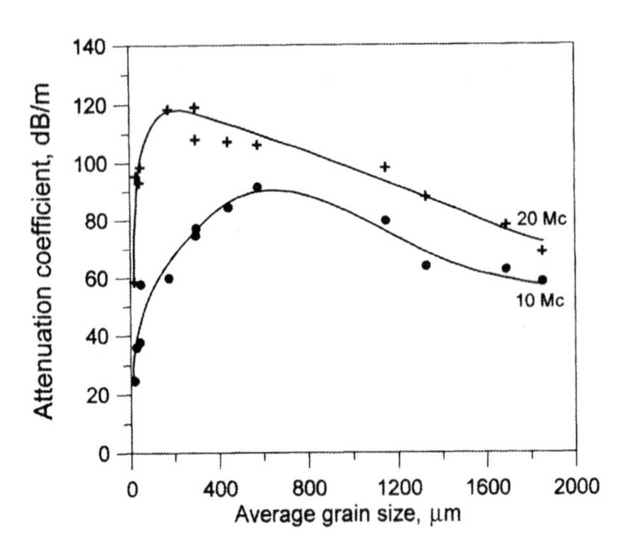

Figure 124. Dependence of ultrasound attenuation in Mg on grain size. Longitudinal waves.

dislocations is explained by their being pinned by atoms or disperse particles of impurities. So, in the alloys, the contribution from this mechanism to the whole attenuation should also be insignificant in the well annealed state. However, dislocations can become free from pins when the alloys are being deformed. In this case attenuation connected with oscillation of dislocations should increase [405–406].

For the acoustical aspect, scattering on the grain boundaries of a metal solid solution is caused by elastic anisotropy of separate grains. The magnesium crystal lattice is characterised by small elastic anisotropy and it is one of the causes of low ultrasound attenuation in magnesium alloys [391]. Taking into consideration general physical concepts it should depend on grain size according to a curve with a maximum, because theoretically diminution of grain size would result in glass at the limit where scattering on the grain boundaries should be absent. On the other hand, increase of grain size would result in a single crystal at the limit where scattering on the grain boundaries should also be absent. Dependence of ultrasound attenuation on grain size in various metals and alloys was observed experimentally and treated theoretically in [386, 387, 393–395, 408, 409]. For unalloyed magnesium it was established for a wide range of grain size in [410] and is presented in Figure 124. The various grain size was obtained by annealing of the specimens at different temperatures. The attenuation maximum on every curve in Figure 124 corresponds approximately to the ultrasound wave length which amounts to about 570 and 285 μm for the longitudinal waves used with frequencies 10 and 20 Mc, respectively. In accordance with this the attenuation maximum is shifted to the lower grain size of magnesium for the higher frequency of 20 Mc. In common magnesium alloys produced in industry the grain size is substantially less than the ultrasound wave length at such frequencies which are used in the delay lines most frequently. Therefore, as a rule, in magnesium alloys the less grain size the less ultrasound attenuation.

The above specified mechanisms of ultrasound attenuation concern continuous media conforming to the metal or alloy pieces without any shrinkage pores, cavities, cracks and other macrodefects which can occur in real materials. All macrodefects can evoke

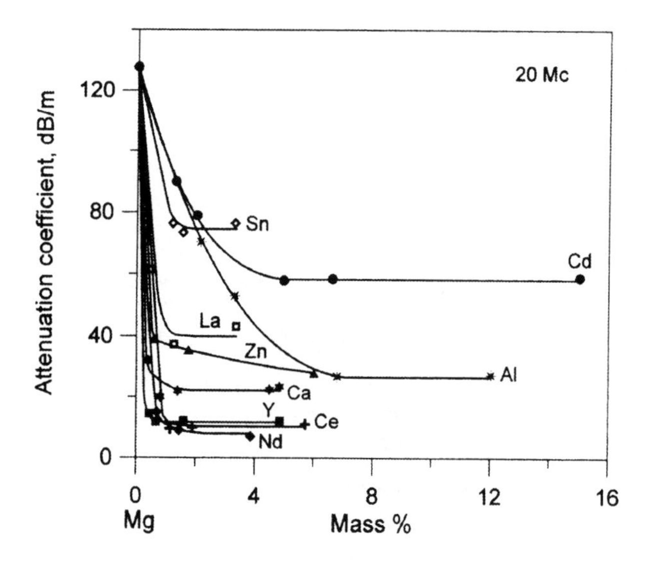

Figure 125. Effect of alloying elements on ultrasound attenuation in Mg. Longitudinal waves, 20 Mc.

significant attenuation of ultrasound and, therefore, should be prevented if low ultrasound attenuation is required [384].

Effect of Composition and Structure on Ultrasound Attenuation in Magnesium Alloys

Because of the enlarged grain size, coarse crystals of the second phases and existence of some shrinkage pores cast magnesium alloys showed high ultrasound attenuation and were not studied in detail [391]. Most of the investigation of ultrasound attenuation of magnesium alloys was carried out, therefore, using wrought materials. Results of the comparative investigations of ultrasound attenuation in binary Mg systems with reasonable contents of alloying elements are presented in Figure 125 [384]. The alloys were prepared by melting, cast into round ingots and hot extruded then into round rods with cross sectional area reduction of about 90%. The extrusion temperature of the alloys (about 400°C) was chosen aiming to ensure completion of recrystallysation and at the same time to prevent excessive grain growth. Ultrasound attenuation was measured by the echo-impulse method using longitudinal waves and frequency 20 Mc.

The data presented in Figure 125 show a decrease of the attenuation coefficient of Mg at the first small contents of all alloying elements used. With further increase of contents, the attenuation coefficient of the binary alloys tends to become approximately constant within the limits of the concentration range considered. At higher concentrations, the attenuation coefficient could increase [384]. The binary Mg alloys with the rare earth metals, cerium and neodymium, show the lowest values of the attenuation coefficient. Quite low values of the attenuation coefficient are also shown by alloys with another rare earth metal, yttrium. The lower attenuation coefficient of the alloys with rare earth metals is caused mainly by smaller grain size. The rare earth metals should also pin dislocations quite effectively.

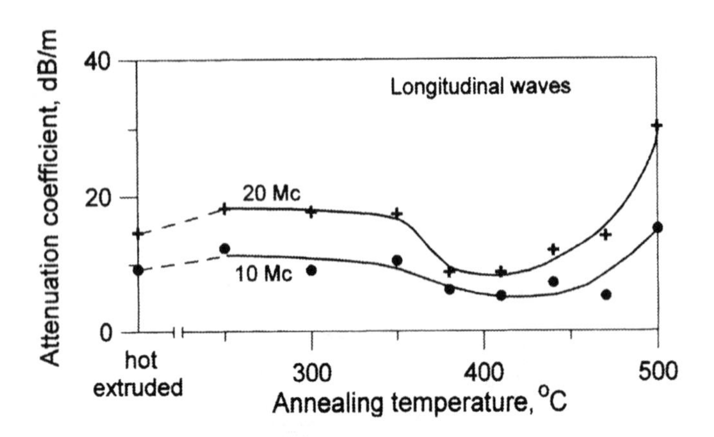

Figure 126. Effect of annealing temperature on ultrasound attenuation in the alloy Mg–1.7 mass % MM.

Grain size is the most important structural factor for ultrasound attenuation in Mg–RE alloys. For low attenuation the grain size should be as small as possible. The lowest attenuation coefficient was observed in the alloys with a structure where recrystallisation after hot working was completed without further grain growth. Recrystallisation replaces grains with lattice imperfections by grains with more perfect lattice and pinned dislocations. Besides, more coarse grains are replaced during recrystallisation by the finer ones. Growth of the recrystallised grains is accompanied by increase of ultrasound attenuation. Such a connection between the structure of the Mg–RE alloys and ultrasound attenuation is revealed during annealing of the hot worked products. A typical example of it is demonstrated by the results of the attenuation measurements shown in Figure 126 [384]. In this case the alloy Mg–1.7 mass % MM (mishmetal) was studied. The rods of the alloy were obtained by hot extrusion, but the extrusion temperature was chosen to be too low so that its structure was not recrystallised completely. Specimens of the alloy were annealed at successive increased temperatures. Every next temperature was higher by 30–50°C than the previous one. Annealing time at every temperature was 8 h. After each annealing, the structure of the alloy was investigated. The attenuation coefficient was measured at longitudinal waves with frequencies 10 and 20 Mc.

Figure 126 shows an insignificant change of attenuation coefficient of the alloy Mg–1.7 mass % MM up to annealing temperature 350°C. Also no visible change in the structure of the alloy at these annealing temperatures was observed. Further increase of annealing temperature was accompanied by substantial decrease of attenuation coefficient and observation of the structure showed the beginning and continuation of the recrystallisation process. Beginning with annealing temperature 440°C the attenuation coefficient rises successively with increasing temperature in accordance with the observation of the growth of the Mg solid solution grains after completion of recrystallisation.

The described dependencies of the ultrasound attenuation on composition and structure lead to the conclusion that magnesium alloys with cerium and neodymium are the most suitable for development of materials with the least ultrasound attenuation. Hot working, or hot working followed by annealing of the materials, should result in as small as possible grain size in them.

Corrosion Resistance of Magnesium Alloys with Rare Earth Metals

Corrosion resistance is of great importance for practical applications of magnesium alloys. In general, magnesium alloys are quite resistant against corrosion attack in rural and some industrial environments. In general, the corrosion rate of magnesium alloys is between that of aluminium alloys and low carbon steels. Nevertheless, the corrosion behaviour of magnesium alloys may be quite different in different environments, and some environments can induce significant corrosion damage. Corrosion behaviour of magnesium and its alloys depend upon alloying elements, their contents and the structure formed. It changes also significantly with the presence of "detrimental" impurities. The main ones are Ni, Fe and Cu. Forming galvanic couples, detrimental impurities in magnesium and its alloys increase their corrosion rate. A number of protective coatings have been developed for magnesium alloys, such that, in general, corrosion of all of them can be prevented effectively even in quite corrosive environments [373, 411, 412].

Corrosion Resistance of Binary Mg–RE Alloys

Insofar as corrosion behaviour of alloys depends upon the acting environment and impurities in the alloys, it is reasonable to assess the influence of separate alloying elements on corrosion based on results of comparative tests conducted in the same conditions. Results of such tests for some binary magnesium systems including Mg–RE were presented in [344, 373, 411]. The alloys were prepared with the main impurity contents in limits Fe $\leq$ 0.003%, Ni $\leq$0.0005%, Cu $\leq$0.003%, except the systems Mg–Ca, Mg–Sc and Mg–Ce where Fe content was 0.007%, 0.02% and 0.008%, respectively. The alloying element contents in the alloys were within the limits of Mg solid solution and some more. The alloys were tested in 3% NaCl aqueous solution and in humid air. There were some differences between results of the tests in both environment for the separate alloys, but, in general, the influence of the rare earth metals on corrosion behaviour as compared with other alloying elements turned out to be the same. Only a part of the rare earth metals were used for preparation of the alloys. They were La, Ce, Pr, Nd, Sm, Y, Sc. However these rare earth metals make it possible to assess the influence of the other rare earths on corrosion of magnesium alloys.

The results of the corrosion tests in 3% aqueous solution are presented in Figure 127 [411]. They show an increase of corrosion rate of the alloys with increasing rare earth contents. This regularity takes place for all the rare earth metals used, except Sm which shows some decrease of corrosion rate with increasing content up to 5.4 mass %. In general, the corrosion rate of the alloys with the rare earth metals is substantially less than that with the known detrimental elements, such as Cu, Ag, Bi. At high contents of Y and Sc the corrosion rate is high, but less than that of the alloys with Cu, Ag and Bi. On the other hand, the alloys with such alloying elements as Mn, Zr, Cd, showed certainly less corrosion rate than that of alloys with the rare earth elements in general. Corrosion rate of the alloys with well known alloying elements for magnesium alloys, Al, Zn and Si, turned out to be close to that of magnesium alloys with rare earth metals. It is reasonable to note one more feature of alloys with the rare earth metals. At small contents they showed lower corrosion rates than the initial magnesium. This fact may be explained by interaction of the rare earth metals in small contents with impurities in initial magnesium resulting in suppression of

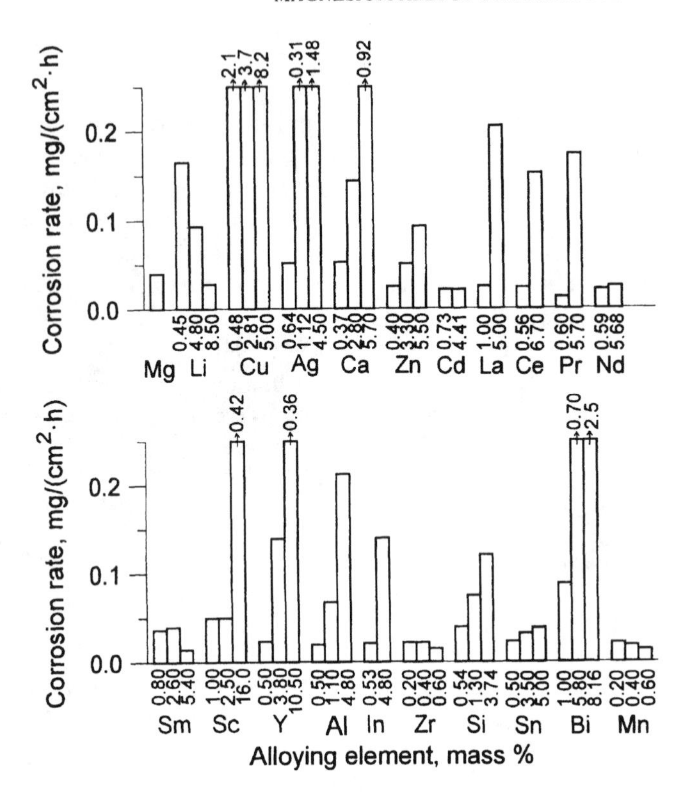

Figure 127. Corrosion rate of binary Mg alloys in 3% NaCl aqueous solution.

their detrimental influence on corrosion resistance.

The alloys containing 8–9 mass % Y inclined to corrosion cracking as with the Mg–Al alloys. Such corrosion behaviour depends on heat treatment [373].

Decrease of corrosion resistance of the Mg–Y alloys with increase of Y contents is explained by formation in their structure of the equilibrium compound $Mg_{24}Y_5$ which is supposed to be an effective cathode for the solid solution [344, 373].

Corrosion behaviour of the binary magnesium alloys considered leads to the conclusion that the rare earth metals can damage corrosion resistance of magnesium, but these alloys should retain enough corrosion resistance for commercial usage, as at least they should be corrosion resistant at the same level as the majority of the common commercial magnesium alloys with aluminium and zinc as the main alloying elements. Corrosion resistance of binary magnesium alloys with the heavy rare earth metals was not studied in [344, 373, 411], but it can be reasonably supposed to be close to that of the alloys with Y and Sc, taking into account the closeness of the alloys with rare earth metals of the yttrium subgroup in chemical properties.

Influence of Additional Alloying Elements on Corrosion Resistance of Mg–RE Alloys

In a few investigations the effect of additional alloying elements on corrosion resistance of Mg–Y alloys was studied [413–416]. According to [413], additions of Mn up to about 1.1

mass % and Zr up to about 0.6 mass % to the alloy Mg–8 mass % Y improve its corrosion resistance in the 0.005N solution of NaCl in water. On the other hand, addition of Cd up to 3 mass % decrease corrosion resistance of Mg–8 mass % in the same aqueous solution. Meanwhile, in [414] where corrosion of the ternary Mg–Y–Cd alloys was studied, results contradictory to [413] were obtained. Corrosion resistance in [414] was investigated in artificial river water containing $NaHCO_3$ – 300 mg/l, $CaCl_2$ – 50 mg/l, $MgSO_4$ – 50 mg/l. Such a water composition corresponded by total salt content (400 mg/l), electrical resistivity (20 Ohm·m) and pH = 7.5–7.8 to the water characteristics of Volga river and soil waters of central Russia. The alloys used contained up to 6 mass % Cd and 15 mass % Y. In these conditions [414] Cd practically did not change the corrosion rate of the alloys, but Y tended to decrease it.

In the book [373] some information about the influence of Zn on corrosion of the binary Mg–Y alloys with 7–8 mass % Y was given. According to [373], addition of up to 2 mass % Zn significantly increased corrosion resistance of Mg–Y alloys. At such contents Zn was dissolved completely in Mg solid solution. Higher contents of Zn resulted in formation of intermetallic phases in the structure and, as a result, decreased corrosion resistance of the Mg–Y alloys. Addition of 0.1–0.2 mass % Zr increased corrosion resistance of the alloy Mg – 9% Y – 2% Zn. This effect of Zr is explained by diminution of the Fe content in the alloy taking into account the results of chemical analysis. The beneficial effect of Mn on corrosion resistance of the Mg–Y alloys was explained in [373] by enrichment of the alloy surface by Mn during corrosion process and the formation of a quite stable oxide film with Mn.

In [416] stress corrosion cracking of the Mg–Y alloys containing additionally Zn, Cd and Zr was studied. The alloys were used in the form of hot rolled sheets 2 mm in thickness. Before rolling, the alloys were homogenized at 480°C for 24 h. The rolling temperature was 460°C and a number of intermediate heatings were used. The tests of stress corrosion cracking were performed by alternate immersion into 0.001% NaCl solution in water at a

Table 47. Stress corrosion cracking of the Mg–Y–base alloys [416].

Alloy composition, mass %	Time before corrosion cracking, days
Mg – 1.5 Y	>365
Mg – 4.5 Y	>365
Mg – 6.0 Y	>365
Mg – 8.4 Y	>365
Mg – 2.5 Y – 2.2 Zn	168
Mg – 7.3 Y – 2.3 Zn	126
Mg – 9.4 Y – 0.5 Zn	>365
Mg – 9.1 Y – 2.0 Zn	236
Mg – 9.3 Y – 2.1 Zn – 0.43 Zr	55
Mg – 9.3 Y – 2.0 Zn – 0.70 Zr – 1.5 Cd	120
Mg – 6.7 Y – 0.5 Zn – 0.40 Zr – 1.5 Cd	62

stress amounting to 90% of yield strength. Results of the tests are presented in Table 47. They showed high resistance against stress corrosion cracking of the binary Mg–Y alloys at all yttrium contents even at the highest of them when coarse crystals of $Mg_{24}Y_5$ can appear in structure. Addition of zinc to the Mg–Y alloys resulted in decrease of the stress corrosion resistance. The detrimental effect of zinc became stronger as its content increased. In the opinion of the authors [416], inclination to stress corrosion cracking of the Mg–Y alloys with zinc addition resulted from the appearance of the ternary phase X in the structure. Zirconium decreased sharply the stress corrosion resistance of the Mg–Y–Zn alloys as a result of formation of its hydrides in this case. The fourth alloying element Cd promoted some corrosion resistance of the Mg–Zn–Zr alloys. This conclusion was supported by comparison of the alloys with the same Y content 9.3 mass %.

Chapter 5

Commercial Magnesium Alloys with Rare Earth Metals, their Composition and Properties

A number of commercial magnesium alloys containing rare earth metals have been developed and recommended for practical applications. Some of them have been produced and used in industry for a long time. Commercial magnesium alloys contain mainly mishmetal consisting substantially of cerium and lanthanum (most of the alloys developed), neodymium and yttrium. The alloys containing yttrium are the latest of the commercial alloys developed for industry. Meanwhile, there is a tendency for development of alloys also containing other rare earth metals, for example gadolinium, as main additives. Other rare earth metals are also present in commercial magnesium alloys as subsidiary constituents, for example praseodymium in the alloys where neodymium is the main alloying element and heavy rare earth metals in the alloys with yttrium.

Although there are many similar features between commercial magnesium alloys with rare earth metals developed in the West and Russia, there are also certain differences between them. Therefore, it is reasonable to describe both groups of the commercial alloys separately.

Magnesium Alloys of the West

The alloys developed and used in industry of the Western World for a long time are fully described in the ASTM Standards (USA) [417, 418]. Their compositions are shown in Table 48. All the alloys were developed and used for applications as light structural materials. Most of them are intended for castings. They are those for sand castings with numbers EZ33A, ZE41A, ZE63A, EQ21A, QE22A, WE54A, QH21A [417]. Only one wrought alloy containing rare earth metals, ZE10A for sheets and plates [418] is in the ASTM Standards. In general, the specified impurity limits in the alloys containing rare earth metals are the same as in the alloys without rare earth metals. The total permissible impurity content is specified to be 0.30 mass %.

Cast Alloys

Cast alloys are characterized by different mechanical and cast properties. There are substantial differences in their cost connected with the kind of rare earth metals in their composition. Minimum values of the tensile properties of the alloys at room temperature are presented in Table 49 according to ASTM Standard [417] in comparison with the strongest cast magnesium alloys of other important systems. As one can see in Table 49 no alloy with rare earth metals surpasses the strongest cast alloy ZK61A of the system Mg–Zn–Zr in

Table 48. Chemical composition of the West's commercial magnesium alloys containing rare earth metals for castings [417] and for wrought products (sheet and plate) [418].

Alloy number	Composition (main components), mass %						
	Mg	RE	Th	Zn	Ag	Cu	Zr
EZ33A	remainder	2.5–4.0	–	2.0–3.1	–	≤0.10	0.50–1.0
ZE41A	remainder	0.75–1.75	–	3.5–5.0	–	≤0.10	0.40–1.0
ZE63A	remainder	2.1–3.0	–	5.5–6.0	–	≤0.10	0.40–1.0
EQ21A	remainder	1.5–3.0[a]	–	–	1.3–1.7	0.05–0.10	0.40–1.0
QE22A	remainder	1.8–2.5[a]	–	–	2.0–3.0	≤0.10	0.40–1.0
WE54A[b]	remainder	1.5–4.0[c]	–	≤0.20	–	≤0.03	0.40–1.0
QH21A	remainder	0.6–1.5[a,d]	0.6–1.6[d]	≤0.20	–	≤0.10	0.40–1.0
ZE10A	remainder	0.12–0.22	–	1.0–15	–	–	–

[a]Rare earth elements are in the form of didymium, with not less than 70% neodymium, and the remainder substantially praseodymium.
[b]Yttrium content for alloy WE54A shall be 4.75 to 5.5%.
[c]Rare earth are 1.5–2.0% neodymium, the remainder being heavy rare earth.
[d]Thorium and didymium total is 1.5–2.4%.

Table 49. Tensile properties of the West's commercial Mg alloys for castings at room temperature [417].

Alloy number	Temper	TS, min, MPa	TYS, min, MPa	E, min, %
EZ33A	T5	138	96	2
ZE41A	T5	200	133	2.5
ZE63A	T6	276	186	5
EQ21A	T6	234	172	2
QE22A	T6	241	172	2
WE54A	T6	255	179	2
QH21A	T6	241	186	2
ZK61A (Mg–Zn–Zr)	T6	276	179	5
AZ92A (Mg–Al–Zn–Mn)	T6	234	124	1
HK31A (Mg–Th–Zr)	T6	186	89	4

strength properties at room temperature. But alloys with rare earth metals have substantially higher strength properties at elevated temperatures as compared with ZK61A. Meanwhile, some strength characteristics of alloys with rare earth metals at elevated temperatures are less than those of alloys with the radioactive element thorium.

The alloy EZ33A (ZRE1 by UK designation) is characterized by good strength properties and high creep resistance up to 200°C combined with good castability. It is weldable and enables high hermetic parts to be cast [294, 419]. Some typical strength properties of the alloy EZ33A at elevated temperatures are shown in Table 50.

Table 50. Typical strength properties of the alloys EZ33A and QE22A for castings at elevated temperatures [420].

Alloy number	Temper	Tensile strength, MPa		Tensile yield strength, MPa		Creep strength, 0.5% in 100 h, MPa		
		204°C	316°C	204°C	316°C	204°C	260°C	316°C
EZ33A	T5	145	83	83	55	69	28	10
QE22A	T6	193	103	172	83	90	34	10

The alloy ZE41A (RZ5 by UK designation) shows higher strength properties than the alloy EZ33A. The alloy is also weldable and hermetic. It is characterized by good castability with high crack resistance and low inclination to micropore formation during casting [294, 419].

The alloy ZE63A is distinguished by high strength properties at room temperature which are close to those of the strongest alloy ZK61A of the system Mg–Zn–Zr. Besides, the alloy ZE63A shows good casting properties and high fatigue resistance [417, 419]. In [419] the version ZE63B was reported. The version ZE63B differed from ZE63A by addition of 0.75–1.25 mass % Ag. The main characteristics of ZE63B were high strength and plasticity, weldability and good casting properties [419].

The alloy QE22A (MSR by UK designation) is distinguished by high strength properties, especially tensile yield strength at elevated temperatures. The alloy contains didymium consisting mainly of neodymium and, therefore, its strength properties at elevated temperatures surpass the alloys with mishmetal. Some typical strength properties of the alloy at elevated temperatures are shown in Table 50. It is recommended to be used up to 250–290°C. Other important characteristics of the alloy QE22A are weldability and ability to obtain hermetic parts. In the United Kingdom two versions of the alloy of this type were suggested. The versions differ from each other by didymium contents at the same concentrations of silver and zirconium. The version MSR–A is distinguished by less didymium which amounts to about 1.7 mass % and is actually outside of the alloy QE22A composition registered in the ASTM Standard [417]. The version MSR–B contains more didymium amounting to about 2.5 mass % and is close to the QE22A composition. As compared with the version MSR–B, the version MSR–A shows lower strength properties combined with higher plasticity. The alloy QE22A is used in T6 condition (solution treatment + ageing) [294, 419, 420].

The alloy EQ21A shows actually the strength properties as the alloy QE22A, and can be used at temperatures higher than 200°C. EQ21A is distinguished by good castability. As compared with QE22A, it contains less expensive silver (about 1.5 mass %) and about 0.08 mass % Cu additionally [376, 417].

The alloy WE54A is one of the latest Mg cast alloys. Its main feature is the presence of yttrium as the dominant alloying element. Apart from this, WE54A contains Nd and heavy rare earth metals, mainly Dy, Er, Yb, Gd [376]. The main advantage of the alloy WE54A is its high strength properties at elevated temperatures. In this aspect it surpasses the most well known alloy QE22A. The alloy WE54A is suggested to be used after solution treatment by quenching from 525°C for 8 h in hot water or polymers followed by ageing at 250°C for 16 h. The alloy shows good weldability and castability, and good corrosion resistance. It can be cast without substantial difficulties into thin sections [376]. Typical mechanical properties of the alloy are at room temperature TS = 255 MPa, TYS = 185 MPa, E = 3% and at 250°C TS = 225 MPa, TYS ≅ 160 MPa, E = 6% [421]. WE54A shows higher creep resistance and higher resistance against overageing at elevated temperatures than the cast magnesium alloys for high temperature applications QE22 and QH21A. The alloy WE54A is recommended for use up to 300°C [422].

Along with the standard alloy WE54A, several cast magnesium alloys of the same base system Mg–Y–Nd were developed and tried for industrial applications. One of them is the alloy WE43 containing about 4 mass % Y and 3 mass % Nd [423]. Typical mechanical properties of WE43 were reported to be at room temperature TS = 265 MPa, TYS = 185 MPa, E = 7–5%, at 250°C TS = 210–230 MPa, TYS = 150–170 MPa, E = 15–20%, at 300°C TS = 150–170 MPa, TYS = 110–130 MPa, E = 30–50%. The alloy was recommended for use up to 300°C [423].

The alloy QH21A may be considered as similar to the alloy QE22A with the rare earth metal didymium replaced by thorium. As a result of the Th presence the alloy QH21A shows higher strength at elevated temperatures as compared with the alloy QE22A, especially higher creep resistance. During creep tests the alloy shows extension of 0.2% at 250°C for 100 h under stress of about 40 MPa. Required mechanical properties of the alloy are obtained after solution treatment followed by ageing (T6 condition). The recommended regime for solution treatment consists of quenching from 525°C for 10 h in hot water (60°C), and that for ageing is 200°C for 16 h. The alloy can be successfully welded by the argon-arc method in cast and heat treated conditions. Its castability is similar to that of the alloy QE22A [424].

In addition to the above alloys, AE42X1 was developed for castings [425]. The alloy contains (in mass %) Al 3.5–4.5, RE 2.0–3.0, Mn ~0.25, Zn <0.20, Fe <0.004, Ni <0.004, Cu <0.004, Be 0.0004–0.001, other impurities <0.01 (each), Mg remainder. Its main advantage is retention of sufficiently high strength properties at moderately elevated temperatures combined with good castability. The alloy AE42X1 can be used at temperatures up to 175°C. It shows high corrosion resistance [425].

There are various applications of the above mentioned alloys, firstly for aircraft and missile components [63, 294, 422]. Thus, castings of the alloy EZ33A (ZRE1) were used in aircraft engines for diffuser and compressor cases, covers of combustion chambers and other parts which could be heated when engines are working. Cast parts of EZ33A (ZRE1) were used also in the research rockets "Skylark" and the missile "Bomark" [294]. One of the possible applications of the alloy AE42X1 are parts of automatic transmissions [425].

Wrought Alloys

The only wrought magnesium alloy containing rare earth metals included in the ASTM Standard of USA [418] is ZE10A. Its composition is shown in Table 48. The alloy is used for production of rolled semiproducts such as sheets and plates. Its main characteristics are the ability to be rolled significantly without difficulties, moderate strength, weldability and low cost [419]. Tensile strength properties of the rolled alloy ZE10A are shown in Table 51.

Table 51. Mechanical properties of sheets and plates of the alloy ZE10A [418].

Temper*	Thickness, mm	TS, min, MPa	TYS, min, MPa	E, min, %
O	0.40–1.60	207	124	15
O	1.60–6.30	207	103	15
O	6.30–12.50	200	83	12
H24	0.40–3.20	248	172	6
H24	3.20–5.00	234	152	6
H24	5.00–6.30	214	138	6

*O means full annealing.
H24 means work hardening + partial annealing.

In [426] the wrought semiproducts of the alloy MSR (QE22A) were studied. The alloy was homogenized at 525°C followed by cooling either by quenching in water or in a furnace. Afterwards hot extrusion with cross area reduction of 70% and ageing were carried out. Extrusion temperature was chosen in the range 100–400°C. Ageing was conducted at 200°C with various exposures from 1 to 48 h. Hot extrusion increased significantly the strength properties of the alloy. Thus, in the case of quenching in water and extrusion at 100°C the alloy showed TS = 400 MPa, TYS = 360 MPa at E = 2.5%. The strength properties decreased and plasticity increased with rising extrusion temperature, so that they amounted to TS = 330 MPa, TYS = 300 MPa, E = 15%, when the alloy was hot worked at 400°C.

In general, corrosion resistance of both cast and wrought commercial magnesium alloys with rare earth metals is similar to that of the rest of commercial magnesium alloys. All of them are characterized as well resistant to atmosphere, but attacked by salt water if special surface treatment improving the corrosion resistance is not applied [63].

Magnesium Alloys of Russia

As in the West a number of commercial magnesium alloys containing various rare earth metals were developed for structural applications and used in the former USSR and then in Russia. Most of them were included in the State Standards [427, 428]. In addition to them, some new commercial magnesium alloys containing rare earth metals were developed and suggested for industry. They have not yet been included in the State Standards.

Cast Alloys

Commercial magnesium alloys with rare earth metals for shape castings are shown in Table 52. Those are the alloys which are included in the State Standard [427] and their

Table 52. Chemical composition of the Russian commercial magnesium alloys containing rare earth metals for castings [427].

Alloy number	Composition (main components), mass %							
	Mg	MM*	Nd	Y	La	In	Zn	Zr
ML11	remainder	2.5–4.0	–	–	–		0.2–0.7	0.4–1.0
ML9	remainder	–	1.9–2.6	–	–	0.2–0.8	–	0.4–1.0
ML10	remainder	–	2.2–2.8	–	–	–	0.1–0.7	0.4–1.0
ML19	remainder	–	1.6–2.3	1.4–2.2	–	–	0.1–0.6	0.4–1.0
ML15	remainder	–	–	–	0.6–1.2	–	4.0–5.0	0.7–1.1

*MM is mishmetal containing not less than 45% Ce.

Table 53. Tensile properties of the Russian commercial Mg alloys for castings at room temperature [427].

Alloy number	Temper*	TS, min, MPa	TYS, min, MPa	E, min, %
ML11	cast	117.5	–	1.5
	T2	117.5	–	1.5
	T4	137.0	83.3	3.0
	T6	137.0	98.0	2.0
ML9	T6	226.0	108.0	4
ML10	T6	226.0	137.0	3
	T61	235.0	137.0	3
ML19	T6	216.0	118.0	3
ML15	T1	206.0	127.5	3
ML5 (7.5–9.0% Al, 0.2–0.8% Zn, 0.15–0.5% Mn)	T4	226.0	83.3	5
ML12 (4.0–5.0% Zn, 0.6–1.1% Zr)	T1	226.0	127.5	5
ML8 (5.5–6.6% Zn, 0.2–0.8% Cd, 0.7–1.1% Zr)	T61	275.0	176.5	4

*T1 means ageing.
T4 means solution treatment with quenching in air.
T6 means solution treatment with quenching in air + ageing.
T61means solution treatment with quenching in water + ageing.

compositions are shown according to this Standard. Table 53 shows the minimum tensile properties of these alloys at room temperature according to the same Standard [427] which are required to be obtained using separately cast specimens. Table 54 shows typical mechanical properties of the alloys at elevated temperatures [429–431]. Also, in Tables 53 and 54, the strength properties of the widely used commercial Mg alloys without rare earth metals are presented for comparison.

The alloys with rare earth metals contain mishmetal consisting mainly of cerium, lanthanum, neodymium and yttrium. Other rare earth metals are not added specially to the commercial magnesium alloys for shape castings, although some of them are present as

Table 54. Typical mechanical properties of the Russian magnesium alloys for castings at elevated temperatures [429–431].

Alloy number	Temper*	Temperature, °C	TS, MPa	TYS, MPa	E, %	σ_{100}, MPa**	$\sigma_{0.2/100}$, MPa**
ML11	T6	200	140	80	6	115	65
(Mg–MM–Zn–Zr)		250	130	75	8.5	55	30
		300	105	60	30	22	–
ML9	T6	200	210	135	5	–	95
(Mg–Nd–In–Zr)		250	170	120	5	–	45
		300	120	100	20	18	–
ML10	T6	200	190	140	8	130	110
(Mg–Nd–Zn–Zr)		250	165	130	13	70	38
		300	135	110	17	25	–
ML19	T6	200	215	120	3	–	–
(Mg–Nd–Y–Zn–Zr)		250	200	110	7	–	–
		300	150	100	12	–	–
		350	110	80	15	–	–
ML15	T1	150	145	105	5	100	65
(Mg–Zn–La–Zr)		200	125	85	13	65	40
		250	100	65	16	40	20
		300	75	50	16		–
ML5	T4	150	185	60	12	85	25
(Mg–Al–Zn–Mn)		200	155	50	15	45	8
		250	120	40	15	25	–
		300	90	–	–	–	–
ML12	T1	150	160	110	8	80	40
(Mg–Zn–Zr)		200	140	85	10	50	25
		250	100	60	12	20	–
ML8	T6	150	155	110	15	85	45
(Mg–Zn–Cd–Zr)		200	120	90	15	–	–

*T1, T4, T6 mean the same tempers as in the previous Table 53.
**σ_{100} is the stress resulting in rupture of material in the 100 h exposure at an elevated temperature. $\sigma_{0.2/100}$ is the stress resulting in creep of 0.2% in 100 h.

minor constituents of mishmetal or inevitable impurities of neodymium (the main one of them is praseodymium) and yttrium. Unlike the West's alloys, the Russian alloys do not contain silver and thorium, but contain other minor alloying elements, for example indium in the alloy ML9. Meanwhile, all Russian alloys contain zirconium for modification of grains as those of the West.

The alloys ML11, ML9, ML10 and ML19 have been developed for applications at elevated temperatures and are distinguished by their higher strength properties and more retention of strength after heating than other Mg structural alloys of the base systems Mg–Al–Zn–Mn and Mg–Zn–Zr. Table 54 demonstrates this fact.

The alloy ML11 contains mishmetal and shows lower strength properties than ML9, ML10, ML19 both at room and at elevated temperatures, but ML11 is the cheapest one amongst them. The alloys ML9 and ML10 are developed on the basis of the system Mg–Nd–Zr. They surpass the alloy ML11 by strength at room and elevated temperatures, but are more expensive because they contain neodymium. As compared with ML10, the

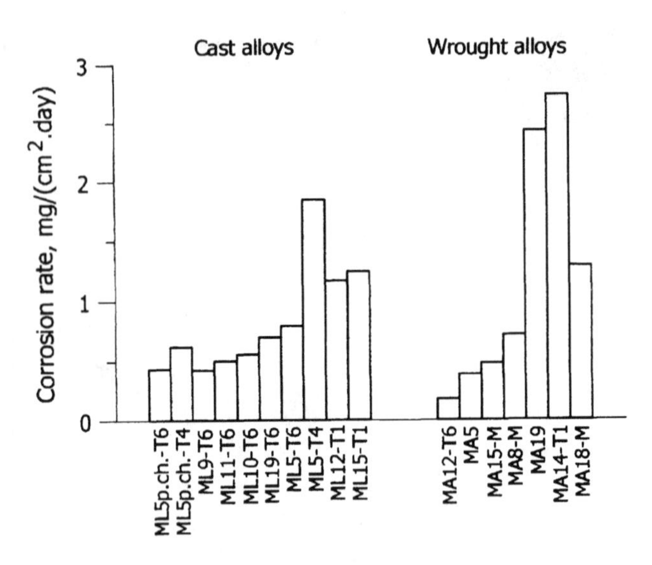

Figure 128. Corrosion rate of the Russian commercial alloys in 3% NaCl aqueous solution. T1 –ageing without preliminary quenching, T4 – quenching, T6 – quenching + ageing, M – annealing.

alloy ML9 shows higher tensile strength at 200 and 250°C, but less tensile yield strength both at room and at elevated temperatures. The alloy ML9 retains also better strength after heating at 200 and 250°C [430]. The alloy ML19 shows higher tensile strength at elevated temperatures as compared with alloys ML9 and ML10 owing to the presence of 1.4–2.2 Y. Besides, ML19 retains its strength after heating at 200 and 250°C better than the alloys ML9 and ML10.

The Russian alloy ML11 is like the West's alloy EZ33A, although differs from it by significantly less Zn. The Russian alloys ML9 and ML10 are like the West's alloy QE22A, but differ from it by the absence of silver and the presence of small contents of indium (ML9) or zinc (Ml10). The Russian alloy ML19 differs substantially from any known Western alloys. It belongs to the same base system Mg–Y–Nd as the Western alloy WE54A, but contains significantly less Y.

The alloy ML15 should be considered as that belonging to the system Mg–Zn–Zr where La is added as a subsidiary alloying element. Additional alloying by La results in some increase of strength at the highest temperatures, density of castings and weldability. Nevertheless, strength and plasticity at room temperature decrease with La [431].

Another Russian alloy of the base system Mg–Zn–Zr with small RE addition was named ML17. This alloy has not yet been included in the State Standard [427] and, therefore, is not shown in Tables 52 and 53. It contains (in mass %) Zn 7.8–9.2, Nd 0.03–0.3, Cd 0.2–1.2, Zr 0.7–1.1, Mg and impurities being the remainder. The alloy is reported to show quite high strength properties resulting from high Zn content and presence of 0.03–0.3 mass % Nd [431]. Typical mechanical properties of the alloy ML17–T61 at room temperature are TS = 305 MPa, TYS = 205 MPa, E = 5.0%, CYS = 205 MPa [431].

The corrosion behaviour of cast magnesium alloys containing rare earth metals may be characterized by the results of corrosion tests reported in [373, 411]. They are shown in Figure 128 for full immersion into 3% NaCl aqueous solution. They show a low corrosion rate for the alloys ML9 and ML10 containing neodymium as the main additive and ML11 with mishmetal as the main additive. It is of the same order as compared with the corrosion rate of the alloy ML5p.ch. which is a version of the alloy ML5 (Mg–Al–Zn–Mn) with less impurity contents for better corrosion resistance. On the other hand, the alloy ML5 with the normal level of impurities showed significantly higher corrosion rate than the alloys ML9, ML10 and ML11. The alloy ML19 containing neodymium and yttrium as main additives showed higher corrosion rate than that shown by ML9, ML10 and ML11, but less than corrosion rate shown by ML5 with standard impurities. All the mentioned alloys ML9, ML10, ML11 and ML19 showed significantly less corrosion rate in the 3% NaCl solution as compared with the high strength alloy ML12 of the Mg–Zn–Zr system. The alloy ML15 of the Mg–Zn–Zr–La system showed the corrosion rate being higher than that of ML12, but ML15 contains the same contents of Zn and Zr with a comparatively small addition of La, so that its corrosion is caused mainly by Zn and Zr, but not by La.

The data from the corrosion tests [373, 411] lead to the conclusion that corrosion resistance of the cast commercial magnesium alloys containing rare earth metals is within the limits of that of the common cast commercial alloys without rare earth metals.

As with the cast alloys of the West, the Russian cast alloys are intended for making castings for various parts of machines where weight saving is of great importance. The main applications are aircraft, helicopters and the like. The alloys with rare earth metals as the main components are preferable because of their higher strength properties at elevated temperatures. They are, therefore, intended for parts which can be heated during application.

Table 55. Chemical composition of the Russian commercial wrought alloys containing rare earth metals [428].

Alloy number	Composition (main components), mass %										
	Mg	Ce*	Nd	La	Al	Mn	Zn	Ni	Cd	Li	Zr
MA11	r**	–	2.5–3.5	–	–	1.5–2.5	–	0.1–0.22	–	–	–
MA12	r	–	2.5–3.5	–	–	–	–	–	–	–	0.3–0.8
MA19	r	–	1.4–2.0	–	–	–	5.5–7.0	–	0.2–1.0	–	0.5–0.9
MA8	r	0.15–0.35	–	–	–	1.3–2.2	–	–	–	–	–
MA15	r	–	–	0.7–1.1	–	–	2.5–3.5	–	1.2–2.0	–	0.45–0.9
MA20	r	0.12–0.25	–	–	–	–	1.0–1.5	–	–	–	0.05–0.12
MA18	r	0.15–0.35	–	–	0.5–1.0	0.1–0.4	2.0–2.5	–	–	10.0–11.5	–
MA21	r	0.00–0.15	–	–	4.3–5.3	0.0–0.1	1.0–2.0	–	4.0–5.0	7.5–9.0	–

*As a rule, Ce is used as mishmetal with Ce being a main constituent.
**r = remainder.

Wrought Alloys

The Russian wrought magnesium alloys containing rare earth metals are listed in Table 55 according to the State Standard [428]. They were developed for applications as light structural materials and are produced as sheets, plates, round and rectangular rods, tubes, profiles with complicated cross sections and forgings. Components are made from them by machining, stamping, welding and other technological operations. The alloys are used, in

Table 56. Typical mechanical properties of the Russian wrought magnesium alloys with rare earth metals at room temperature [429, 432–435].

Alloy number (mass %)	Semiproduct	TS, MPa	TYS, MPa	Elongation, %	Elasticity E, GPa	Toughness, kJ/m^2	FS[a], MPa
MA11	extruded products[b]	260	130	5	43	35	85
MA12	sheets[b],	280	140	4	42	–	70
	extruded products[b]	260	130	5	42	130	90
MA19	extruded products,	390	345	6.5	42	50	90
	forgings	340	320	8	42	50	90
MA8	sheets[c],	250	145	16	41	70	70
	sheets,	250	170	11	–	–	–
	extruded products,	270	135	9.5	41	–	–
	forgings	250	180	10	41	–	–
MA15	sheets[c],	280	210	11	41	60	110
	extruded products,	310	255	10	43	50	120
	forgings	300	230	6	42	–	–
MA20	sheets[c],	260	185	24	41	100	100
	extruded products	270	160	18	41	100	80
MA18	sheets,	205	160	16	45	240	70
	forgings	165	140	34		250	–
MA21	sheets,	275	240	10	–	70	100
	forgings	215	160	13.5	44.5	60	–
MA2–1 (3.8–5.0 Al, 0.8–1.5 Zn, 0.3–0.7 Mn)	sheets[c],	280	180	12	41	60	105
	extruded products,	270	190	11	42	–	–
	forgings	270	160	10	41	–	–
MA5 (7.8–9.2 Al, 0.2–0.8 Zn, 0.15–0.5 Mn)	extruded products,	310	220	8	43	–	130
	forgings	300	220	8	43	–	–
MA14 (5.0–6.0 Zn, 0.3–0.9 Zr)	extruded products,	345	295	9.5	43	60	120
	forgings	310	250	12	43	–	95

[a]FS means fatigue stress.
[b]solution treatment + ageing.
[c]annealing.

general, in the same industrial areas as cast magnesium alloys with rare earth metals.

The alloys listed in Table 55 can be divided into three groups taking into consideration the importance of rare earth additives in them. The first group includes alloys where rare earth metals are the most important alloying elements which are mainly responsible for the strength level and possible applications. These alloys are MA11, MA12, MA19. They are distinguished by the presence of comparatively expensive Nd and, as a result, show the

Table 57. Typical mechanical properties of the Russian wrought magnesium alloys at elevated temperatures [429, 432–435].

Alloy	Semiproduct	Temperature, °C	TS, MPa	TYS, MPa	Elongation, %	σ_{100} [a], MPa	$\sigma_{0.2/100}$ [a], MPa
MA11	extruded products[b]	200	210	110	13	150	80
		250	180	90	15	90	35
		300	140	80	19	15	–
MA12	extruded products[b]	200	170	120	10	110	70
		250	140	100	16	–	–
		300	100	80	20	–	–
MA19	rods	150	280	220	15	100	50
		200	220	150	35	60	–
		250	130	90	50	35	–
MA8	sheets[c]	150	150	70	30	110	–
		200	130	60	32	50	8
		250	110	50	34	20	–
	rods	150	140	100	36	120	60
		200	120	100	34	75	31
		250	100	70	32	35	–
MA15	sheets[c]	150	160	100	20	70	25
		200	120	50	35	–	–
	rods	150	190	140	25	100	30
		200	160	90	30	–	–
		250	110	50	45	–	–
MA20	sheets[c]	150	120	100	65	60	20
		200	90	80	70	–	–
MA2–1	sheets[c]	150	140	100	32	50	28
		200	130	100	34	–	–
		250	70	50	60	–	–
	plates	150	190	100	25	–	–
		200	140	70	25	–	–
		250	90	50	30	–	–
MA5	extruded products	150	170	100	30	–	–
		200	125	70	38	–	–
		250	85	55	45	–	–
MA14	rods[d]	150	210	140	28	36	17.5
		200	150	100	55	–	–
		250	100	–	59	–	–

[a]σ_{100} and $\sigma_{0.2/100}$ mean the same as in Table 54.
[b]solution treatment + ageing.
[c]annealing.
[d]ageing.

highest strength properties at elevated temperatures. The second group includes the alloys MA8, MA15, MA20 where importance of rare earth metals is comparable with that of other main alloying additives such as Mn in MA8 and Zn with Zr in MA15 and MA20. The third group includes two alloys MA21 and MA18 where rare earth additives are minor ones. The main alloying element in the alloys MA21 and MA18 is lithium which makes them attractive as materials with especially low density. The rare earth metals in the alloys MA21 and MA18 somewhat improve their properties, but their role is less than that of other additional alloying elements such as aluminium, zinc and cadmium. Typical mechanical properties of the alloys at room and elevated temperatures are shown in Tables 56 and 57 together with the properties of the widely used magnesium alloys having high strength without rare earth metals [429, 432–435].

The alloys MA11 and MA12 are characterized by higher strength properties at elevated temperatures as compared with those of other wrought magnesium alloys of the State Standard [422]. They can be used for rolled, extruded and forged products. The alloy MA11 surpasses the alloy MA12 at elevated temperatures owing to the presence of Ni. Meanwhile, Ni in the alloy MA11 reduces its corrosion resistance, weldability and behaviour during hot working. The alloy MA12 shows lower strength properties at elevated temperatures than the alloy MA11, but can be easily hot worked, is weldable and shows higher corrosion resistance.

The alloy MA19 shows quite high strength properties at room temperature combined with adequate high strength properties at elevated temperatures, up to about 200°C. At higher temperatures and long-term tests the alloy MA19 shows less strength than the alloys MA11 and MA12. Along with Nd the alloy MA19 contains enough Zn to be considered as an intermediary variant between MA12 and the high strength Russian alloy MA14 of the system Mg–Zn–Zr which is similar to the West's ZK61A.

The alloy MA8 is one of the oldest Russian magnesium alloys. Its main advantages are excellent technological characteristics and improved strength properties at elevated temperatures. The alloy may be used for production of rolled products, such as sheets and plates, extrusion products and forgings. It can be easily welded and hot worked, and shows good corrosion resistance. In general, the alloy reveals lower strength properties at elevated temperatures than alloys MA11 and MA12 because it contains cerium, but not neodymium. Meanwhile, MA8 is significantly cheaper than MA11 and MA12 and this fact results in its wider application.

The alloy MA15 is characterized by high strength properties at room and near room temperatures with good technological characteristics. Its behaviour and properties are conditioned mainly by the medium range Zn content of 2.5–3.5 mass % as compared with other alloys of the base system Mg–Zn–Zr. Addition of lanthanum to the alloy MA15 promotes increase of its strength properties at room and elevated temperatures, in particular tensile yield strength and compressive yield strength. Besides, La in this alloy improves its weldability [432].

The alloy MA20 was developed mostly as a material combining moderate strength with quite high plasticity. It is quite convenient for producing thin sheets and working them. A small amount of rare earth metals in the alloy MA20 improves its strength and plasticity.

The alloys MA18 and MA21 show moderate strength properties at room and near room temperatures and low strength properties at elevated temperatures. The main advantage of these alloys is lower density than that of pure magnesium connected with the presence of sufficient lithium. As compared with MA21, the alloy MA18 contains more lithium and is lighter, but shows lower strength properties.

Corrosion resistance of standard wrought alloys containing rare earth metals is characterized by data of the comparative corrosion tests reported in [373, 411]. They were obtained by measuring the weight change of samples immersed in solution of 3% NaCl in water. Results of the tests are presented in Figure 128. They show a low corrosion rate of the alloy MA12 of the Mg–Nd–Zr system with Nd as the main alloying element. The alloys MA15 of the Mg–Zn–Cd–Zr–La system and MA8 of the Mg–Mn–Ce(mishmetal) system showed corrosion rates which were higher than that of MA5 (Mg–Al–Zn–Mn) and significantly lower than that of MA14 (Mg–Zn–Zr). The alloy MA19 containing Nd showed a high corrosion rate which turned out to be close to that of MA14, but MA19 contains a high Zn like MA14.

According to [373], the alloys MA11, MA12, MA8 are not in practice inclined to stress corrosion cracking surpassing in this aspect the alloys of the Mg–Al–Zn–Mn system. The alloy MA19 derived actually from the alloy MA14 of Mg–Zn–Zr by addition of 0.2–1.0 mass % Cd and 1.4–2.0 mass % Nd shows insignificant inclination to stress corrosion cracking being less than that of MA14.

In general, corrosion resistance of wrought commercial magnesium alloys containing rare earth metals may be considered to be within the limits of that typical of commercial magnesium alloys of other systems. Certainly, the presence of rare earth metals in commercial magnesium alloys does not make their corrosion resistance significantly worse. This conclusion corresponds to the conclusion made on cast commercial magnesium alloys.

Apart from the alloys of the State Standard [428] a number of new wrought magnesium alloys containing rare earth metals were suggested in Russia for commercial applications. They are named IMV5, IMV6, IMV7, IMV8 [106]. Compositions of the alloys and their properties are listed in Table 58. All alloys contain yttrium as the main alloying element which is present together with other rare earth metals, such as scandium, cerium, gadolinium and neodymium. As a result of the Y presence the alloys show higher strength properties than the standard alloys MA11 and MA12 where Nd is the main alloying element. The alloys with Y also show quite high strength properties at room temperature.

Table 58. Composition and typical tensile properties of the Russian wrought commercial alloys containing Y as the main additive [106].

Alloy number	Composition, mass %	Product	Temper	Temperature, °C	TS, MPa	TYS, MPa	Elongation, %
IMV5	5–7 Y, 8–11 Sc, 0.3–1.0 Mn, rest Mg	Rod, $d = 20$ mm	as extruded	room	314	238	16
				250	283	202	20
				300	230	135	69
IMV6	7–11 Y, 0.05–0.15 Ce, 0.15–1.5 Mn, 0.05–1.0 Al, 0.1–3.0 Cd, rest Mg	Sheet*, $t = 2$ mm	Ageing at 200°C for 100 h	room	392	294	9
				250	285	234	14
				300	190	148	40
IMV7	4–6 Y, 8–10 Gd, 0.3–1.0 Mn, rest Mg	Rod, $d = 20$ mm	Ageing at 200°C for 24 h	room	417	363	4
				250	369	290	7
				300	277	232	20
IMV8	8–11 Y, 0.5–1.2 Nd, 0.3–1.2 Zn, 0.1–0.6 Zr, rest Mg	Rod, $d = 10.5$ mm	Ageing at 210°C for 36 h	room	373	304	11
				250	304	235	16
				300	196	147	69

*Longitudinal direction.

The alloy IMV5 contains significant contents of Sc apart from Y. In accordance with high solubility in solid magnesium all Sc is in solution in it increasing strength properties. The high strength properties of this alloy are obtained without heat treatment resulting in solid solution decomposition. The dependence of the strength properties of the alloy IMV5 on the Y and Sc contents are described in [436, 437].

The alloy IMV6 contains sufficiently high Y and small contents of Ce. The highest strength properties of this alloy are achieved after ageing at 200°C for 100 h without preliminary quenching. The alloy can be used for production of sheets by rolling [106, 289].

The alloy IMV7 is distinguished by the highest strength properties at room and elevated temperatures owing to the presence of Gd. The alloy requires ageing at 200°C for high strength, as with the alloy IMV6, but for a shorter time, 24 h. It can be used for production of various extruded products including rods and plates [280, 351]. In addition to IMV7 the alloy IMV7-1, belonging to the same base system Mg–Y–Gd, was developed [438]. The alloy IMV7-1 contains less expensive Gd and more Y. Besides, Mn in IMV7 is replaced by Zr in order to obtain a fine grain in ingots. The alloy IMV7-1 includes (in mass %) 5–6.5 Y, 3.5–5.5 Gd, 0.15–0.7 Zr, remainder Mg and inevitable impurities. Properties of IMV7-1 are close to those of IMV7 [438].

The alloy IMV8 combines high strength properties with lower cost because it does not contain the quite expensive scandium and gadolinium as do IMV5 and IMV7, respectively. This alloy is intended for production of extruded and forged products [106, 289].

In addition to the above mentioned alloys the alloy VMD10 was developed on the basis of the Mg–Y–Zn system [289, 439–442]. The alloy has an advantage in low anisotropy of mechanical properties after hot working and high compressive yield strength (CYS) which is close to tensile yield strength (TYS). The alloy is convenient for production of high loaded forgings where high strength properties including CYS and low anisotropy are required. The typical strength properties of forgings produced from the alloy VMD10 are shown in Table 59. For comparison the strength properties of forgings produced from the alloy MA14 with high strength at room temperature and the alloy MA12 with high strength at elevated temperatures without yttrium are presented also [442]. The advantage of VMD10 as compared with MA14 and MA12 is evident from Table 59. The alloy VMD10 also has high strength properties at elevated temperatures. For forgings at 200°C along fibres these were TS = 280 MPa, TYS 200 = MPa, creep strength $\sigma_{0.2/100}$ = 90 MPa [442].

Table 59. Mechanical properties of forgings (up to 100 kg) produced from some magnesium alloys [442].

Alloy number	System	Temper	Direction to fibers	TS, MPa	TYS, MPa	CYS, MPa	Elongation, %	Density, g/cm^3
VMD10	Mg–Y–Zn	Annealing	Longitudinal	340	270	290	10	1.89
			Transverse	320	240	270	12	
MA14	Mg–Zn–Zr	Ageing	Longitudinal	320	250	140	6	1.78
			Transverse	240	170	80	6	
MA12	Mg–Nd–Zr	ST* +	Longitudinal	260	150	120	4	178
		Ageing	Transverse	210	120	–	4	

*ST means solution treatment.

In addition to the above alloys for structural applications, special wrought magnesium alloys for ultrasound guides of ultrasonic delay lines were developed in Russia. The alloys contain rare earth metals as the main alloying elements. One of them, MA17, is characterized by especially low ultrasound attenuation. It is produced in the form of hot rolled plates with a thickness of about 25–35 mm. Hot rolling is carried out in conditions chosen to obtain the fully recrystallized structure in the plates, but without superfluous growth of grain size. As a rule, the grain size of the plates turned out to be less than 10 μm. Composition of the alloy MA17 according to the State Standard [428] is 0.7–1.5 mass % Ce, 0.2–0.7 mass % Mn, rest Mg and impurities. Cerium in MA17 is used commonly in the form of mishmetal enriched in cerium. The main properties of the MA17 plates are shown in Table 60. Along with low attenuation coefficient of ultrasound the alloy MA17

Table 60. Properties of hot rolled plates from the alloy MA17 [384].

Property	Value
Attenuation coefficient of ultrasound for longitudinal waves, α, dB/m	
at 10 Mc	2–3
at 15 Mc	3–5
at 20 Mc	5–8
Velocity of longitudinal waves, c_l, m/c	5760–5840
Velocity of transverse waves, c_t, m/c	3040–3150
Thermal coefficient of delay for longitudinal waves, γ_l, 10^{-4}/°C	1.3–1.6
Thermal coefficient of delay for transverse waves, γ_t, 10^{-4}/°C	2.2–2.4
Density, d, g/cm^3	1.754–1.763
Electrical resistance, ρ, μOhm·cm	5.32–5.34
Tensile strength, σ_b, MPa	190–210
Tensile yield strength, $\sigma_{0.2}$, MPa	110–140
Elongation, δ, %	12–14
Reduction area, ψ, %	14–22
Hardness, HB	40–52

shows a good combination of strength properties and plasticity. The alloy was developed based on the results of investigations of the binary alloys where the least ultrasound attenuation in the alloys of the Mg–Ce system was revealed [384].

In addition to the alloy MA17 another alloy with low ultrasound attenuation was developed. This magnesium alloy was named MDZ-2. Its composition includes 2–4 mass % Nd, 1.0–1.5 mass % Ca, 0.15–0.4 mass % Mn, rest Mg and impurities [391]. The main alloying element in MDZ-2 is neodymium which also promotes an abrupt decrease of ultrasound attenuation in magnesium as does cerium [384]. The alloy MDZ-2 was not used actually in industry for ultrasonic delay lines because it did not show significant advantages as compared with MA17, and the presence of neodymium made it substantially more expensive than MA17.

Conclusion

The results reviewed of investigations of magnesium alloys with various rare earth metals show successive change of their characteristics with the position of RE atoms in the Periodic Table of the Elements. This regularity is based on the regular change of the physico-chemical interaction between magnesium and separate rare earth metals, in particular solubility of them in solid magnesium. Solubility of separate rare earth metals in solid magnesium changes within quite wide limits and, in accordance with this, mechanical properties of the alloys can change in quite wide limits at common and elevated temperatures.

Despite the substantial difference in constitution and properties, magnesium alloys with separate rare earth metals show similar features. All rare earth metals result in improvement of strength when they are added to magnesium. Their favourable influence begins even at small contents when grain refinement and increase of plasticity are observed. Mechanical properties of the alloys depend to a great extent on the possibility of solid solution decomposition in the alloys resulting in formation of dispersed particles of RE-rich phases in the structure. Quantity and size of the particles are responsible for general level of strength of the alloys. Kinetics and phase transformations of the solid solution decomposition depend regularly on the place of the rare earth metals in the Periodic Table.

Regularities in constitution and properties of magnesium alloys with rare earth metals are caused eventually by successive change of the electron constitution of the rare earth atoms with changing atomic numbers and atomic radii.

High strength properties of most magnesium alloys with rare earth metals make them attractive for commercial applications in modern technology. Rare earth metals are especially effective for creating magnesium alloys with high strength properties at elevated temperatures. Some of the commercial alloys with rare earth metals are already in use successfully. A number of the alloys with higher mechanical properties have been suggested to industry over the last years, although they are still not in use. Meanwhile, opportunities for use of separate rare earth metals as alloying elements for magnesium alloys can not be considered as exhausted. Only a few rare earth metals have been used for alloying magnesium alloys so far. They are actually only lanthanum, cerium, neodymium and yttrium, although many other rare earth metals showed themselves also as quite effective alloying elements. Further investigations of magnesium alloys with rare earth metals are likely to result in the development of new commercial materials with improved properties both at common and elevated temperatures.

The wide use of separate rare earth metals as alloying elements in magnesium alloys is restrained by their cost which still remains rather high for some of them. Nevertheless, the cost of rare earth metals tends to decrease, and almost all separate rare earth metals gradually will become acceptable as commercial products for applications in industry.

References

1. G.A. Mellor and R. W. Ridley, *J. Inst. Metals*, 75, 679–692 (1949).
2. G.A. Mellor and R. W. Ridley, *J. Inst. Metals*, **81**, 245–269 (1952–53).
3. T.E . Leontis, *J. Metals*, **1**(12), 968–983 (1949).
4. T.E. Leontis, *J. Metals*, **3**(11), 987–993 (1951).
5. T.E. Leontis, *J.* Metals, **4**(3), 287–294 (1952).
6. T.E. Leontis, *J. Metals*, **4**(6), 633–643 (1952).
7. B.E. Bockrath, *Modern Metals*, **14**(8), 52–58 (1958).
8. R.V. London, R.E. Edelman and H. Markus, *Trans. ASM*, **59**(2), 250–261 (1966).
9. R.S. Busk, *Modern Metals*, **24**(6), 43–46 (1968).
10. Nomenclature of Inorganic Chemistry, *J. Amer. Chem. Soc.*, **82**(21), 5523–5544 (1960).
11. H.W. King, in *Binary Alloy Phase Diagrams,* Ed. Th.B. Massalski, American Society for Metals, Metals Parc Ohio, 2179–2181(1986).
12. W. Hume-Rothery, *Elements of Structural Metallurgy*, The Institute of Metals, London (1961).
13. Th. Moeller, in *The Rare Earth*, John Wiley and Sons Inc., New York–London, 9–28 (1961).
14. K.A. Gschneidner, Jr., *Rare Earth Alloys. A Critical Review of the Alloy Systems of the Rare Earth, Scandium and Yttrium Metals*, D. Van Nostrand Co, Inc., Princeton, New Jersey–New York–Toronto–London (1961).
15. B.J. Beaudry, K.A. Gschneidner, Jr., in *Handbook on the Physics and Chemistry of Rare Earth*, North-Holland Publishing Co, Amsterdam–New York–Oxford, **1**, 173–232 (1978).
16. W.B. Pearson, *The Crystal Chemistry and Physics of Metals and Alloys*, Wiley–Interscience, A Division of John Wiley and Sons, Inc., New York–London–Sydney–Toronto (1972).
17. H. Templeton and C.H. Dauben, *J. Amer. Chem. Soc.*, **76**(20), 5237–5239 (1954)
18. R.D. Shannon and C.T. Prewitt, *Acta Crystallogr.*, **B25**(5), 925–946 (1969).
19. A. Iandelli and A. Palenzona, *J. Less-Common Metals*, **9**(1), 1–6 (1965).
20. J.H.N. van Vucht and K.H.J. Buschov, *J. Less-Common Metals*, **10**(1), 98–107 (1965).
21. W. Gordy, *Phys. Rev.*, **69**(11/12), 604–607 (1946).
22. D.C. Koskenmaki, in *Handbook on the Physics and Chemistry of Rare Earth*, North-Holland Publishing Co., Amsterdam–New York–Oxford, **1**, 337–377 (1978).
23. A.E. Miller and A.H. Daane, *Trans. Metallurg. Soc. AIME*, **230**(4), 568–572 (1964).
24. E.M. Savitsky and V.F. Terekhova, *Metals Science of the Rare-Earth Metals*, Nauka, Moscow, (1975) (in Russian).
25. G.V. Raynor, *The Physical Metallurgy of Magnesium and Its Alloys*, Pergamon Press, London–New York–Paris–Los Angeles (1959).
26. I.V. Shakhno, Z.N. Shevtsova, P.I. Fedorov and S S. Korovin, *Khimiya i Tekhnologiya Redkikh i Rasseyanykh Elementov*, Part 2, Vysshaya Shkola, Moscow (1976) (in Russian).
27. F.E. Block and T.T. Campbell, in *The Rare Earth*, ed. F.H. Spedding and A.H. Daane, John Wiley and Sons, Inc., New York–London, 89–101 (1961).
28. V.I. Mikheeva, M.E. Kost and A.I. Konstantinova, in *Khimiya Metallicheskikh Splavov,* 154–158 (1973) (in Russian).
29. J. Dexpert-Chys, C. Loier, Ch.H. la Blanchetais and P.E. Caro, *J. Less-Common Metals*, **41**(1), 105–113 (1975).
30. H.A. Eick, in *Rare Earth Research*, The Macmillan Company, New York, 297–305 (1961).
31. D.H. Templeton and G.F. Carter, *J. Phys. Chem.*, **58**(11), 940–944 (1954).
32. H. Bommer, *Z. anorg. allgem. Chem.*, **241**(2/3), 273–280 (1939).

33. A. Iandelli, in *Rare Earth Research*, The Macmillan Company, New York, 135–141 (1961).
34. E.I. Gladyshevsky, *Crystal Chemistry of Silicides and Gemanides*, Metallurgiya, Moscow, (1971) (in Russian).
35. G.V. Samsonov, T.Ya. Kosolapova, M.D. Lyutaya and G.N. Makarenko, in *Redkozemelnye Elementy (Rare-Earth Elements)*, Izdat. Akad. Nauk SSSR, Moscow, 8–21 (1963) (in Russian).
36. C.L. Huffine and J.M. Williams, in *The Rare Earth*, ed. F.H. Spedding and A.H. Daane, John Wiley and Sons, Inc., New York–London, 145–162 (1961).
37. C.E. Lundin, in *The Rare Earth*, ed. F.H. Spedding and A.H. Daane, John Wiley and Sons, Inc., New York–London, 224–385 (1961).
38. K.A. Gschneidner, Jr. and F.W. Calderwood, in *Handbook on the Physics and Chemistry of Rare Earths*, North-Holland Physics Publishing Co., Amsterdam–New York–Oxford–Tokyo, **8,** Chapt. 54, 1–161 (1986)
39. F.H. Spedding, B. Sandeen and B.J. Beaudry, *J. Less-Common Metals*, **31**(1), 1–13 (1973).
40. H. Okamoto, *J. Phase Equilibria*, **12**(1), 118 (1991).
41. E.I. Gladyshevsky and O.I. Bodak, *Crystal Chemistry of the Rare-Earth Metal Intermetallic Compounds*, Vyscha Shkola, Lvov, (1982).
42. P. Chiotti and J.T. Mason, *Trans. Amer. Soc. AIME*, **233**(4), 786–795 (1965).
43. A. Iandelli and A. Palenzona, *J. Less-Common Metals*, **18**(3), 221–227 (1965).
44. K.H.J. Buschow, and A.S. Goot, *J. Less-Common Metals*, **19**(3), 153–158 (1969).
45. K.H.J. Buschow, *Philips Res. Rep.*, **26**(1), 49–64 (1971).
46. R. Vogel and Th. Heumann, *Z. Metallkd.* **35**(2) 29–42 (1943).
47. F. Gaume-Mahn, in *Progress in the Science and Technology of the Rare Earth,* Vol. **1**, Pergamon Press, Oxford–London–New York–Paris, 259–309 (1964).
48. E.F. Westrum, in *Progress in the Science and Technology of the Rare Earth*, Vol. **1**, Pergamon Press, Oxford–London–New York–Paris, 310–350 (1964).
49. D.A. Mineev, *Lantanoids in Rare-Earth Ores and Complex Deposits*, Nauka, Moscow, (1974) (in Russian).
50. A.N. Zelikman and G.A. Meerson, *Metallurgy of Rare Metals*, Metallurgiya, Moscow, (1973) (in Russian).
51. L.H. Ahrens, in *Progress in the Science and Technology of the Rare Earth*, Vol. **1**, Pergamon Press, Oxford–London–New York–Paris, 1–29 (1964).
52. R.M. Healy and H.E. Kremers, in *The Rare Earth*, ed. F.H. Spedding and A.H. Daane, John Wiley and Sons, Inc., New York–London, 29–37 (1961).
53. D.F. Peppard, in *The Rare Earth*, ed. F.H. Spedding and A.H. Daane, John Wiley and Sons, Inc., New York–London, 38–54 (1961).
54. J.E. Powell, in *The Rare Earth*, ed. F.H. Spedding and A.H. Daane, John Wiley and Sons, Inc., New York–London, 55–76 (1961).
55. F.E. Block and T.T. Campbell, in *The Rare Earth*, ed. F.H. Spedding and A.H. Daane, John Wiley and Sons, Inc., New York–London, 89–101 (1961).
56. A.H. Daane, in *The Rare Earth*, ed. F.H. Spedding and A.H. Daane, John Wiley and Sons, Inc., New York–London, 102–112 (1961).
57. O.N. Carlson and F.A. Schmidt, in *The Rare Earth*, ed. F.H. Spedding and A.H. Daane, John Wiley and Sons, Inc., New York–London, 113–125 (1961).
58. B.L. Porozhenko, *The Reference Book on Technical Conditions on the Rare and Small Metals and Their Compounds*, Part 3, Metallurgiya, Moscow (1969) (in Russian).
59. *Collection of Technical Conditions on the Rare-Earth Metals and Their Oxides*, Giredmet, ONTIP, Moscow (1974) (in Russian).
60. Price List No 02-01. *Wholesale Prices of Nonferrous Metals, Alloys and Powders*, Preiskurantizdat, Moscow, 1980 (in Russian).
61. J.C. Barker, in *Precious and Rare Metal Technologies*, Proceedings of a Symposium on Precious and Rare Metals, Albuquerque, NM, USA, April 6–8, 1988, Elsevier, Amsterdam–Oxford–New York–Tokyo, 415–434 (1989).
62. *Standard Specification for Magnesium Alloys in Ingot Form for Sand Castings, Permanent Mold Castings, and Die Castings*. American Society for Testing and Materials. Designation: B 93/B 93M - 91.
63. *Materials Engineering*, **103**(12), 95–97 (1986).

64. W. Unsworth, *Metals and Mater.*, **4**(2), 83–86 (1988).
65. W. Unsworth, in *1st Int. SAMPE Metals and Metals Process Conf., Cherry Hill, N.Y. Aug. 18–20, 1987*, Vol. 1, Covina (Calif.) (1987).
66. R. Vogel, *Z. Anorg. Chem.*, **91**(4), 277–298 (1915).
67. G. Canneri, *Metallurgia Ital.*, **23**, 803–823 (1931).
68. G. Canneri, *Metallurgia Ital.*, **25**, 250–252 (1933).
69. J.L. Haughton and T.H. Schofield, *J. Inst. Metals*, **60**(1), 339–347 (1937).
70. R. Vogel and Th. Heumann, *Metallforschung*, **2**(1), 1–8 (1947).
71. F. Weibke and W. Schmidt, *Z. Elektrochem. und Angew. Phys. Chem.*, **46**(6), 357–364 (1940).
72. J.J. Park andL.L. Wyman, *Nat. Bur. Stand. Rep.*, WADS-TR-57-504 (1957).
73. E.M. Savitsky, V.F. Terekhova and I.A. Novikova, *Zhurn. Neorg. Khimii*, **3**(9), 2138–2142 (1958) (in Russian).
74. E.D. Gibson and O.N. Carlson, *Trans. ASM*, **52**, 1084–1096 (1960).
75. V.F. Terekhova, I.A. Markova and E.M. Savitsky, *Zh. Neorgan. Khim.*, **5**(1), 235–236 (1960) (in Russian).
76. D. Mizer and J.B. Clark, *Trans. Metallurg. Soc. AIME*, **221**(2), 207–208 (1961).
77. E.M. Savitsky, V.F. Terekhova, I.V. Burov and I.A. Markova, *Zh. Neorg. Khim.*, **6**(7), 1734–1737 (1961) (in Russian).
78. O.D. McMasters and K.A. Gschneidner, *J. Less-Common Metals*, **8**(5), 289–298 (1965).
79. W. Muelpfordt and W. Klemm, *J. Less-Common Metals*, **17**(1), 127–129 (1969).
80. L.N. Komissarova and B.I. Pokrovsky, *Zh. Neorg. Khim.*, **9**(10), 2277–2278 (1964) (in Russian).
81. B.J. Beaudry and A.H. Daane, *J. Less-Common Metals*, **18**(3), 305–308 (1969).
82. A.E. Miller and A.H. Daane, *Trans. Metallurg. Soc. AIME*, **230**(4), 568–572 (1964).
83. R.R. Joseph and K.A. Gschneidner, *Trans. Metallurg. Soc. AIME*, **233**(12), 2063–2069 (1965).
84. A. Saccone, S. Delfino, G. Borzone and R. Ferro, *J. Less-Common Metals*, **154**(1), 47–60 (1989).
85. A. Saccone, S. Delfino, D. Maccio and R. Ferro, *Z. Metallkd.* **82**(7), 568–569 (1991).
86. A. Saccone, S. Delfina, D. Maccio and R. Ferro, *Metallurg. Trans. A*, **23A**(3), 1005–1012 (1992).
87. A. Saccone, S. Delfino, D. Maccio and R. Ferro, *J. Phase Equilibria*, **14**(3), 280–287 (1993).
88. A. Saccone, S. Delfino, D. Maccio and R. Ferro, *J. Phase Equilibria*, **14**(4), 479–484 (1993).
89. M.E. Drits, Z.A. Sviderskaya and L.L. Rokhlin, in *Metallurgiya, Metallovedenie i Fiziko-Khimicheskie Metody Issledovaniya*, Russian Academy of Sciences Publishers, Moscow, (12), 143–151 (1962) (in Russian).
90. L.L. Rokhlin, *Izv. Akad. Nauk SSSR, Otd. Tekh. Nauk, Met. Toplivo,* (2), 126–130 (1962) (in Russian).
91. Z.A. Sviderskaya and E.M. Padezhnova, *Izv. Akad. Nauk SSSR , Met.*, (6), 183–190 (1968) (in Russian).
92. N.R. Bochvar, in *Structure and Properties of Light Alloys*, Nauka, Moscow, 21–23 (1971) (in Russian).
93. L.L. Rokhlin and N.R. Bochvar, *Izv. Akad. Nauk SSSR, Met.* (2), 193–197 (1972) (in Russian).
94. L.L. Rokhlin and N.R. Bochvar, in *Metalloved. Tsvetn. Met. Splav*, Nauka, Moscow, 58–61 (1972) (in Russian).
95. Z.A. Sviderskaya and N.I. Nikitina, in *Metalloved. Tsvetn. Met. Splav*, Nauka, Moscow, 61–65 (1972) (in Russian).
96. L.L. Rokhlin, E.M. Padezhnova and L.S. Guzey, *Izv. Akad. Nauk SSSR, Met.*, (6), 204–208 (1976) (in Russian).
97. L.L. Rokhlin and L.P. Deeva, *Tekhnol. Legkich Splavov*, (12), 83–84 (1976) (in Russian).
98. M.E. Drits and L.L. Rokhlin, *Izv. Vyss. Uchebn. Zaved., Tsvetn. Metall.*, (1), 169–171 (1977) (in Russian).
99. L.L. Rokhlin and N.I. Nikitina, *Izv. Vyss. Uchebn. Zaved., Tsvetn. Metall.*, (1), 167–168 (1977) (in Russian).
100. L.L. Rokhlin, *Izv. Akad. Nauk SSSR, Met.*, (1), 181–183 (1977) (in Russian).
101. L.L. Rokhlin, *Izv. Vyss. Uchebn. Zaved., Tsvetn. Metall.*, (6), 142–144 (1977) (in Russian).
102. M.E. Drits, L.L. Rokhlin, E.M. Padezhnova and L.S. Guzey, *Metalloved. Term. Obrab. Met.*,

(9), 70–73 (1978) (in Russian).
103. L.L. Rokhlin and L. P. Deeva, *Izv. Akad. Nauk SSSR, Met.*, (5), 219–221 (1978) (in Russian).
104. L.L. Rokhlin, N. I. Nikitina and Z. K. Zolina, *Metalloved. Term. Obrab. Met.*, (7), 15–18 (1978) (in Russian).
105. L.L. Rokhlin, *Izv. Akad. Nauk SSSR, Met.*, (4), 185–187 (1979) (in Russian).
106. L.L. Rokhlin. *Magnesium Alloys Containing Rare Earth Metals*, Nauka, Moscow (1980) (in Russian).
107. L.L. Rokhlin and N.I. Nikitina, in *Legirov. Obrabot. Legkich Splavov*, Nauka, Moscow, 111–114 (1981) (in Russian).
108. L.L. Rokhlin, N.I. Nikitina, *Izv. Vyss. Uchebn. Zaved., Tsvetn. Metall.*, (1), 43–45 (1998) (in Russian).
109. A. Rossi and A. Iandelli, *Atti Reale Accad. Naz. Lincei*, **19**(6), 415–420 (1934).
110. H. Nowotny, *Z. Metallkd*, **34**(11), 247–253 (1942).
111. F. Laves, *Naturwissenschaften*, **31**(7/8), 96 (1943).
112. A. Iandelli, in *The Physical Chemistry of Metallic Solutions and Intermetallic Compounds*, Proceedings of a Symposium held at National Physical Laboratory on 4th, 5th and 6th June, 1958, London, **1**, 3F (1959).
113. A. Iandelli, *Atti Accad. Naz. Lincei*, Classe di Scienze fisiche, matematiche e naturali, Ser. 8, **29**, 62–69 (1960).
114. P.I. Kripyakevich and V.I. Evdokimenko, *Dopov. Akad. Nauk Ukr. RSR*, (11), 1610–1611 (1962) (in Ukrainian).
115. V.I. Evdokimenko and P.I. Kripyakevich, *Kristallografiya*, **8**(2), 186–193 (1963) (in Russian).
116. P.I. Kripyakevich, V.F. Terekhova, O.S. Zarechnyuk and I.V. Burov, *Kristallografiya*, **8**(2), 268 (1963) (in Russian).
117. E.I. Gladyshevsky, P.I. Kripyakevich, E.E. Cherkashin, O.S. Zarechnyuk, I.I. Zalutsky and V.I. Evdokimenko, in *Redkozemelnye Elementy*, Izdat. Akad. Nauk SSSR, Moscow, 67–70 (1963) (in Russian).
118. P.I. Kripyakevich, V.I. Evdokimenko and I.I. Zalutsky, *Dopov. Akad. Nauk Ukr. RSR*, (5), 766–767 (1964) (in Ukrainian).
119. V.I. Evdokimenko and P.I. Kripyakevich, *Kristallografiya*, **9**(4), 554–556 (1964) (in Russian).
120. P.I. Kripyakevich and V.I. Evdokimenko, in *Voprosy Teorii i Primeneniya Redkozemelnykh Metallov*, Nauka, Moscow, 146–148 (1964) (in Russian).
121. W. Klemm, H. Kock and W. Muehlpfordt, *Angew. Chem.*, **76**(20), 862 (1964).
122. N.F. Lashko and G.I. Morozova, *Zavodskaya Laboratoriya*, **30**(10), 1187–1189 (1964) (in Russian).
123. N.F. Lashko and G.I. Morozova, *Kristallografiya*, **9**(2), 269–270 (1964) (in Russian).
124. Q.C. Johnson, G.S. Smith, D.H. Wood and E.M. Cramer, *Nature*, **201** (4919), 600 (1964).
125. A. Iandelli and A. Palenzona, *J. Less-Common Metals*, **9**(1), 1–6 (1965).
126. O. Schob and E. Parthe, *Acta Crystallogr.*, **19**(2), 214–224 (1965).
127. D.H. Wood and E. M. Cramer, *J. Less-Common Metals*, **9**(5), 321–337 (1965).
128. P.I. Krypiakewytsch and V.I. Evdokimenko, *Z. Anorg. Allg. Chem.*, **355**(1/2), 104–112 (1967).
129. Q. Johnson and G.S. Smith, *Acta Crystallogr.*, **22**(3), 360–365 (1967).
130. Q. Johnson and G.S. Smith, *Acta Crystallogr.*, **23**(2), 324–329 (1967).
131. P.I. Kripyakevich and V.I. Evdokimenko, *Visn. Lviv. Univ.*, Ser. Khim., (11), 3–7 (1969) (in Ukrainian).
132. Q. Johnson and G.S. Smith, *Acta Crystallogr.*, **B26**(4), 434–435 (1970).
133. K.H.J. Buschow, *J. Less-Common Metals*, **44**(1), 301–306 (1976).
134. V.V. Kinzhibalo, L.L. Rokhlin and N.P. Abrukina, *Izv. Akad. Nauk SSSR, Met.*, (1), 204–205 (1985) (in Russian).
135. M.L. Fornasini, P. Manfrinetti and K.A. Gschneidner, Jr., *Acta Crystallogr.*, **C42**(2), 138–142 (1986).
136. L.L. Rokhlin, V.V. Kinzhibalo and N.P. Abrukina, *Izv. Akad. Nauk SSSR, Met.*, (5), 119–122 (1988) (in Russian).
137. H.W. Zandbergen, G. van Tendeloo, D.B. de Mooij, K.H.J. Buschow, *J. Less-Common Metals*, **154**(2), 375–380 (1989).
138. *Binary Alloy Phase Diagrams*, Ed.-in-Chief T. B. Massalski, American Society for Metals.

Metals Park. Ohio. Vol. 1–2, (1986–1987).
139. *Binary Alloy Phase Diagrams. Second Edition*. Ed.-in Chief T. B. Massalski, Publisher William W. Scott, Jr., Vol. 1–3, (1990).
140. *Phase Diagrams of the Binary Metal Systems*. Ed.-in Chief N. P. Lyakishev, Mashinostroenie, Moscow, Vol. 1–2, (1996–1997) (in Russian).
141. P. Villars and L. D. Calvert, *Paerson's Handbook of Crystallographic Data for Intermetallic Phases. Second Edition.* The Materials Information Society, Materials Park, OH, Vol. 1–4, (1991).
142. A.A. Nayeb-Hashemi and J.B. Clark, *Phase Diagrams of Binary Magnesium Alloys*, ASM International Metals Park, OH, (1988).
143. H. Okamoto, *J. Phase Equilibria*, **13**(1), 105 (1992).
144. H. Okamoto, *J. Phase Equilibria*, **14**(4), 534–535 (1993).
145. H. Okamoto, *J. Phase Equilibria*, **16**(6), 535 (1995).
146. L.L. Rokhlin, *J. Phase Equilibria*, **16**(6), 504–507 (1995).
147. A.A. Nayeb-Hashemi and J.B. Clark, *Bull. Alloy Phase Diagrams*, **9**(2), 172–178 (1988).
148. P. Manfrinetti and K.A. Gschneidner, Jr, *J. Less-Common Metals*, **123**(1/2), 267–275 (1986).
149. B. Darriet, M. Pezat, A. Hibika and P. Hagenmuller, *Mater. Res. Bull.*, **14**, 377–385 (1979).
150. Chai Liang, Lin Haichun and Yeh Yupu, *Proc. 5th Nat. Symp. Phase Diagr. Wuhan, Nov. 18–21, 1988*, Wuhan, 15 (1988).
151. N.M. Tikhova and L.A. Afanasyeva, *Metalloved. Term. Obrab. Met.* (3), 38–41 (1958).
152. A.A. Nayeb-Hashemi and J.B. Clark, *Bull. Alloy Phase Diagrams*, **9**(2), 162–172 (1988).
153. M.S. Beletsky and E.L. Galperin, *Fiz. Met. Metalloved.*, **11**(5), 698–703 (1961) (in Russian).
154. L.L. Rokhlin, *Investigaton of Structure and Properties of Magnesium Alloys Containing Neodymium*, Dissertation for the Candidate Degree, Baikov Institute of Metallurgy, Moscow, (1963).
155. C.S. Roberts, *J. Metals*, **6**(5), sec. II, 634–640 (1954).
156. Unpublished work of Dow Chemical Company, Midland, Mich., cited from [157].
157. M. Hansen and K. Anderko, *Constitution of Binary Alloys*. McGraw–Hill Book Company, Inc., New York–Toronto–London, (1958).
158. A.A. Nayeb-Hashemi and J.B. Clark, *Bull. Alloy Phase Diagrams*, **10**(1), 23–27 (1989).
159. A. Saccone, A.M. Cardinale, S. Delfino and R. Ferro, *Intermetallics*, **1**, 151–158 (1993).
160. H. Okamoto, *J. Phase Equilibria*, **16**(6), 535 (1995).
161. Chai Liang, Lin Haichun and Yeh Yupu, *Proc. 5th Nat. Symp. Phase Diagr. Wuhan, Nov. 18–21, 1988*, Wuhan, 16 (1988).
162. K.H.J. Buschow, R.C. Sherwood and F.S.L. Hsu, *J. Appl. Phys.*, **49**(3), 1510–1512 (1978).
163. S. Delfino, A. Saccone and R. Ferro, *Metal. Trans.* **21A**(8), 2109–2114 (1990).
164. R. Ferro, S. Delfino, G. Borzone, A. Saccone and G. Cacciamani, *J. Phase Equilibria*, **14**(3), 273–279 (1993).
165. M.E. Drits, L.L. Rokhlin and N.P. Sirchenko, *Izv. Akad. Nauk SSSR, Met.*, (6), 78–82 (1983) (in Russian).
166. R.R. Joseph, and K.A. Gschneidner, Jr., *Trans. AIME*, **233**(12), 2063–2069 (1965).
167. P.I. Kripyakevich, V.I. Evdokimenko, and E.I. Gladyshevsky, *Kristallografiya*, **9**(3), 410–411 (1964).
168. J.F. Smith, D.M. Bailey, D.B. Novotny and J.E. Davison, *Acta Metall.* **13**(8), 889–895 (1965).
169. H. Okamoto, J. *Phase Equilibria*, **13**(1), 105–106 (1992).
170. A.A. Nayeb-Hashemi and J.B. Clark, *Bull. Alloy Phase Diagrams*, **7**(6), 574–577 (1986).
171. P.I. Kripyakevich, E.I. Gladyshevsky, O.S. Zarechnyuk, V.I. Evdokimenko, I.I. Zalutsky and D.P. Frankevich, *Kristallografiya*, **8**(4), 595–599 (1963) (in Russian).
172. J.E. Pahlman and J.F. Smith, *Metal. Trans.*, **3**(9), 2423–2432 (1972).
173. L.L. Rokhlin, *J. Phase Equilibria*, **19**(2), 142–145 (1998).
174. W. Hume-Rothery and G.V. Raynor, *The Structure of Metals and Alloys*, The Institute of Metals, Grosvenor Gardens, London, S.W. (1954).
175. M. Giovannini, A. Saccone, R. Marazza and R. Ferro, *Metall. and Mater. Trans. A*, **26A**(1), 5–10 (1995).
176. M. Giovannini, R. Marraza, A. Saccone and R. Ferro, *J. Alloys Compd.*, **203**, 177–180 (1994).
177. E.M. Padezhnova, T.V. Dobatkina and E.V. Muratova, *Izv. Akad. Nauk SSSR Met.* (4),

194–197 (1983) (in Russian).
178. T.V. Dobatkina, E.V. Muratova, V.V. Kinzhibalo and A.T. Tyvanchuk, *Izv. Akad. Nauk SSSR, Met.*, (1), 211–214 (1984) (in Russian).
179. E.M. Padezhnova, E.V. Melnik, V.V. Kinzhibalo and T.V. Dobatkina, *Izv. Akad. Nauk SSSR, Met.*, (4), 220–223 (1981) (in Russian).
180. M.E. Drits, E.M. Padezhnova, T.V. Dobatkina, E.A. Voitekhova and V.V. Kinzhibalo, *Izv. Akad. Nauk SSSR, Met.*, (6), 206–210 (1981) (in Russian).
181. Z.A. Sviderskaya and E.M. Padezhnova, *Izv. Akad. Nauk SSSR, Met.*, (6), 200–204 (1971) (in Russian).
182. M.E. Drits, L.L. Rokhlin and N.I. Nikitina, *Izv. Akad. Nauk SSSR, Met.*, (5), 215–219 (1983) (in Russian).
183. L.L. Rokhlin, T.V. Dobatkina and N.I. Nikitina, *Metally*, (4), 109–112 (1997) (in Russian).
184. L. L. Rokhlin, N.I. Nikitina and T.V. Dobatkina, *J. Alloys Compd.*, **239**, 209–213 (1996).
185. V.E. Kolesnichenko, V.V. Karonik and O.A. Nesvetaeva, in *Fazovye Ravnovesiya v Metallicheskikh Splavakh*, Nauka, Moscow, 37–42 (1981) (in Russian).
186. T.V. Dobatkina, E.V. Muratova and E.I. Drozdova, *Izv. Akad. Nauk SSSR, Met.*, (1), 205–208 (1987) (in Russian).
187. O.S. Zarechnyuk, V.V. Kinzhibalo, A.T. Tyvanchuk and R.M. Rykhal, *Izv. Akad. Nauk SSSR, Met.* (5), 221–223 (1981) (in Russian).
188. M.E. Drits, E.I. Drozdova, I.G. Korolkova, V.V. Kinzhibalo and A.T. Tyvanchuk, *Izv. Akad. Nauk SSSR, Met.*, (2), 198–200 (1989).
189. Kh.O. Odinaev, I.N. Ganiev and V.V. Kinzhibalo, *Izv. Akad. Nauk SSSR, Met.*, (5), 91–94 (1988) (in Russian).
190. Kh.O. Odinaev, I.N. Ganiev, V.V. Kinzhibalo and A.T. Tyvanchuk, *Izv. Akad. Nauk SSSR Met.*, (4), 94–97 (1988) (in Russian).
191. Z.A. Sviderskaya and E.M. Padezhnova, in *Legkie Splavy i Metody ikh Obrabotki*, Nauka, Moscow, 169–180 (1968) (in Russian).
192. M.E. Drits, Z.A. Sviderskaya and L.L. Rokhlin, *Zh. Neorg. Khim.*, **7**(12), 2771–2777 (1962) (in Russian).
193. M.E. Drits, E.M. Padezhnova and N.R. Bochvar, *Izv. Akad. Nauk SSSR, Met.*, (1), 149–152 (1966) (in Russian).
194. M.E. Drits and N.R. Bochvar, *Dokl. Akad. Nauk SSSR*, **169**(4), 884–886 (1966) (in Russian).
195. V.V. Kinzhibalo, A.T. Tyvanchuk and E.V. Melnik, in *Stabilnye i Metastabilnye Fazovye Ravnovesiya v Metallicheskikh Sistemakh*, Nauka, Moscow, 70–74 (1985) (in Russian).
196. T.V. Dobatkina, E.V. Melnik, A.T. Tyvanchuk and E.V. Muratova, in *Stabilnye i Metastabilnye Fazovye Ravnovesiya v Metallicheskikh Sistemakh*, Nauka, Moscow, 75–79 (1985) (in Russian).
197. M.E. Drits, E.M. Padezhnova and N.V. Miklina, *Izv. Vyss. Uchebn. Zaved. Tsvetn. Metall.*, (4), 104–107 (1971) (in Russian).
198. M.E. Drits, E.M. Padezhnova and N.V. Miklina, *Izv. Akad. Nauk SSSR, Met.*, (3), 225–229 (1974) (in Russian).
199. M.E. Drits, E.M. Padezhnova and L.S. Guzey, *Izv. Akad. Nauk SSSR, Met.*, (1), 218–220 (1978) (in Russian).
200. M.E. Drits, E.M. Padezhnova and T.V. Dobatkina, *Izv. Akad. Nauk SSSR, Met.*, (4), 193–198 (1979) (in Russian).
201. T.V. Dobatkina, *Izv. Akad. Nauk SSSR, Met.*, (2), 211–214 (1979) (in Russian).
202. E.M. Padezhnova, E.V. Melnik, R.A. Milievsky, T.V. Dobatkina and V.V. Kinzhibalo, *Izv. Akad. Nauk SSSR, Met.*, (4), 204–208 (1982) (in Russian).
203. E.M. Padezhnova, E.V. Melnik and T.V. Dobatkina, *Izv. Akad. Nauk SSSR, Met.*, (1), 217–221 (1979) (in Russian).
204. M.E. Drits, E.M. Padezhnova and L.S. Guzey, *Izv. Akad. Nauk SSSR, Met.*, (3), 218–221 (1977) (in Russian).
205. M.E. Drits, L.L. Rokhlin, N.P. Abrukina, V.V. Kinzhibalo and A.T. Tyvanchuk, *Izv. Akad. Nauk SSSR, Met.*, (6), 194–200 (1985) (in Russian).
206. L.L. Rokhlin, N.R. Bochvar and E.V. Lysova, *Metally*, (5), 122–126 (1997) (in Russian).
207. L.L. Rokhlin, N.R. Bochvar and E.V. Lysova, *J. Mater. Sci. Lett.*, **15**, 2077–2079 (1996).

208. L.L. Rokhlin and N.I. Nikitina, *Izv. Akad. Nauk SSSR, Met.*, (3), 213–219 (1992) (in Russian).
209. E.V. Muratova, T.V. Dobatkina and V.V. Kinzhibalo, *Izv. Akad. Nauk SSSR, Neorg. Mater.*, **25**(1), 66–70 (1989) (in Russian).
210. M.E. Drits, E.M. Padezhnova and N.V. Miklina, in *Metallovedenie Tsvetnykh Metallov i Splavov*, Nauka, Moscow, 52–58 (1972) (in Russian).
211. M.E. Drits, E.M. Padezhnova and N.V. Miklina, *Izv. Akad. Nauk SSSR, Met.*, (4), 218–222 (1974) (in Russian).
212. M.B. Altman, N.M. Tikhova, G.I. Morozova, E.M. Nikolskaya, L.P. Nefedova, A.I. Filatova and G.E. Katerinikova, in *Konstruktsionnye i Zharoprochnye Materialy dlya Novoi Tekhniki*, Nauka, Moscow, 256–265 (1978) (in Russian).
213. M.E. Drits and N.R. Bochvar, *Dokl. Akad. Nauk SSSR*, **178**(2), 403–405 (1968) (in Russian).
214. M.E. Drits, E.M. Padezhnova and F.L. Guschina, *Tekhnol. Legkich Splavov*, (10), 20–25 (1979) (in Russian).
215. M.E. Drits, E.M. Padezhnova and F.L. Guschina, *Tekhnol. Legkich Splavov*, (3), 12–14 (1980) (in Russian).
216. M.E. Drits, E.M. Padezhnova and F.L. Guschina, *Izv. Akad. Nauk SSSR, Met.*, (6), 207–211 (1979) (in Russian).
217. E.M. Padezhnova and T.V. Dobatkina, in *Fazovye Ravnovesiya v Metallicheskikh Slavakh*, Nauka, Moscow, 53–57 (1981) (in Russian).
218. M.E. Drits, E.M. Padezhnova and F.L. Guschina, *Izv. Akad. Nauk SSSR, Met.*, (5), 224–227 (1981) (in Russian).
219. M.E. Drits, L.L. Rokhlin and N.P. Abrukina, *Izv. Akad. Nauk SSSR, Met.*, (5), 199–204 (1984) (in Russian).
220. M.E. Drits, L.L. Rokhlin and N.I. Nikitina, *Izv. Akad. Nauk SSSR, Met.*, (4), 217–221 (1976) (in Russian).
221. M.E. Drits, E.M. Padezhnova and T.V. Dobatkina, in *Problemy Metallovedeniya Tsvetnykh Splavov*, Nauka, Moscow, 89–91(1978) (in Russian).
222. V.I. Mikheeva and E.A. Sennikova, *Proceedings of MATI*, (7), 62–81 (1949) (in Russian).
223. D.A. Petrov, M.S. Mirgalovskaya and I.A. Strelnikova, *Proceedings of Baikov Institute of Metallurgy*, (1), Izdat. Akad. Nauk SSSR, 144–147 (1957) (in Russian).
224. S.A. Pogodin and V.I. Mikheeva, *Izv. Sekt. Fizik.-Khim. Analiza*, **14**, 283–297 (1941) (in Russian).
225. O.S. Zarechnyuk and P.I. Kripyakevich, *Izv. Akad. Nauk SSSR, Met.*, (4), 188–190 (1967) (in Russian).
226. A.M. Korolkov and P.Ya. Saldau, *Izv. Sect. Fizik.-Khim. Analiza*, **16**(2), 295–306 (1946) (in Russian).
227. F. Petzold and B. Beier, *Metalloved. Term. Obrab. Met.*, (5), 22–24 (1971) (in Russian).
228. M.E. Drits, E.M. Padezhnova and T.V. Dobatkina, in *Redkie Metally v Tsvetnykh Splavakh*, Nauka, Moscow, 5–11 (1975) (in Russian).
229. M.E. Drits, L.L. Rokhlin and N.P. Abrukina, *Izv. Vyss. Uchebn. Zaved. Tsvetn. Metall.*, (2), 83–87 (1986) (in Russian).
230. E.V. Muratova and T.V. Dobatkina, in *Magnievye Splavy dlya Sovremennoi Tekhniki*, Nauka, Moscow, 63–67 (1992) (in Russian).
231. Drits, Z.A. Sviderskaya, L.L. Rokhlin and N.I. Nikitina, *Tekhnol. Legkich Splavov*, (6), 73–75 (1977) (in Russian).
232. Z.A. Sviderskaya and E.M. Padezhnova, *Izv. Akad. Nauk SSSR, Met.*, (1), 206–210 (1970) (in Russian).
233. M.E. Drits, E.M. Padezhnova and T.V. Dobatkina, *Izv. Akad. Nauk SSSR, Met.*, (3), 223–227 (1979 (in Russian).
234. Kh.O. Odinaev, I.N. Ganiev, V.V. Kinzhibalo and A.T. Kurbanov, *Izv. Vyss. Uchebn. Zaved. Tsvetn. Metall.*, (4), 75–77 (1989) (in Russian).
235. E.M. Padezhnova, *Izv. Akad. Nauk SSSR, Met.*, (3), 196–200 (1969) (in Russian).
236. Kh.O. Odinaev, I.N. Ganiev, V.V. Kinzhibalo and A.T. Tyvanchuk, *Izv. Vyss. Uchebn. Zaved. Tsvetn. Metall.*, (2), 81–85 (1988) (in Russian).
237. Kh.O. Odinaev, I.N. Ganiev and A.V. Vakhobov, *Izv. Akad. Nauk SSSR, Met.*, (4), 195–197 (1991) (in Russian).

238. M.E. Drits, Z.A. Sviderskaya and L.L. Rokhlin, *Izv. Vyss. Uchebn. Zaved. Tsvetn. Metall.*, (3), 117–121 (1962) (in Russian).
239. Z.A. Sviderskaya and L.L. Rokhlin. *Magnesium Alloys Containing Neodymium*. Nauka, Moscow, (1965) (in Russian).
240. M.E. Drits, E.M. Padezhnova and F.L. Guschina, in *Fazovye Ravnovesiya v Metallicheskikh Splavakh*, Nauka, Moscow, 63–69 (1981) (in Russian).
241. S.K. Das, C.F. Chang and D. Raybould, *Light Metal Age*, **44**(11/12), 5–8 (1986).
242. P.J. Meschter, *Metall. Trans.*, **18A**(2), 347–350 (1987).
243. C.F. Chang, S.K. Das, D. Raybould, R.L. Bye and E.V. Limoncrlli, *Light Metal Age*, **47**(9/10), 12–16 (1989).
244. S.K. Das and C.F. Chang, in *Rapidly Solidified Cryst. Alloys: Proc. TMS-AIME Northeast Reg. Meet., Morristown, N.T., May 1–3, 1985*. Warrendale (Pa), 137–156 (1985).
245. G. Nussbaum, H. Gjestland and G. Regassoni, *Light Metal Age*, **46**(7/8), 16–19 (1988).
246. C.F. Chang, S.K. Das and D. Raybould, in *Rapidly Solidified Mater. Proc. Int. Conf., San Diego, Calif., Febr. 3–5, 1985*. Metals Park, Ohio, 129–135 (1985).
247. Kazuo Alkawa and Katsuyuki Kawasaki, Patent of USA No 5,304,260. U.S. Cl. **148/403**, 148/420, Apr. 19, 1994.
248. L.L. Rokhlin, T.V. Dobatkina, I.G. Korolkova and Yu.N. Grin, *Izv. Akad. Nauk SSSR, Met.*, (5), 181–184 (1991) (in Russian).
249. L.L. Rokhlin and I.G. Korolkova, *Kristallografiya*, **37**(5), 1341–1344 (1992) (in Russian).
250. L.L. Rokhlin, T.V. Dobatkina and I.G. Korolkova, *Izv. Akad. Nauk, Met.*, (2), 152–156 (1994) (in Russian).
251. T.V. Dobatkina, M.E Drits, I.G. Korolkova and L.L. Rokhlin, *Izv. Akad. Nauk SSSR, Met.*, (1), 190–194 (1988) (in Russian).
252. F. Hehmann, F. Sommer and B. Predel, *Mater. Sci. and Eng. A.*, **125** (2), 249–265 (1990).
253. M.E. Drits, Z.A. Sviderskaya and L.L. Rokhlin, in *Issledovanie Splavov Tsvetnykh Metallov*, Nauka, Moscow, 68–74 (1962) (in Russian).
254. D. Mizer and B. Peters, *Proc. 2nd Int. Conf. on Strength of Metals and Alloys, Pacific Grove Calif., 1970*, Metals Park, OH, **2**, 669–673 (1970).
255. D. Mizer and B.C. Peters, *Metal. Trans.*, **3**, 3262–3264 (1972).
256. M.E. Drits, Z.A. Sviderskaya and E.M. Padezhnova, *Tekhnol. Legkich Splavov*, (2), 15–21 (1972) (in Russian).
257. T.J. Pike and B. Noble, *J. Less-Common Metals*, **30**(1), 63–74 (1973).
258. Omori Goro and Matsuo Shigeru, *J. Jap. Inst. Metals*, **39**(5), 444–451 (1975).
259. L.L. Rokhlin, in *Problemy Metallovedeniya Tsvetnykh Splavov*, Nauka, Moscow, 59–70 (1978) (in Russian).
260. M.E. Drits and L.L. Rokhlin, in *Konstruktsionye i Zharoprochnye Materialy dlya Novoi Tekhniki*, Nauka, Moscow, 78–91 (1978) (in Russian).
261. L.L. Rokhlin, in *Magnievye Splavy*, Nauka, Moscow, 63–69 (1978) (in Russian).
262. Omori Goro, *Trans. National Research Institute for Metals*, **21**(4), 38–51 (1979).
263. B.N. Ovechkin, L.G. Klimovich and N.N. Kulakov, *Tekhnol. Legkich Splavov*, (1), 13–16 (1978) (in Russian).
264. P.A. Nuttal, T.J. Pike and B. Noble, *Metallography*, **13**(1), 3–20 (1980).
265. L.L. Rokhlin, *Fiz. Met. Metalloved.*, **54**(2), 315–319 (1982) (in Russian).
266. L.L. Rokhlin, *Fiz. Met. Metalloved.*, **55**(4), 733–738 (1983) (in Russian).
267. M.E. Drits, L.L. Rokhlin and I.E. Tarytina, *Izv. Akad. Nauk SSSR, Met.*, (3), 111–116 (1983) (in Russian).
268. L.L. Rokhlin and I.E. Tarytina, *Fiz. Met. Metalloved.*, **59**(6), 1188–1193 (1985) (in Russian).
269. M.E. Drits, L.L. Rokhlin and N.I. Nikitina, in *Metallovedenie Legkikh Splavov*, VILS, Moscow, 133–139 (1985) (in Russian).
270. L.L. Rokhlin and N.P. Abrukina, in *Metallovedenie Legkikh Splavov*, VILS, Moscow, 139–146 (1985) (in Russian).
271. L.L. Rokhlin and N.I. Nikitina, *Fiz. Met. Metalloved.*, **62**(4), 781–786 (1986) (in Russian).
272. L.L. Rokhlin and I.E. Tarytina, *Izv. Vyss. Uchebn. Zaved. Tsvetn. Metall.*, (1), 106–110 (1986) (in Russian).
273. L.L. Rokhlin, *Fiz. Met. Metalloved.*, **63**(1), 146–150 (1987) (in Russian).

274. L.L. Rokhlin, *Metalloved. Term. Obrab. Met.*, (3), 50–52 (1987) (In Russian).
275. P. Vostry, T. Stulikova, B. Smola, M. Cieslar and B.L. Mordike, *Z. Metallkd.*, **79**(5), 340–344 (1988).
276. Sato Tatsuo, Takahashi Isao, Tezuka Hiroyasa and Kamio Akihiko, *J. Jap. Inst. Light Metals*, **42**(12), 804–809 (1992).
277. Kamado Shigeharu, Iwasawa Shigeru, Ohuchi Kiyoaki, Kojima Yo and Ninomiya Ryuuji, *J. Jap. Inst. Light Metals*, **42**(12), 727–733 (1992).
278. N.P. Abrukina, in *Magnievye Splavy dlya Sovremennoi Tekhniki*, Nauka, Moscow, 82–86 (1992) (in Russian).
279. L.L. Rokhlin, N.I. Nikitina and I.E. Tarytina, *Fiz. Met. Metalloved.*, **76**(6), 82–87 (1993) (in Russian).
280. L.L. Rokhlin and N.I. Nikitina, *Z. Metallkd.*, **85**(12), 819–823 (1994).
281. L.Y. Wei, G.L. Dunlop and H. Westengen, *Metall. Mater. Trans. A*, **26A**(7), 1705–1716 (1995).
282. L.L. Rokhlin and N.I. Nikitina, *Fiz. Met. Metalloved.*, **82**(4), 113–118 (1996) (in Russian).
283. S. Kamado, Y. Kojima, R. Ninomiya and K. Kubota, in *Proceedings of the Third International Magnesium Conference*, The Institute of Materials, London, 327–342 (1997).
284. M. Hisa, J.C. Barry and G.L. Dunlop, in *Proceedings of the Third International Magnesium Conference*, The Institute of Materials, London, 369–379 (1997).
285. R.A. Khosrhoshahi, R. Pilkington and G.W. Lorimer, in *Proceedings of the Third International Magnesium Conference*, The Institute of Materials, London, 242–256 (1997).
286. F.J. Humphreys and G.W. Lorimer, in *Advanced Light Alloys and Composites*, Kluwer Academic Publishers, Dordrecht–Boston–London, 343–353 (1998).
287. M.E. Drits and L.L. Rokhlin, in *Legkie i Zharoprochnye Splavy i Ikh Obrabotka*, Nauka, Moscow, 130–134 (1986) (in Russian).
288. L.L. Rokhlin and N.I. Nikitina, *Proceed. Int. Conf. "Light Alloys and Composites", May 13–16, 1999, Zakopane, Poland*, Wykonano w Akademickim Centrum Graficzno-Marketinggowym SA. LODART, Lodz, 205–211 (1999).
289. M.E. Drits, L.L. Rokhlin, E.M. Padezhnova, I.I. Gurev, N.V. Miklina, T.V. Dobatkina and A.A. Oreshkina. *Magnesium Alloys with Yttrium*. Nauka, Moscow (1979) (in Russian).
290. D.S. Gencheva, A.A. Katsnelson, L.L. Rokhlin, V.M. Silonov and F.A. Khavaja, *Fiz. Met. Metalloved.*, **51**(4), 788–793 (1981) (in Russian).
291. L.Y. Wei, G.L. Dunlop and H. Westengen, *Metall. Mater. Trans. A*, **26A**(8), 1974–1985 (1995).
292. L.L. Rokhlin, N.I. Nikitina and A.A. Oreshkina, in *Tekhnologiya obrabotki legkikh i spetsialnykh splavov*, Metallurgiya, Moscow, 115–124 (1994) (in Russian).
293. L.L. Rokhlin and N.I. Nikitina, *J. Alloys Compd.*, **279**, 166–170 (1998).
294. E.F. Emley, *Principles of Magnesium Technology*, Metallurgiya, Moscow (1972) (Russian translation).
295. E.M. Savitsky, *Plastic Properties of Magnesium and Some Its Alloys*, Izdat. Akad. Nauk SSSR, Moscow–Leningrad, (1941) (in Russian).
296. Th. Ernst and F. Laves, *Z. Metallkd.*, **40** (1), 1–12 (1949) (in German).
297. J. W. Suiter and W. A. Wood, *J. Inst. Metals*, **81** (4), 181–184 (1952).
298. M.E. Drits, L.L. Rokhlin and N.I. Nikitina, *Tekhnol. Legkich Splavov*, (1), 22–26 (1976) (in Russian).
299. M.E. Drits, Z.A. Sviderskaya and L.L. Rokhlin, *Izv. Akad. Nauk SSSR, Metallurg. Toplivo.*, (5), 191–196 (1962) (in Russian).
300. L.L. Rokhlin, Z.A. Sviderskaya and R.P. Volchkova, in *Metallurgiya, Metallovedenie, Fiziko-Khimicheskie Metody Issledovaniya. Trudy In-ta Metallurgii im. A.A. Baikova*, Izdat. Akad. Nauk SSSR, Moscow, (12), 161–165 (1962) (in Russian).
301. Z.A. Sviderskaya and A.A. Oreshkina, in *Splavy Tsvetnykh Metallov*, Nauka, Moscow, 161–166 (1972) (in Russian).
302. M.E. Drits, L.L. Rokhlin and A.A. Oreshkina, in *Obrabotka Metallov i Splavov Davleniem*, VILS, Moscow, 39–46 (1976) (in Russian).
303. M.E. Drits, I.I. Gurev, N.I. Vasileva and A.E. Ansyutina, in *Metallovedenie Legkikh Splavov*, Nauka, Moscow, 217–225 (1965) (in Russian).
304. Z.A. Sviderskaya and A.A. Oreshkina, in *Splavy Tsvetnykh Metallov*, Nauka, Moscow, 161–166 (1972) (in Russian).

305. M.E. Drits, Z.A. Sviderskaya and N.I. Nikitina, in *Metallovedenie i Tekhnologiya Legkikh Splavov*, Nauka, Moscow, 105–109 (1976) (in Russian).
306. J.B. Newkirk, in *Starenie Splavov* (Translation of the book "Precipitation from Solid Solution"), Metallurgizdat, Moscow, 12–142 (1962) (in Russian).
307. E.C. Burke and W.R. Hibbard, *Trans. AIME*, **194**, 295–303 (1952).
308. F.E. Hauser, C.D. Starr, L. Tietz and J.E. Dorn, *Trans. Am. Soc. Met.*, **47**, 102–134 (1955).
309. R.E. Reed-Hill and W.D. Robertson, *Acta Metall.*, **5**(12), 717–727 (1957).
310. R.E. Reed-Hill and W.D. Robertson, *Acta Metall.*, **5**(12), 728–737 (1957).
311. R.E. Reed-Hill and W.D. Robertson, *J. Metals*, **9**(4), sec. 2, 496–502 (1957).
312. H. Asada and H. Yoshinaga, *Nippon Kinzoku Gakkai-Si*, **23**(11), 649–652 (1959).
313. J.D. Mote and J.E. Dorn, *Trans. Metall. Soc. AIME*, **218**(3), 491–497 (1960).
314. R.E. Reed-Hill, *Trans. Metall. Soc. AIME*, **218**(3), 554–558 (1960).
315. H. Yoshinaga, T. Obara and S. Merozumi, *Mater. Sci. Eng.*, **12**(5/6), 255–264 (1973).
316. M.E. Drits, Z.A. Sviderskaya, I.I. Gurev, L.L. Rokhlin and A.A. Oreshkina, in *Metallovedenie Legkikh Splavov*, Nauka, Moscow, 125–134 (1965) (in Russian).
317. K.H. Eckelmeyer and R.W. Herzberg, *Metall. Trans.*, **1**(12), 3411–3414 (1970).
318. E.S. Kadaner and A.A. Oreshkina, in *Metallurgiya, Metallovedenie, Fiziko-Khimicheskiye Metody Issledovaniya. Proced. Baikov Institute of Metallurgy*, USSR Acad. Sci. Publishers, Moscow, (14), 130–138 (1963) (in Russian).
319. M.E. Drits and E.S. Kadaner, in *Issledovaniye Splavov Tsvetnykh Metallov*, USSR Acad. Sci. Publishers, Moscow, (4), 211–223 (1963) (in Russian).
320. S.B. Felgina, *Izv. Akad. Nauk SSSR, Metallurgiya i Gornoye Delo*, (6), 137–141 (1964) (in Russian).
321. M.E. Drits and S.B. Felgina, *Izv. Akad. Nauk SSSR, Met.*, (4), 75–83 (1964) (in Russian).
322. Omori Garo, *J. Jap Inst. Metals*, **39**(5), 451–459 (1975).
323. L.L. Rokhlin, *Fiz. Met. Metalloved.*, **16**(5), 703–709 (1963) (in Russian).
324. M. E. Drits and S.B. Felgina, in *Legkie Splavy i Metody Ikh Obrabotki*, Nauka, Moscow, 184–190 (1968) (in Russian).
325. M.E. Drits, Z.A. Sviderskaya, I.I. Gurev, L.L. Rokhlin and A.A. Oreshkina, in *Metallovedenie Legkikh Splavov*, Nauka, Moscow, 125–134 (1965) (in Russian).
326. C.S. Roberts, *J. Metals*, **5**(9),1121–1126 (1953).
327. S.L. Couling and C.S. Roberts, *J. Metals*, **9**(10), 1252–1256 (1957).
328. J.W. Suiter and W.A. Wood, *J. Inst. Metals*, **81**(4), 181–184 (1952).
329. A.R. Chaudhuri, H.C. Chang and N.J. Grant, *J. Metals*, **7**(5), 682–688 (1955).
330. A.R. Chaudhuri, N.J. Grant and J.T. Norton, *J. Metals*, **5**(5), sec. II, 712–716 (1953).
331. L.L. Rokhlin, in *Legirovaniye i Obrabotka Legkikh Splavov*, Nauka, Moscow, 96–103 (1981) (in Russian).
332. L.L. Rokhlin and A.A. Oreshkina, *Fiz. Met. Metalloved.*, **22**(3), 420–423 (1966) (in Russian).
333. M.E. Drits, Z.A. Sviderskaya, L.L. Rokhlin and A.A. Oreshkina, *Metalloved. Term. Obrab. Met.*, (3), 55–58 (1968) (in Russian).
334. M.E. Drits, Z.A. Sviderskaya and A.A. Oreshkina, in *Struktura i Svoistava Legkikh Splavov*, Nauka, Moscow, 127–136 (1971) (in Russian).
335. M.E. Drits, A.A. Oreshkina and Z.A. Sviderskaya, *Fizika i Chim. Obrab. Mater.*, (1), 112–114 (1968) (in Russian).
336. Z.A. Sviderskaya and A.A. Oreshkina, in *Redkie Metally v Tsvetnykh Splavakh*, Nauka, Moscow, 147–151 (1975) (in Russian).
337. T.V. Lebedeva, I.G. Kovalev and O.V. Emelyanova, in *Redkie Metally v Tsvetnykh Splavakh*, Metallurgizdat, Moscow, 209–218 (1960) (in Russian).
338. M.E. Drits, Z.A. Sviderskaya and N.I. Turkina, *Izv. Akad. Nauk SSSR, Metallurg. Toplivo*, (4), 111–119 (1960) (in Russian).
339. M.E. Drits, *Magnesium Alloys for Application at Elevated Temperatures*, Nauka, Moscow, 39–48 (1964) (in Russian).
340. M.E. Drits, E.M. Padezhnova and E.V. Muratova, *Izv. Akad. Nauk SSSR, Metally*, (1), 195–199 (1984) (in Russian).
341. M.E. Drits, L.L. Rokhlin and N.P. Abrukina, *Metalloved. Term. Obrab. Met.*, (7), 27–29 (1985) (in Russian).

342. L.L. Rokhlin, *Proc. XV Physical Metallurgy and Materials Science Conf. on Advanced Materials and Technologies, AMT'98, Krakow-Krynica, Poland 17–21 May, 1998; Inzynieria Materialowa*, **19**(3), 146–152 (1998).
343. S.S. Gorelik, *Recrystallization of Metals and Alloys*, Metallurgiya, Moscow, (1967) (in Russian).
344. M.A. Timonova, A.A. Blyablin, E.F. Chirkova, E.I. Smirnova, A.I. Kutaisteva and F.L. Gurevich, in *Metallovedenie i Tekhnologiya Legkikh Splavov*, Nauka, Moscow, 113–118 (1976) (in Russian).
345. M.E. Drits, L.L. Rokhlin, A.A. Oreshkina and N.I. Nikitina, *Izv. Akad. Nauk SSSR, Metally*, (5), 98–103 (1982) (in Russian).
346. L.L. Rokhlin and N.I. Nikitina, *Metalloved. Term. Obrab. Met.*, (6), 37–39 (1999) (in Russian).
347. M.E. Drits, Z.A. Sviderskaya and N.I. Nikitina, in *Redkie Metally v Tsvetnykh Splavakh*, Nauka, Moscow, 49–54 (1975) (in Russian).
348. L.L. Rokhlin and A.A. Oreshkina, in *Problemy Bolshoy Metallurgii i Fizicheskoy Khimii Novykh Splavov*, Nauka, Moscow, 235–240 (1965) (in Russian).
349. M.E. Drits, E.M. Padezhnova and N.V. Miklina, *Tekhnol. Legkich Splavov*, (5), 42–45 (1978) (in Russian).
350. M.E. Drits, I.I. Gurev, E.M. Padezhnova, N.V. Miklina and T.V. Dobatkina, in *Magnievye Splavy*, Nauka, Moscow, 78–82 (1978) (in Russian).
351. M.E. Drits, Z.A. Sviderskaya, L.L. Rokhlin and N.I. Nikitina, *Metalloved. Term. Obrab. Met.*, (11), 62–64 (1979) (in Russian).
352. M.E. Drits, Z.A. Sviderskaya and N.I. Nikitina, in *Metallovedenie Tsvetnykh Metallov i Splavov*, Nauka, Moscow, 106–110 (1972) (in Russian).
353. K. Grube, J.A. Davis and L.W. Eastwood, *Proc. ASTM*, **50**, 965–988 (1950).
354. K. Grube and L.W. Eastwood, *Proc. ASTM*, **50**, 989–1012 (1950).
355. A.I. Murphy and R.J.M. Payne, *J. Inst. Metals*, **73**(3), 105–127 (1946).
356. R.J.M. Payne and N. Bailey, *J. Inst. Metals*, **88**(6), 417–427 (1960).
357. J.F. Pashak and T.E. Leontis, *Trans. Metallurg. Soc. AIME*, **218**(1), 102–107 (1960).
358. T.E. Leontis and D.H. Feisel, *J. Metals*, **9**(10), sec. 2, 1245–1252 (1957).
359. N.M. Tikhova, V.A. Blokhina and L.A. Afanasyeva, in *Redkie Metally i Splavy*, Metallurgizdat, Moscow, 219–226 (1960) (in Russian).
360. I.I. Gurev, N.N. Kulakov, N.I. Vasileva and S.V. Bukhman, in *Redkie Metally v Tsvetnykh Splavakh*, Nauka, Moscow, 77–81 (1975) (in Russian).
361. Z.A. Sviderskaya, in *Redkie Metally v Tsvetnykh Splavakh*, Nauka, Moscow, 152–159 (1975) (in Russian).
362. M.E. Drits, Z.A. Sviderskaya and L.L. Rokhlin, in *Metallurgiya, Metallovedenie i Fiziko-Khimicheskie Metody Issledovaniya*, Russian Acad. Sci. Publishers, Moscow, (14), 120–129 (1963) (in Russian).
363. M.E. Drits and N.R. Bochvar, *Izv. Vyss. Uchebn. Zaved. Tsvetn. Metall.*, (4), 99–103 (1969) (in Russian).
364. N.M. Tikhova, G.G. Solovyeva, V.A. Blokhina, A.P. Antipova and T.P. Vasileva, in *Redkozemelnye Metally i Splavy*, Nauka, Moscow, 99–103 (1971) (in Russian).
365. M.V. Chukhrov and Z.N. Khrisanova, *Izv. Akad. Nauk SSSR, Metally*, (6), 149–153 (1979) (in Russian).
366. M.E. Drits, E.M. Padezhnova and T.V. Dobatkina, in *Legirovaniye i Obrabotka Legkikh Splavov*, Nauka, Moscow, 91–96 (1981) (in Russian).
367. J.E. Morgan, B.L. Mordike, "*Strength Metals and Alloys (ICSMA6)". Proc. 6th Int. Conf., Melbourn, 16–20 Aug., 1982*, **2**, Oxford e.a., 643–648 (1982).
368. J.E. Morgan and B.L. Mordike, *Metall. Trans.*, **12A**(9), 1581–1585 (1981).
369. F. Sauerwald, *Z. Metallkd.*, **45**(5), 257–269 (1954).
370. F. Sauerwald, *Z. Anorg. Allg. Chem.*, **255**(1/3), 212–220 (1947/1948).
371. I.M. Vesey and H.J. Bray, *J. Inst. Metals*, **92**(11), 383–384 (1964).
372. G. Mellor, *J. Inst. Metals*, **77**(2), 163–174 (1950).
373. M.A. Timonova, *Protection of Magnesium Alloys against Corrosion*, Metallurgiya, Moscow, (1977) (in Russian).
374. T.E. Leontis, *Metal Progr.*, **76**(5), 82–87 (1959).

375. R.J.M. Payne and N. Bailey, *Metallurgia*, (346), 67–68 (1958).
376. A. Stevenson, *J. Metals*, **39**(5), 16–19 (1987).
377. N.M. Tikhova, V.A. Blokhina, A.P. Antipova and T.P. Vasileva, *Liteinoe Proizvodstvo*, (7), 20–21 (1970) (in Russian).
378. P.J. Meschter and J.E. O'Neal, *Metall. Trans.*, **15A**(1), 237–240 (1984).
379. Sung Gyoo Kim, Akihisa Inoue and Tsuyoshi Masumoto, *Mater. Trans., JIM*, **31**(11), 929–934 (1990).
380. H. Gjestland, G. Nussbaum and G. Regazzoni, *Light Metal Age*, **47**(7/8), 18–21 (1989).
381. T.V. Dobatkina, L.L. Rokhlin and I.G. Korolkova, in *Advanced Light Alloys and Composites*, Kluwer Academic Publishers, Dordrecht–Boston–London, 461–466 (1998).
382. L.L. Rokhlin, T.V. Dobatkina, I.G. Korolkova and V.V. Kuleshov, in *Metallovedenie, Litye i Obrabotka Splavov*, VILS, Moscow, 101–105 (1995).
383. A.G. Sokolinsky and Yu.M. Sukharevsky, *Magnesium Ultrasonic Delay Lines*. Sovetskoye radio, Moscow (1966) (in Russian).
384. M.E Drits and L.L. Rokhlin, *Magnesium Alloys with Special Acoustical Properties*. Metallurgiya, Moscow (1983).
385. F.A. Metz and W.M.A. Anderson, *Electronics*, **22**(7), 96–99 (1949).
386. W. Roth, *J. Appl. Phys.*, **19**(10), 901–940 (1948).
387. W.P. Mason and H.J. McSkimin, *J. Appl. Phys.*, **19**(10), 940–946 (1948).
388. C.T. Brockelsby, J.S. Palfreeman and R.W. Gibson, *Ultrasonic Delay Lines*. Hiffe Books Ltd, (1963).
389. R.W. Gibson, *Ultrasonics*, **3**(2), 49–61 (1965).
390. J.S. Palfreeman, *Ultrasonics*, **3**(1), 1–8 (1965).
391. L.L. Rokhlin, *Acoustical Properties of Light Alloys*, Nauka, Moscow (1974) (in Russian).
392. W.P. Mason, *Physical Acoustics and the Properties of Solids*, D. Van Nostrand Company, Princeton (N.J.), (1958).
393. W.P. Mason and H.J. McSkimin, *J. Acoust. Soc. Amer.*, **19**(3), 464–473 (1947).
394. E.P. Papadakis, *J. Acoust. Soc. Amer.*, **43**(4), 876–879 (1968).
395. L.G. Merkulov, *Zh. Tekhn. Fiz.* **26**(1), 64–75 (1956) (in Russian).
396. C.F. Ying and R. Truell, *J. Appl. Phys.*, **27**(9), 1086–1097 (1956).
397. G. Johnson and R. Truell, *J. Appl. Phys.*, **36**(11), 3466–3475 (1965).
398. A. Granato and K. Lucke, *J. Appl. Phys.*, **27**(6), 583–593 (1956).
399. K. Lucke, *J. Appl. Phys.* **27**(12), 1433–1438 (1956).
400. L.L. Rokhlin, in *Legkie Splavy I Metody ikh Obrabotki*, Nauka, Moscow, 212–225 (1968) (in Russian).
401. N.F. Fiore and Ch.L. Bauer, *J. Appl. Phys.*, **35**(7), 2242–2247 (1964).
402. T. Hinton and J.G. Rider, *J. Appl. Phys.*, **32**(2), 283–284 (1961).
403. S. Weing and E.S. Machlin, *J. Appl. Phys.*, **27**(7), 734–738 (1956).
404. S. Takanashi, *J. Phys. Soc. Japan*, **11**(12), 1253–1272 (1956).
405. G. Alers and D. Tomson, *J. Appl. Phys.*, **32**(2), 283–284 (1961).
406. A. Hikata, B. Chick, C. Elbaum and R. Truell, *Acta Metall.*, **10**(4), 423–429 (1962).
407. A. Hikata, R. Truell, A. Granato, B. Chick and K. Lucke, *J. Appl. Phys.*, **27**(4), 396–404 (1956).
408. L.L. Rokhlin, *Akust. Zhurn.*, **18**(1), 90–95 (1972) (in Russian).
409. I.M. Lifshits and G.D. Parkhamovsky, *Zh. Eksp. Teor. Fiz.*, **20**(2), 175–182 (1950) (in Russian).
410. M.E. Drits, L.L. Rokhlin and L.L. Zusman, *Tekhnol. Legkikh Splavov*, (2), 50–53 (1969) (in Russian).
411. M.A. Timonova, in *Magnievye Splavy. Sprovochnik (Magnesium Alloys. A Reference Book)*. **1**. Metallurgiya, Moscow, 175–204 (1978) (in Russian).
412. E. Ghali, in *Magnesium Alloys 2000. Proc. of the First Nagaoka International Workshop on Magnesium Platform Science and Technology 2000, Nagaoka City, Japan, 27–29, July 2000*. Materials Science Forum, **350–351**, Trans Tech Publications Ltd, Switzerland, 261–272 (2000).
413. M.A. Timonova, in *Magnievye Splavy (Magnesium Alloys)*. Nauka, Moscow, 23–27 (1978).
414. V.V. Krasnoyarsky, E.M. Padezhgnova, T.V. Dobatkina and V.P. Manokhin, in *Magnievye Splavy (Magnesium Alloys)*, Nauka, Moscow, 85–89 (1978).

415. V.V. Krasnoyarsky and L.M. Petrova, in *Magnievye Splavy dlya Sovremennoi Tekhniki*, Nauka, Moscow, 37–41 (1992).
416. E.F. Volkova, L.I. Dyachenko and A.A. Blyablin, in *Magnievye Splavy dlya Sovremennoi Tekhniki*, Nauka, Moscow, 149–154 (1992).
417. *Standard Specification for Magnesium Alloy Sand Castings*. American Society for Testing and Materials. Designation: B 80–91.
418. *Standard Specification for Magnesium-Alloy Sheet and Plate [Metric]*. American Society for Testing and Materials. Designation: B 90M–90.
419. A. Raman, *Z. Metallkd.*, **68**(3), 163–172 (1977).
420. *Materials Engineering*, Dec., 89–91 (1992).
421. W. Unsworth, *Metals and Mater.* **4**(2), 83–86 (1988).
422. *Des. Eng. (Gr. Brit.)*, Dec., 30 (1985).
423. *Mater. + Manuf.*, **6**(9), 5 (1989).
424. W.H. Unsworth, *Light Metal Age*, **35**(5/6), 14, 16 (1977).
425. *Light Metal Age*, **49**(7/8), 30 (1991).
426. K. Buehler, *Mater. and Manuf. Processes*, **4**(4), 603–616 (1989).
427. Casting magnesium alloys. Grades. *State Standard of USSR* , No 2856–79.
428. Strained magnesium alloys. Grades. *State Standard of USSR*, No 14957–76.
429. B.N. Arzamasov, ed. *Structural Materials*. Mashinostroenie, Moscow, 272–291 (1990).
430. N.M. Tikhova, V.A. Blokhina – A.P. Antipova, in *Magnievye Splavy. Sprovochnik (Magnesium Alloys. Reference Book)* . **1.** Metallurgiya, Moscow, 87–103 (1978) (in Russian).
431. V.V. Krymov, in *Magnievye Splavy. Sprovochnik (Magnesium Alloys. Reference Book).* **1.** Metallurgiya, Moscow, 68–87 (1978) (in Russian).
432. A.A. Blyablin and E.I. Smirnova, in *Magnievye Splavy. Sprovochnik (Magnesium Alloys. Reference Book).* **1**. Metallurgiya, Moscow, 103–123 (1978) (in Russian).
433. A.A. Kazakov, in *Magnievye Splavy. Sprovochnik (Magnesium Alloys. Reference Book).* **1**. Metallurgiya, Moscow, 123–132 (1978) (in Russian).
434. M.E Drits, F.M. Elkin, I.I. Gurev, B.I. Bondarev, V.F. Trokhova, A.D. Sergievskaya and T.N. Osokina, *Magnesium–Lithium Alloys*, Metallurgiya, Moscow (1980) (in Russian).
435. A.A. Blyablin, in *Magnievye Splavy*, Nauka, Moscow, 17–22 (1978) (in Russian).
436. M.E. Drits, Z.A. Sviderskaya and N.I. Nikitina, in *Splavy Tsvetnykh Metallov*, Nauka, Moscow, 193–197 (1972) (in Russian).
437. M.E. Drits, Z.A. Sviderskaya and N.I. Nikitina, in *Metallovedenie Tsvetnykh Metallov i Splavov*, Nauka, Moscow, 106–110 (1972) (in Russian).
438. L.L. Rokhlin, T.V. Dobatkina, D.G. Eskin, N.I. Nikitina and M.L. Charakterova, in *Institutu Metallurgii i Materialovedeniya im. A.A. Baikova 60 let*, ELIS, Moscow, 235–242 (1998) (in Russian).
439. T.M. Moskovchenko, B.I. Bondarev, O.S. Bochvar and M.S. Kogan, *Tekhnol. Legkikh Splavov*, (5), 62–63 (1981) (in Russian).
440. I.I. Gurev, A.A. Blyablin, A.E. Ansyutina, A.A. Kazakov and E.I. Smirnova, *Tekhnol. Legkikh Splavov*, (3), 10–14 (1982) (in Russian).
441. B.I. Bondarev, E.V. Ekhina, T.M. Kunyavskaya and T.M. Moskovchenko, in *Legkie i Zharoprochnye Splavy i Ikh Obrabotka*, Nauka, Moscow, 171–175 (1986) (in Russian).
442. E.F. Volkova, V.M. Lebedev, F.L. Gurevich and Z.N. Khristova, in *Metallovedenie, Lityo i Obrabotka Splavov*, VILS, Moscow, 106–112 (1995) (in Russian).

Index